SUSTAINABLE ENERGY TECHNOLOGIES

Sustainable Energy Technologies

Options and Prospects

Edited by

K. HANJALIĆ

Delft University of Technology and
Marie Curie Chair, Department of Mechanics and Aeronautics,
University of Rome "La Sapienza", Rome, Italy

R. VAN DE KROL

Department DelftChemTech / Energy,
Delft University of Technology,
Delft, The Netherlands

and

A. LEKIĆ

Faculty of Mechanical Engineering,
University of Sarajevo,
Sarajevo, Bosnia and Herzegovina

Springer

A C.I.P. Catalogue record for this book is available from the Library of Congress.

ISBN 978-94-007-9214-2
ISBN 978-1-4020-6724-2 (eBook)

Published by Springer,
P.O. Box 17, 3300 AA Dordrecht, The Netherlands.

www.springer.com

Printed on acid-free paper

Table of Contents

Preface

In pursuit of feasible near-future sustainable and environmentally acceptable energy solutions to specific needs and based on the available – often specific and limited – resources as pertinent to many smaller transitional countries, the Academy of Sciences and Arts of Bosnia and Herzegovina in cooperation with Dubrovnik Inter-University Centre invited a panel of experts to present and discuss the state-of-the-art in the development of new energy technologies at a conference held in Dubrovnik, Croatia, on 23–25 September 2006. Up-to-date reviews of the status and prospects of different options in energy conversion and storage technologies were presented by some of the world leading authorities – members of the national and international academies of sciences, directors of institutes, directors of national and international (EU) energy programmes and academics from Belgium, France, Germany, Italy, Japan, the Netherlands, Portugal and Switzerland. The articles covered new clean and zero-emission coal technologies, solar, wind, nuclear, fuel cells, hydrogen and hybrid technologies, accompanied by treatises on the challenge of increasing global energy needs and consumption, issues of sustainability, and on efficient production and use of energy based on modern rationing technologies. This overview was complemented by several regional surveys of needs, resources and priorities, as well as specific initiatives towards meeting future energy objectives, pursued in several countries in South-Eastern Europe.

This volume has grown out of (but is not confined to) extended and revised invited articles presented at the Dubrovnik conference under the auspices of the Academy of Sciences and Arts of Bosnia and Herzegovina. The book aims at providing information and guidance to engineers and planners in energy sector, employees in energy utility companies, various levels of governmental organizations and offices. It is also intended to serve as a graduate-level textbook to meet growing demand for new courses in alternative, renewable and sustainable energy technologies at technical and general universities.

The Editors would like to thank the contributors for their effort to provide up-to-date revised and extended manuscripts for this volume. Thanks are also due to the Academy of Sciences and Arts of Bosnia and Herzegovina, for initiating, sponsoring and organizing the conference in Dubrovnik in September 2006, and to the Inter-University Centre in Dubrovnik for its support. We thank Professor Joop Schoonman from the Delft University of Technology for providing invaluable advice in selecting and inviting experts in renewable energy technologies. Our appreciations go also to Amra Avdagić and Danica Radić for the skilful organisation of the Conference, and to Gina Landor for her assistance in the editing of this volume.

June 2007
The Editors

List of Contributors

NAIM HAMDIA AFGAN
Member of the Academy of Sciences and Arts of Bosnia and Herzegovina, UNESCO Chair Holder, Technical University of Lisbon, Lisbon, Portugal,
nafgan@ist.utl.pt

SERGEY V. ALEKSEENKO
Corresponding Member of the Russian Academy of Sciences, Director of the Institute of Thermophysics, Siberian Branch of RAS, 1 Lavrentyev Av., Novosibirsk 630090, Russia
aleks@itp.nsc.ru

JEAN-FRANÇOIS BONNET
Laboratoire TREFLE, ENSAM, University of Bordeaux 1, Bordeaux, France
jean-francois.bonnet@u-bordeaux1.fr

MICHEL COMBARNOUS
Corresponding Member of the French Academy of Sciences, Laboratoire TREFLE, ENSAM, University of Bordeaux 1, Bordeaux, France/University of Gabes, Tunisia
michel.combarnous@bordeaux.ensam.fr

DAN GABRIEL CACUCI
Honorary Member of the European Academy of Arts and Sciences and of the Romanian Academy, University of Karlsruhe, Germany/Commissariat à l'Energie Atomique (CEA), Direction de l'Energie Nucléaire (DEN) CEA Saclay, Bât. 121, 91191 Gif sur Yvette Cedex, France
dan-gabriel.cacuci@cea.fr

MAURIZIO CUMO
Member of the National Italian Academy of Sciences ("the Forty"), Department of Nuclear Engineering and Energy Conversions, University of Rome "La Sapienza", Corso Vittorio Emanuele II 244, 00186 Rome, Italy
maurizio.cumo@uniroma1.it

BOZIN DONEVSKI
Faculty of Technical Sciences, University St. Kliment Ohridski, P.O. Box 99, 7000 Bitola, Macedonia
bozin.donevski@uklo.edu.mk

AGGELOS DOUKELIS
Laboratory of Steam Boilers and Thermal Plants, National Technical University of Athens, Heroon Polytechniou 9, 15780 Athens, Greece
adoukel@central.ntua.gr

DIONYSIOS GIANNAKOPOULOS
Laboratory of Steam Boilers and Thermal Plants, National Technical University of Athens, Heroon Polytechniou 9, 15780 Athens, Greece
dgia@central.ntua.gr

MICHAEL GRÄTZEL
Honorary Member of the Société Vaudoise des Sciences Naturelles, Laboratory for Photonics & Interfaces, Faculty of Basic Sciences, Ecole Polytechnique Fédérale de Lausanne, CH-1015 Lausanne, Switzerland
michael.graetzel@epfl.ch

KEMAL HANJALIĆ
Fellow of the Royal Academy of Engineering (UK), Member of the Academy of Sciences and Arts of Bosnia and Herzegovina, Emeritus Professor of Delft University of Technology, The Netherlands, and Marie Curie Chair, Department of Mechanics and Aeronautics, University of Rome "La Sapienza", Via Eudossiana 18, 00161 Rome, Italy
hanjalic@dma.ing.uniroma1.it

RALPH HENDRIKS
Delft University Wind Energy Research Institute DUWIND, Kluyverweg 1, 2629 HS Delft, The Netherlands
r.l.hendriks@tudelft.nl

HIROSHI IWAI
Department of Aeronautics and Astronautics, Kyoto University, Kyoto 606-8501, Japan
iwai@mbox.kudpc.kyoto-u.ac.jp

EMMANOUIL KAKARAS
Laboratory of Steam Boilers and Thermal Plants, National Technical University of Athens, Heroon Polytechniou 9, 15780 Athens, Greece
ekak@central.ntua.gr

ANTONIOS KOUMANAKOS
Laboratory of Steam Boilers and Thermal Plants, National Technical University of Athens, Heroon Polytechniou 9, 15780 Athens, Greece
akouman@central.ntua.gr

EVALD KRANJČEVIČ
Energy Efficiency Centre, Jožef Stefan Institute, Jamova 39, 1000 Ljubljana, Slovenia
evald.kranjcevic@ijs.si

URŠKA LAVRENČIČ ŠTANGAR
Laboratory for Environmental Research, University of Nova Gorica, Vipavska 13, 5000 Nova Gorica, Slovenia
urska.lavrencic@p-ng.si

ALIJA LEKIĆ
Member of the Academy of Sciences and Arts of Bosnia and Herzegovina, Faculty of Mechanical Engineering, University of Sarajevo, Vilsonovo 9, 71000 Sarajevo, Bosnia and Herzegovina
lekic@mef.unsa.ba; alilekic@yahoo.com

PHILIPPE MATHIEU
Department of Aeronautics and Mechanical Engineering, Power Generation, University of Liège, Chemin des Chevreuils, 1, B52/3, 4000 Liège, Belgium
pmathieu@ulg.ac.be

AUGUSTIN J. MCEVOY
Laboratory for Photonics & Interfaces, Faculty of Basic Sciences, Ecole Polytechnique Fédérale de Lausanne, CH-1015 Lausanne, Switzerland
mcevoy@bluewin.ch

DIMITAR MLADENOV
President, Energy Utilisation Association, 33 Bakston Blv., Sofia, Bulgaria
d_mladenov@abv.bg

SIMEON N. OKA
Director of the National Energy Efficiency Program of the Ministry of Science of Republic of Serbia, Former Scientific Advisor of the Institute for Nuclear Sciences Vinča, Belgrade, Serbia, Gandijeva 49b/11, 11070 Novi Beograd, Serbia
okasn@afrodita.rcub.bg.ac.yu

KEN OKAZAKI
Department of Mechanical and Control Engineering, Tokyo Institute of Technology (Tokyo Tech) O-okayama, Meguro-ku, Tokyo 152-8552, Japan
okazakik@mech.titech.ac.jp

JOOP SCHOONMAN
Department DelftChemTech / Energy, Delft University of Technology, P.O. Box 5045, 2600 GA Delft, The Netherlands
j.schoonman@tudelft.nl

HARTMUT SPLIETHOFF
Chair for Energy Systems, Technische Universität München, Germany
spliethoff@tum.de

KRASIMIR STANCHEV
Executive Manager, ERATO Holding PLC, 10 Poruchik Nedelcho Bonchev Blv., Sofia, Bulgaria
k_stanchev@erato.bg

PETER STANKOV
Center for Research and Design in Human Comfort, Energy and Environment, CERDECEN, Technical University of Sofia, 8 Kliment Ohridski Blv., 1000 Sofia, Bulgaria
pstankov@tu-sofia.bg

KENJIRO SUZUKI[1]
Department of Machinery and Control Systems, Shibaura Institute of Technology, 307 Fukasaku, Saitama 337-8570, Japan

BART UMMELS
Delft University Wind Energy Research Institute DUWIND, Kluyverweg 1, 2629 HS Delft, The Netherlands
b.c.ummels@tudelft.nl

ROEL VAN DE KROL
Department DelftChemTech / Energy, Delft University of Technology, P.O. Box 5045, 2600 GA Delft, The Netherlands
R.vandeKrol@tudelft.nl

GIJS VAN KUIK
Delft University Wind Energy Research Institute DUWIND, Kluyverweg 1, 2629 HS Delft, The Netherlands
g.a.m.vankuik@tudelft.nl

HIDEO YOSHIDA
Department of Aeronautics and Astronautics, Kyoto University, Kyoto 606-8501, Japan
yoshida@mbox.kudpc.kyoto-u.ac.jp

[1] Deceased 25 April 2007.

Abbreviations and Acronyms

Common Abbreviations and Acronyms

ASU	Air Separation Unit
AZEP	Advanced Zero Emissions Power Plant
BWR	Boiling Water (nuclear) Reactor
BRICs	Large populous countries: Brazil, Russia, India and China
CCP	Combined Cycle Plant
CCS	(CO_2) Capture and Storage Technologies
CCT	Clean Coal Technology
CFB	Circulating Fluidised Bed
COE	Cost of Electricity Production
CPP	Condensation Power Plant
CCUJ	Center for Coal Utilisation, Japan
DOE	Department of Energy (USA)
EAA	European Atomic Agency
ENBIPRO	Energie-Billanz-Programme
EOR	Enhanced Oil Recovery
EPR	European Pressurised (nuclear) Reactor
ESBWR	European/Economic Simplified Boiling Water Reactor
ESOA	"Earth Surface, Oceans and Atmosphere" system
EUR	European Utility Requirements
FBC	Fluidised Bed Combustion
FBR	Fast Breeder (nuclear) Reactor
FC	Fuel Cell
FGD	Flue Gas Desulphurisation
FLOX	FLameless OXidation
GCV	Gross Calorific Value
GDP	Gross Domestic Product
GeoPP	Geothermal Power Plant
GTU	Gas-Turbine Unit

GHG	Green House Gases
HPP	Heat Power Plant
IEA	International Energy Agency
IGCC	Integrated Gasification Combined Cycle
IGFC	Integrated Coal Gasification Fuel Cell Combined Power Cycle
IPCC	Intergovernmental Panel on Climate Change
LHV	(fuel) Low Heat Value
LWR	Light Water (nuclear) Reactor
MCFC	Molten Carbonate Fuel Cell
MEA	Monoethanolamine
MGT	Micro Gas Turbine
NCV	Net Caloric Value (idem LHV, low heat value)
NEDO	New Energy and Industrial Technology Development Organisation
NEEP	National Energy Efficiency Programme (Serbia)
NGCC	Natural Gas Combined Cycle
NPP	Nuclear Power Plant
OECD	Organisation for Economic Co-operation and Development
OPEC	Organisation of Oil Exporting Countries
O&M	Operation and Maintenance
PC	Pulverised Coal
PEFC	Polymer Electrolite Fuel Cell
PR	Progress Ratio
PV-TRAC	Photovoltaic Technology Advisory Council (EU)
PWR	Pressurised Water (nuclear) Reactor
RAO UES	Russian Power Grid Joint Stock Company
RAS	Russian Academy of Sciences
R&D	Research and Development
R&D&D	Research, Development and Demonstration
ReNEP	Resolution on the National Energy Programme (Slovenia)
RES	Renewable Energy Sources
RTD	Research and Technological Development
SMW	Solid Municipal Wastes
SOFC	Solid Oxide Fuel Cell
STU	Steam-Turbine Unit
TIT	Turbine Inlet Temperature
TSO	Transmission System Operator
UNFCCC	UN Framework Convention on Climate Change
USAID	US Agency for International Development
WEC	World Energy Council
ZEP	Zero Emission Power Plant

Abbreviations for Measures

G	prefix for "giga", 10^9
goe (g.e.f)	gramme of oil equivalent (gramme of equivalent fuel)
kWh	kilowatt-hour
M	prefix for "mega", 10^6
MW	megawatt
MWe	power unit: megawatt in the form of electricity
MWth	power unit: megawatt in the form of heat
ppm	parts per million (mass)
ppmv	parts per million volume
T	prefix for "tera", 10^{12}
t	tonne
toe (t.e.f.)	tonne of oil equivalent (tonne of equivalent fuel)

Introduction

Kemal Hanjalić

Recent record increases in crude oil prices have once again brought the energy issue into the prime focus of world politics and public awareness. Developing sustainable and renewable energy sources is again seen as the top priority in many countries, especially those that are highly developed. Thus, it is not the alarming observation of melting polar ice caps or other incontestable evidences of climate change, nor Kyoto, post-Kyoto or similar protocols and declarations that compel us to search for new energy alternatives. It is rather the issues of national security, independence from imports, the immediate and near-future prospects for our economies and possible undesired social and societal implications, that provide the prime motivation for extensive research on new renewable as well as conventional energy sources, conversion and storage technologies. It should be no surprise that the issue of security of energy supplies and "absolute necessity" for new energy efficient technologies appeared as primary items on the agenda of the recent Annual Meeting of the World Economic Forum in Davos, Switzerland.

However, the problems and challenges in different parts of the globe are very different. And so are the motivations, public attitudes, priorities and official policies. The highly developed countries, the USA, EU, Japan, see the issues from a different perspective than the highly populated countries such as Brazil, China, India, Indonesia, and very different still from smaller developing "transitional" countries, such as those in South Eastern Europe. "Hydrogen has the future", "Hydrogen is the answer to energy question", claims the European Commissioner for Research. Wind, solar, tidal, etc., have the potential to satisfy all our needs, claim others. Yet, according to some (unconfirmed) evidence, there are currently over 850 large fossil-fuel plants under construction or in the design stage worldwide, most in China, India, Indonesia, Russia, South America, and also in the USA.

While every tonne of fossil fuel replaced by a renewable source should be welcome, and the research into and development of novel renewable and sustainable technologies should be stimulated even at a larger scale than hitherto (see e.g. the recent EU Energy Technology Plan), we should ask ourselves:

- Which of the proposed options in alternative and renewable energy technologies provides the fastest route towards meeting the challenges and fulfilling promises by providing efficient, safe, environmentally acceptable and reliable solutions?
- What are the feasible near-future options, in particular for specific countries and regions?
- The total installed capacity of renewable energy sources, apart from hydroelectric power and biomass, is small, despite extensive research and large investments. Will it remain small, and will these sources need to be subsidised for a long time to come?
- Is the role of renewable energy overestimated, especially in the increased liberalisation of the energy market? For example, which (combination of) renewable sources can currently replace the above mentioned 850 fossil-fuel plants? To put this into perspective, consider that a single plant with an average capacity of 500 MW is equivalent to 200–500 giant wind mills, or 1/10th of the worldwide total installed PV capacity.
- How real are the perspectives for the revival of nuclear energy? Is it time to reconsider existing bans and previously made decisions in many countries to phase out nuclear plants? How can we assure ourselves and the public that nuclear power can be safe?
- Or, is the solution for many countries to be found in so-called "clean fossil fuel" technology, which combines efficiency improvements with CO_2 sequestration?

The widespread consensus among experts is that the solution is not in one single energy source or technology, but that all options need to be pursued and explored. The 2007 Report on Energy and Climate Change by the World Energy Council (WEC) emphasizes that no single energy source can meet the energy needs of the world and its emission goals at the same time, nor can any single policy or measure provide the full solution. This statement can be projected to a single country or even a region: each country should develop its own strategy to climate change and to energy needs. The three "A" criteria of the WEC (Accessibility to affordable energy, environmental Acceptability of the energy sources, and reliable and secure Availability) can serve as guidance for assessing possible technological options and policy decision.

This volume presents an overview of recent developments in sustainable energy technologies (excluding vehicle propulsion) and provides a platform for answering some of the above-posed and other questions when projected to specific situations. It also aims at stimulating activities in selecting, researching and pursuing specific technologies that are feasible, economical and realisable with special emphasis on the needs and potential of countries in South Eastern Europe and their counterparts in other regions of the world. A panel of internationally renowned experts present recent research advances and prospects for different options. These are complemented with a series of short "regional" overviews of initiatives, trends and activities in some of the countries of South Eastern Europe.

*****

The material is organised in two parts. Part I begins with discourses of some global energy issues, such as the challenge of increasing global energy needs and consumption, the concept of sustainability, and potential improvement of efficiency in production and use of energy. Different novel conversion and storage technologies, their development status and perspectives of implementation are then presented. Part II contains overviews of current activities in some of the countries in South Eastern Europe. In addition to giving an insight into current initiatives, policies and actions, these chapters provide examples of possible measures towards short- and mid-term mitigation of energy shortage in small and medium-sized countries as a precursor for major undertakings.

The first three chapters deal with different topics, but their common denominator is in their global concern and significance. In Chapter 1, M. Combarnous and J.-F. Bonnet review the world energy situation, current status and future trends. A distinction is made between "explicit" energy sources and main energy "vectors", and their correlation with world population evolution, emission of green-house gases and climate change. Using the average individual energy consumption as a key parameter, different future energy scenarios are considered and possible solutions are discussed. Recognising the enormous complexity of the energy systems, their size and inertia to implementing technological novelties, and rigorous safety issues, the authors conclude that all feasible measures should be pursued. They especially emphasize energy saving, the development of all forms of energy production, developing technologies for carbon capture, and the need for a more balanced geographical distribution of resources.

Overall sustainable growth cannot be ensured if sustainability is not accounted for in all major planning exercises and undertakings that involve future use of energy, water and other natural resources. N.H. Afgan introduces in Chapter 2 the sustainability concept and measures of sustainability, defined in terms of a sustainability index. Examples of multi-criteria assessment of an energy system for several different scenarios illustrate the feasibility of the approach.

In Chapter 3, S.V. Alekseenko gives a Russian perspective on efficiency of production and use of energy, based on modern rationing technologies. Presented are novel developments in both conventional and renewable conversion and storage technologies, new types of energy sources, as well as some interesting innovative methods for efficient use of energy, such as powerful light sources based on a plasmatron. Modernisation of existing and introduction of new environmentally acceptable technologies based primarily on gas, coal and nuclear fuel, together with energy rationing, are seen as priorities, though admittedly all require immense investments. Some of the technologies discussed are specifically targeted towards the Russian "landscape", which is characterised by a harsh climate, huge space and distances, and widely dispersed consumers. Nevertheless, many of the issues, solutions and novelties discussed are ubiquitous and relevant to broader communities worldwide.

The next nine chapters deal with specific prospective technologies of energy conversion and storage, some already matured and ready for large-scale implementation, and some foreseen as feasible solutions in the mid-term period of the next two or three decades. Considered are wind, solar, nuclear and hydrogen energy, fuel cells and their hybrid systems, clean and zero-emission coal conversion and combustion technologies. In Chapter 4, G. van Kuik, B. Ummels and R. Hendriks present the state-of-the-art and perspectives of wind energy, emphasising its successful development and spreading over the last three decades. Issues considered include market and costs perspectives, technological development of wind turbines and wind power plants with a specific account of offshore applications, grid integration, local and global impacts and research agenda. The chapter closes with optimistic forecast on wind energy as a soon-to-become mainstream electricity source in a number of countries.

Chapter 5 presents a review of the development, technology and applications of photovoltaic cells and their role in a renewable energy scenario. Reiterating that solar radiation is the most abundant and the most fundamental sustainable energy source, and its direct conversion into electricity the most attractive and cleanest way of electricity production, A.J. McEvoy and M. Grätzel discuss various options of photovoltaic semiconductor materials, cell designs and their performances. Projected decreases in the production costs and current market trends (>30% growth per year) are expected to work in favour of PV. It is estimated that in 2030, PV could generate 4% of electricity worldwide.

R. van de Krol and J. Schoonman go a step further in Chapter 6 and consider the direct conversion of solar energy into hydrogen. The potential of using metal oxide semiconductors that can split water into oxygen and hydrogen is considered as a challenging alternative to photovoltaic cells, since hydrogen can be stored and used as a clean fuel for a variety of purposes, including vehicle propulsion. Interested readers can learn about the principles of operation, design concepts and their efficiencies, material requirements, main impediments and research challenges, as well as recent innovations based on nanotechnology that offer new promises in the development of this exciting new technology.

Another new concept, researched in Japan, is presented by K. Suzuki, H. Yoshida and H. Iwai in Chapter 7. Anticipating that massive production of hydrogen from renewable energy sources will not reach a wide commercial appeal anytime soon, distributed energy generation is considered using the solid-oxide fuel cell, which can be combined with a micro gas turbine. This hybrid concept, based on mature modern technologies, is claimed to match the efficiency of most advanced oil-fired large-scale combined-cycle electricity generation plants. Distributed energy systems of this kind are seen as very promising, providing a low-emission, interim solution especially in areas with high concentration of population, business and industry, although, admittedly, high cost and durability of fuel cells still pose a serious challenge.

The next two chapters deal with nuclear energy, the revival of which is viewed by many as the most effective interim solution for stopping the ominous climate change before it is too late. In Chapter 8, D.G. Cacuci reviews the evolution of "generations" I to IV of nuclear fission reactor designs. Based on thousands of years of cumulated operation, convincing arguments are provided that matured industrial technologies are available today for all stages of fuel cycles, including spent-fuel treatment and the conditioning and storage of waste. Yet, in order to receive a wider public acceptance and to establish itself as a viable long-term energy option, a number of technological and societal prerequisites are still to be fulfilled. Current international initiatives and activities towards meeting the acceptance criteria are reviewed, leading to the conclusion that novel generation IV nuclear fission reactors will fulfil the criteria for sustainable nuclear energy. In Chapter 9, M. Cumo provides further support to optimistic prospects of nuclear energy by presenting a new fission reactor concept called MARS (Multi-purpose Advanced Reactor, inherently Safe), developed at the University of Rome "La Sapienza". Based on a "passive" safety concept, the MARS reactor meets very high safety standards and thus should be suitable for electricity production, district heating, water desalination and other applications even in highly populated areas.

The last three chapters in Part I consider technologies for clean conversion of fossil fuels and related CO_2 mitigation and carbon-capturing methods. Recognising that fossil fuels will for long remain the dominant primary energy source in many countries, P. Mathieu discusses in Chapter 10 three possible options for capturing CO_2 during power generation: post-combustion removal of CO_2 from flue gases, pre-combustion carbon removal from fuel and oxy-fuel combustion systems. The impact of each technology on plant performances, costs and pollutant emissions is considered. It is argued that all three technologies are proven and mature though associated with high costs. Each has its own pros and cons, but their scale up to large power systems and market penetration depends on political will and substantial financial incentives.

In Chapter 11, K. Okazaki presents a new CO_2 coal utilisation technology researched in Japan, which is based on system integration of clean coal combustion, CO_2 recovery and sequestration, and hydrogen production from coal and its utilisation. It is argued that such combined systems could play an important role in suppressing CO_2 emissions in the interim period up to 2030–2050 when fossil fuels could be to a large degree eliminated by widespread use of renewable energy.

Part I closes with Chapter 12 in which H. Spliethoff also gives a brief overview of possible CO_2 separation methods, but focusing on coal-fired power plants. Oxy-fuel processes are then considered in more detail, followed by a feasibility analysis of steam generator designs suited for coal combustion with pure oxygen. Different methods for limiting the maximum temperature are discussed. Among several schemes, an innovative controlled fuel/oxygen staged combustion with rich/lean burners, researched at the Technical University of Munich, is seen as a promising technology for coal utilisation in power production.

Part II, entitled "Initiatives in South-East Europe", begins with Chapter 13 in which E. Kakaras, A. Doukelis, D. Giannakopoulos and A. Koumanakos present an analysis of possible retrofitting of domestic low-quality coal-fired power plants by oxy-fuel combustion and flue gas decarbonisation by amine scrubbing. The study, based on thermodynamic simulations, indicates the potential to reduce CO_2 emission by 70–85%, though at considerable energy penalties and associated reduction in plant efficiency. Nevertheless, the study shows that retrofitting could provide electricity at costs comparable to natural gas units.

Chapter 14 by U. Lavrenčič-Štangar and E. Kranjčevič gives a survey of renewable energy in Slovenia. Despite a relatively significant share of 11% in the country's total energy budget, it is recognised that the major contribution comes from the conventional hydro-plants and that major efforts are needed to utilise the relatively large potential of other renewable sources. This exemplary analysis of the current and potential participation of renewable energy closes with a short description of several new initiatives for promoting the exploitation of renewable energy sources.

The contribution from Bulgaria, presented by P. Stankov, D. Mladenov and K. Stanchev in Chapter 15, deals with energy wood production and use in the framework of the country's existing laws and regulations, the energy market and entrepreneurship. Considered is the present status of the wood processing industry, the structure of wood biomass and its potential for energy production This is followed by an outlook of the perspectives for further development of wood biomass industry and its increased role in providing a portion of Bulgaria's energy needs. Most countries in the area have a similar climate and orography with large percentages of the land covered by forest and agriculture. Thus, sufficient bio-resources for sustainable development of biomass energy should be available.

The last three chapters in Part II present overviews of the current status and plans for future developments in the energy sectors in Serbia, Macedonia and Bosnia and Herzegovina. Comprehensive information on the energy status and prospects in smaller and transitional countries is scarce in the open literature, in contrast to the detailed surveys available for the large, developed and populous states. Although the reviews presented here may not be exemplary, some features are common to a number of countries of a similar size, level of development and structure of the energy sector.

In Chapter 16, S. Oka gives a review of the available energy resources, present and projected future consumption, and the state of the current energy industry in Serbia, and identifies energy efficiency as the most critical issue. The problems are rooted in outdated technology, energy-intensive industry, inadequate management, inappropriate energy strategy and pricing policy, and the lack or neglect of standards and regulations. It is argued that apart from technology modernisation, which requires large investments, most other causes of inefficiency could be removed or diminished in a relatively short time. The National Energy Efficiency Programme offers plausible solutions that could significantly improve the country's energy situation, but is still awaiting political decision and, admittedly, a more favourable economic environment.

Because of modest primary energy resources in Macedonia, B. Donevski in Chapter 17 advocates the country's future economic developments to be based on low-energy-consuming industries that should bring significant reduction in energy consumption and demand. Another prospect is seen in expanding the grid connections with neighbouring countries and taking the advantage of the favourable geostrategic location of Macedonia to become a major energy hub in the region.

In contrast, A. Lekić in Chapter 18 considers that substantial primary energy resources in Bosnia and Herzegovina offer prospects for a profitable energy industry. Subject to political consolidation of the country, substantial foreign investments are expected both for modernisation of the existing and erection of new coal-fired and hydropower plants, which could make Bosnia and Herzegovina a significant energy exporter in the region.

Part I

New and Sustainable Energy Conversion and Storage Technologies

Chapter 1
World Thirst for Energy: How to Face the Challenge

Michel Combarnous and Jean-François Bonnet

Abstract We review the global energy situation from an anthropogenic point of view. What are the world's energy needs? How can they be satisfied in the mid-term as the world population continues to grow? What kind of balance can be promoted between various sources of energy? How can we progress gradually towards a necessarily limited system, the "Earth Surface, Oceans and Atmosphere" (ESOA) system? Can we define guidelines for energy planning at local, regional and global levels?

Key words: Biomass resources, energy conversion, energy saving, fossil fuels, precautionary principles, primary energy consumption, renewable sources of energy.

1.1 Introductory Remarks

Before we begin with the presentation, it is useful to make some general remarks and introduce some definitions that are related to energy problems on a global scale:

- We do not consider "implicit forms" of energy such as the solar energy required for plant growth, agriculture, forestry, breeding, or the natural drying of agricultural products. Even though these forms of energy are important, particularly for food production and nutrition [1, 2], they are not included in this discussion, which deals primarily with what we call "explicit forms" of energy.
- Regarding quantitative information in the field of energy, one must make the distinction between energy sources and energy vectors when accumulating data. The clearest case is electricity, which is primarily a vector for energy rather than an energy source. When integrating all possible sources of electricity, we must consider the various rates of efficiency of the conversion of original sources into electricity: hydraulic and nuclear conversion, and coal, fuel oil and gas combustion.

K. Hanjalić et al., Sustainable Energy Technologies: Options and Prospects, 3–24.

- Global annual energy consumption can be seen as a mean power use (energy use/time period). Two units of power are generally used, the watt and the tonne of oil equivalent (toe) per year (toe/yr). Without taking into account the various types of energy vectors or the question of output in energy production, the equivalence between those two units, based on the standard heat of combustion for oil, is relatively simple: 1 toe/yr equals approximately 1.4 kW (gross calorific value, GCV) or 1.33 kW (net calorific value, NCV, excluding water evaporation latent heat).
- Related to the above, the international equivalences can be stated thus as: oil products 1 t = 1 toe, coal 1 t = 0.667 toe, gas 1,000 m^3 = 0.900 toe NCV = 1 toe GCV, heat or electrical energy 1 MWh = 0.086 toe (NCV). In equivalent heat content, 1 toe = 10 Gcal · 42 GJ NCV · 44.6 GJ GCV. This last conversion factor relies on a weighted mean value of the following thermal GCV/NCV ratios: 1.11 for natural gas, 1.04 for coal, 1.05 for oil.
- In cases in which a balance is established between various sources before consolidating the data, equivalence rules must be chosen in good congruence with the problem under study. For instance, if we look at the contribution of gas to the production of electricity, one can either take the equivalence as mentioned before (1,000 m^3 = 0.900 toe NCV = 10.5 MWh), or include in the computation the efficiency of the conversion from gas to electricity (for instance through a gas turbine with 40% efficiency: 1,000 m^3 is equivalent to an electrical energy output of 0.4×0.9 toe = 4.2 MWh).

1.2 World Population Evolution

The growth rate of the world population is the first element to be considered when looking at present and future energy consumption. The population growth in recent decades can be classified as "exponential", implying that it cannot be controlled. Today we are clearly in what scientists in the field call a "demographic transition" [3, 4]. Because the death rate is currently decreasing more rapidly than the birth rate, the overall population is increasing in this period of transition.

Whereas the birth rate and death rate used to be equal and relatively high (more than 3% per year), a new situation has emerged in the last decades in which the population is likely to stabilise, but with relatively lower birth rates and death rates (approximately 2% per year).

The current situation corresponds roughly to the inflexion point of the "S" curve describing the evolution of the world population *versus* time.

The world population is currently growing at a rate of 200,000 inhabitants per day (the balance between births and deaths). As the world population is now around 6.5 billion (6.5×10^9 = 6.5 Giga-Inhabitants, mid-year 2005), it will soon be between 8 and 10 billion (8 to 10×10^9) [5].)

1.3 Explicit World Energy Consumption

This section gives an overview of the "explicit" sources (or vectors) of energy. These include primary energy vectors, i.e. fossil fuels (coal, oil, gas), nuclear and hydropower, as well as less concentrated and renewable forms of energy such as wind, sea and tide, solar (thermal and photovoltaic energies), geothermal and biomass (wood, other combustion products, biogas, bio-fuels, etc.).

The quality of the data regarding these two groups of energy differs significantly. The collection of data concerning the main energy vectors has been uniform and consistent for a relatively long time [6] in contrast with a lack of consistency and longevity in the collection of data for renewable energy forms. Despite this disparity we can conclude from available data that energy from renewable forms is comparatively stable and amounts to roughly 10 to 20% of total explicit consumption.

1.3.1 Overview: The Five Major Vectors

The five major vectors (coal, oil, gas, nuclear and hydropower) provide most of the energy consumed on the global scale (Figure 1.1). For the year 2005, the total is equivalent to the continuous production of about 15,000 "virtual plants" of 1,000 MW each (such a "virtual plant" differs from an electric power plant, which converts primary energy to electric energy with a low efficiency). The participation of major vectors is:

- Fossil fuels with 88% of total supply are the main source of energy, resulting from long-term storage of solar energy (oil 37%, coal 28% and gas 23%).
- Electricity from nuclear energy is equivalent to 6% of total supply and is the vector with the highest "exergetic" value. The equivalence used for primary sources of electricity is given in Figure 1.1.
- Electricity from hydropower, also equivalent to 6% of total supply, is a form of renewable solar energy (electricity again being the vector with the highest "exergetic" value).

Complementary observations about the future of fossil fuels are as follows:

- Of coal, oil and gas, coal has by far the largest estimated reserves. Including lignite, these reserves are estimated as equivalent to about 600 Gtoe [6, 7], which amounts to several centuries of production at the same level as today [7].
- World reserves of gas are estimated to be 160 Gtoe (180 Tm^3) [6], i.e. equivalent to more than half a century at a consumption rate equal to that of the years 2000 (ultimate reserves might in fact be three to four times larger). The use of gas has increased at a constant rate over recent years (Figure 1.2), due to its overall flexibility.
- The proven recoverable reserves of oil of about 190–250 Gtoe, including tar sands [6, 7], are estimated to last approximately the same period as the gas re-

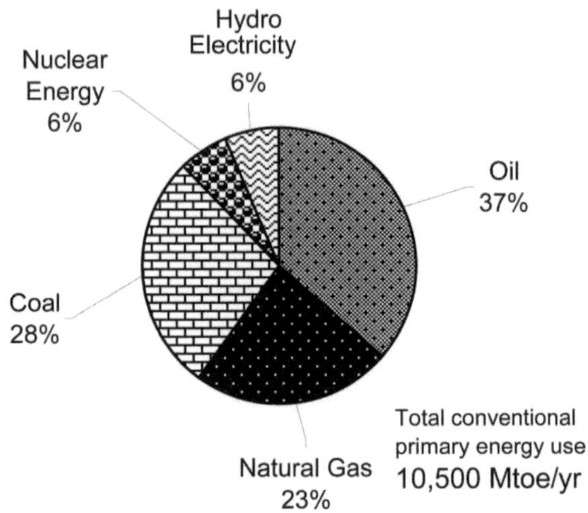

Fig. 1.1 World energy consumption (for the year 2005, corresponding to 6.5×10^9 inhabitants). Primary sources of electricity are converted into heat equivalent, with the assumption that the efficiency rate for the conversion of heat to electricity is 38% [6].

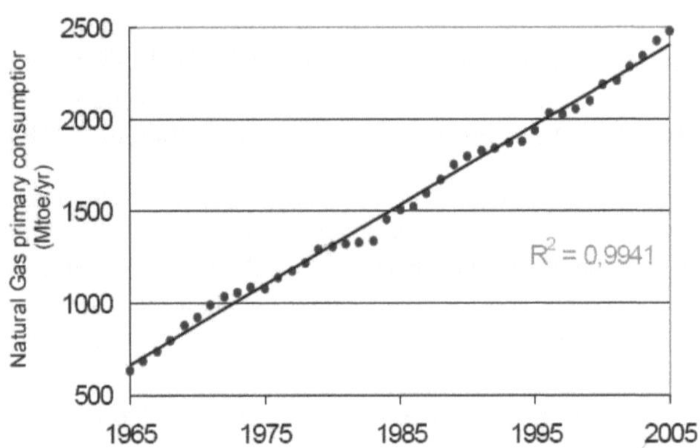

Fig. 1.2 Evolution of world gas consumption.

serves (at the current rate of consumption), i.e., about 40 to 50 years. Increasingly, oil will be used mainly for transportation. It must be stressed, however, that the development of enhanced oil recovery techniques plays, in the same way as the price by the barrel, a significant role in the estimation of recoverable quantities, which now include very heavy types of oil [8]. Each year they are re-evaluated as being higher than in previous years [9]. That is also the case for the so-called "peak oil" problem. Concerning the peak oil, the existence of which cannot be suspected, one can notice that it is partially postponed as long as reserves are re-evaluated.

Three further points must be highlighted regarding fossil fuels [12]:

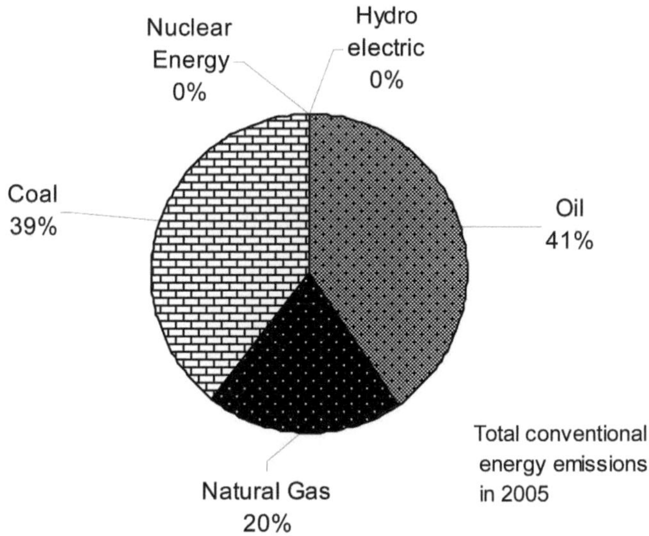

Fig. 1.3 Sources of annual 8 Gt of carbon emissions from fossil fuels (2005), or in CO_2 equivalent $(44/12) \times 8 \cong 30$ Gt/yr.

- Coal, oil and gas are the main contributors to the anthropogenic production of carbon dioxide (Figure 1.3). Wood and other products from contemporaneous biomass also produce CO_2 through their combustion, but this constitutes part of an exchange process in the "Earth Surface, oceans and Atmosphere" (ESOA) system and is therefore not considered as net production (i.e., biomass is replanted after harvest).
- The role of oil as a privileged source for fuels used in transportation is partly due to its suitability and convenience for propelling vehicles, associated with its liquid state and the very high volumetric energy density of oil fractions (Figure 1.3).
- It is not surprising then to observe the significant contribution of transportation to total carbon dioxide emissions, as well as among the diffuse sources of CO_2 (Figure 1.4).

1.3.2 Wood and Other Biomasses

Vegetal biomasses were the first sources of energy used by human beings. The various features of biomass must first be detailed, as we have briefly done with wood. Despite, or because of, the fact that they are very disparate in type and in use, biomass resources are often either forgotten or underestimated in world balances.

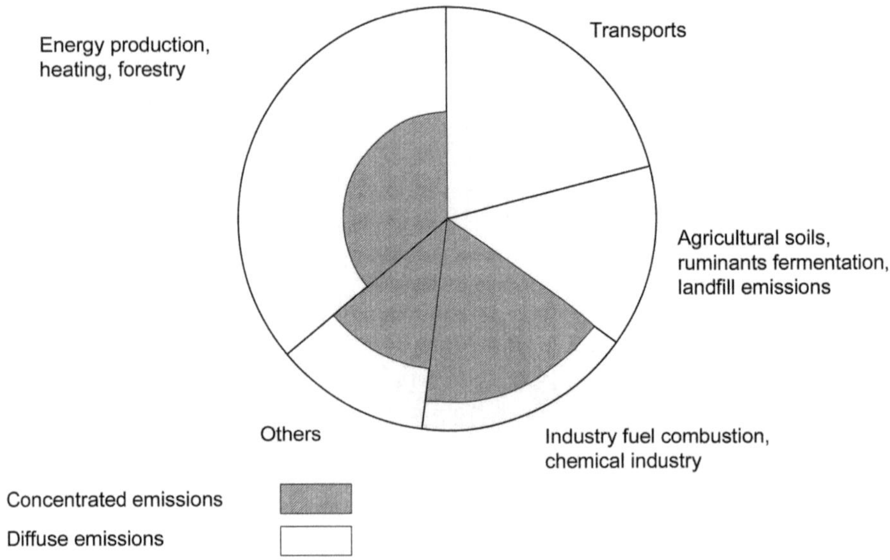

Fig. 1.4 CO_2 distribution between concentrated and diffuse sources [12].

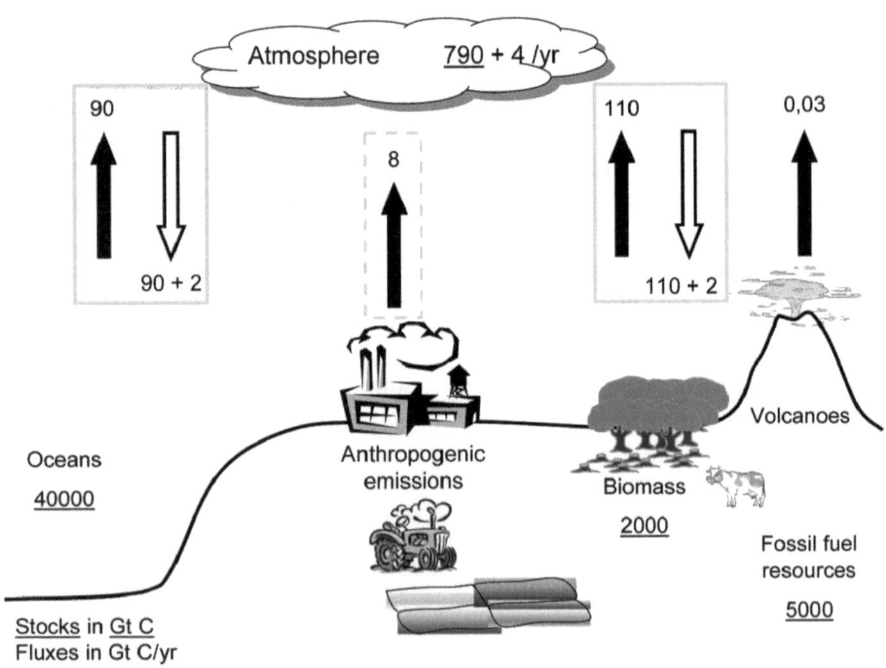

Fig. 1.5 The global carbon cycles: average fluxes and stocks (2005) (Gt C = gigatonnes of carbon, i.e., 10^9 tonnes of carbon).

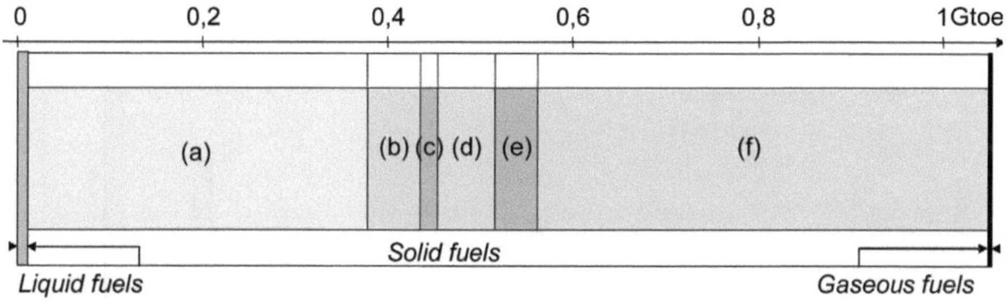

Fig. 1.6 Use of energy biomass in the year 2000: Solid – (a) wood, (b) charcoal, (c) wood residues, (d) black liquor, (e) bagasse and (f) other (animal waste etc.) – Liquid and Gaseous fuels.

Vegetal terrestrial biomass forms a significant stock of carbon worldwide (about 500 Gt C of the total 2000 Gt terrestrial biomass, [13] and Figure 1.5). Important fluxes of carbon are exchanged between vegetal biomass and the atmosphere (about 110 Gt C/yr) through implicit consumption of solar energy. Fluxes exchanged between the atmosphere and oceans are of similar magnitude (about 90 Gt C/yr). Compared to the fluxes exchanged between these components, the anthropogenic CO_2 emission flux is much lower, but unlike the other exchanges mentioned, it is not compensated by an inverse equivalent flux.

As an energy source, wood makes a significant contribution to the gross world energy supply, roughly in the same range as nuclear or hydropower. This contribution is not quantitatively as easy to determine as other major contributions. That is one of the reasons it is not always integrated in local, regional or global balances. In some developing countries, wood contributes up to 80% of the energy consumed for domestic use in rural areas, and from 40 to 80% (in more extreme cases) in the national balance between primary energy sources [14].

In Figure 1.6, all types of biomasses are presented roughly, with the special mention of liquid fuels (on the left) and gaseous fuels (on the right). Although they currently represent relatively small amounts of energy, they look very promising in the following cases: (a) biogas for heat and electricity production and (b) bio-fuels, which are a result of developing agricultural programmes throughout the world.

1.3.3 General Survey: A Large Variety of Other Sources

Other sources will now be briefly mentioned, some of which show potential for the future:

- Wind energy. From a historical perspective, just as the use of wood and other vegetal biomasses as sources of energy, wind was also used very early as a source of energy, most significantly through the development of windmills over the last centuries. After a decline at the beginning of the industrial era, which especially in Europe and North America was characterised by the development of massive

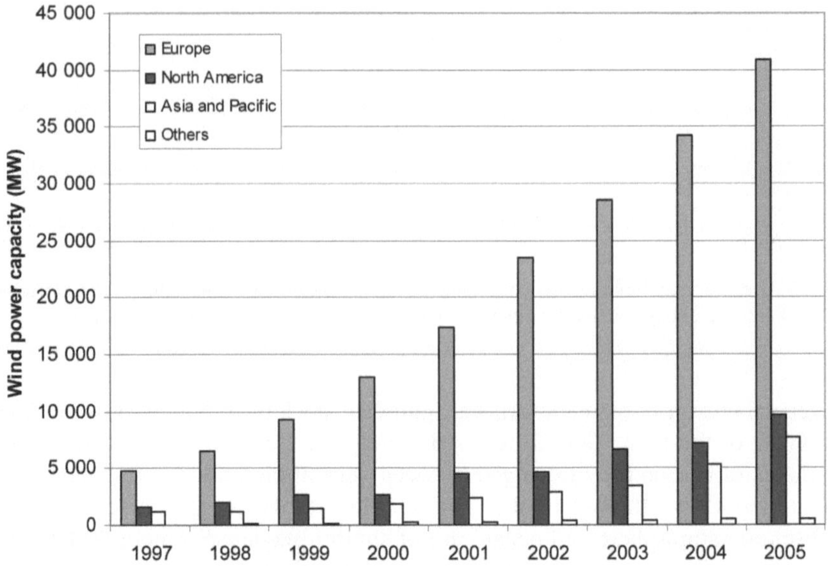

Fig. 1.7 Evolution of wind energy capacities in the last decade.

sources of energy, wind energy is now coming back into use, prompted by the combined effect of energy crises and progress in technology, especially in the material sciences. This has resulted in a new generation of windmills, the extension of which over the last years grew significantly, primarily in Europe (Figure 1.7).

- Direct solar energy. The importance of the heat that reaches the Earth's surface is universally recognised. Its value, 1.7×10^{17} W (170,000 TW), is about 10,000 times higher than the anthropogenic world energy consumption in the year 2005 (Figure 1.1). Direct thermal solar energy has been used since the early days of agriculture, particularly for drying products, either in diffused or more concentrated ways. It should be emphasised that the more the original source flux is dispersed, the harder it is to collect significant data on the processes involved.
- Solar energy for the production of electricity. This has been widely explored through photovoltaic cells of various types and through thermodynamic cycles (Stirling engines, thermo-chemical cycles), all of which are now gaining new attention.
- Hydroelectricity. Water represents a major contribution to the production of energy through the exploitation of potential elevation differences (i.e., differences in altitudes). Water and its kinetic energy offer possibilities which are now actively explored: currents, waves, various types of swells. Tidal energy is a special case, with a prototype 250 MW plant at la Rance, France, launched in 1966, which produced 16 TWh in the first 30 years.
- Geothermal energy. Although the average heat flux flowing through the terrestrial crust is very low compared to the heat flux coming from the sun, there are many sites where the heat contained in the upper layers of the crust can be efficiently

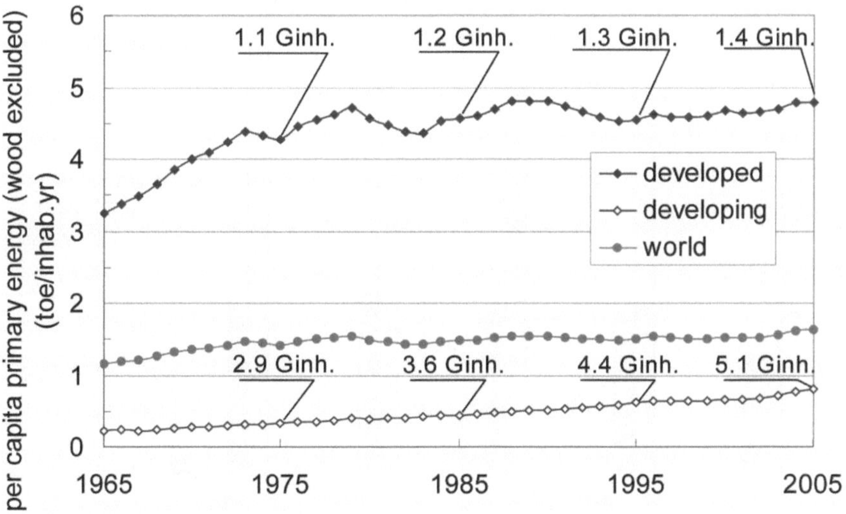

Fig. 1.8 World, developing countries and developed countries. Individual conventional energy consumption (data BP and FAO) [12].

recovered. Some of these projects are based on the injection of water into deep fractured "dry" rocks (e.g. granites), but most of these industrial plants have systems that use in situ water as the vector to transfer underground energy to the surface. Depending on the thermodynamic conditions on the surface, the energy produced is referred to as low, medium or high energy potential. "High energy" (surface temperature of more than 150°C) can be used for the production of electricity. Overall world production at the beginning of the 1990s was about 7,000 MWe (with the US first, then the Philippines, Mexico, Italy, Japan, etc.). "Medium energy" water (90–150°C) produced a world total of about 500 MWe. "Low energy" (25–90°C) is widely used in buildings (in China, the US, Iceland, France, Japan, etc.), with a total power generation of more than 10,000 MWth. The predictions for 2010 are roughly 15,000 MWe and 35,000 MWth.

With the exception of tidal and geothermal energy (the renewable character of which is questionable) all types of energy presented in this section originate from the impact of the sun on the "Earth Surface, Oceans and Atmosphere" system. These renewable energy sources can be considered as various forms of solar energy.

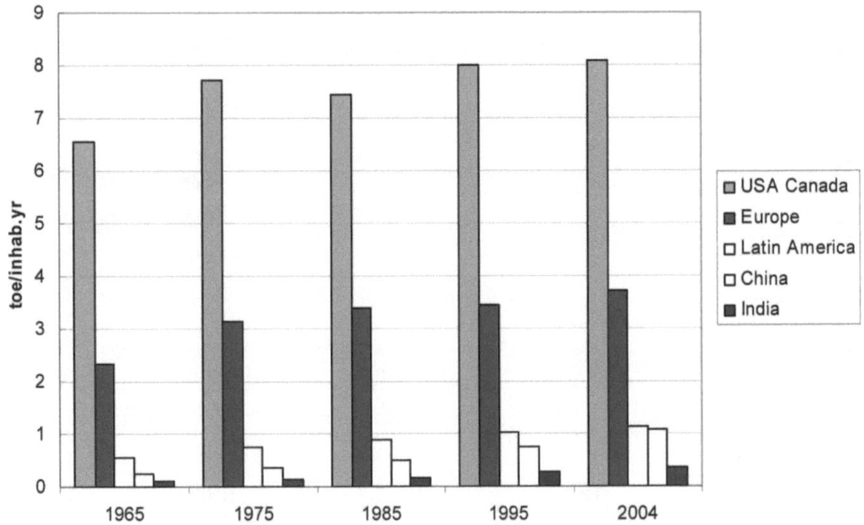

Fig. 1.9 Individual conventional energy consumption.

1.4 Short and Middle-Term Evolutions

1.4.1 The Approach via a "Key Parameter": Average Individual Consumption

Let us consider the average individual energy consumption, i.e. per inhabitant worldwide (Figure 1.8) [12]. Over the past 30 years, the world individual energy consumption, per inhabitant and per year, has been constant, with an average value of about 1.5 toe/(inh.yr) – another value, 1.6 toe/(inh.yr), which includes all forms of explicit energy sources, is also frequently considered. This surprising observation is the result of two contradictory effects, which counterbalance each other:

- Countries with a high individual consumption rate did not significantly modify their consumption during this period, but their relative weight, in terms of population, decreased over time: 1.1 Ginh in 1975 and 1.4 in 2005, compared to 2.9 Ginh and 5.1 Ginh respectively for developing countries.
- The individual energy consumption rate of less developed countries, the populations of which are relatively numerous, grew significantly.

These general observations of two groups of energy consumers should be complemented and re-evaluated by more precise observations taking into account the heterogeneous characteristics of each group. For instance, Figure 1.9 emphasises the high individual energy consumption in North America (more than 5 times the world average), while India and China have low individual energy consumptions, but significant increases over the last 40 years.

This relative stability of the world average can be used as a basis for a robust estimate of future short-term energy needs. As the population increases at a rate of 200,000 inhabitants per day, and as each inhabitant uses approximately 1.5 toe/yr $\approx$ 2 kW (NCV) · 2.12 kW (GCV), world energy consumption will grow at a *daily rate* of about +300,000 toe/yr (or +110 Mtoe/yr each year). In terms of equivalent energy production capacity, this leads to a need for a capacity increment of *about 400 MW added each day*. This analysis demonstrates the magnitude of the task ahead. Some conclusions can immediately be drawn from these observations.

1.4.2 Fossil Fuels, a Major Part of Future Energy Consumption

From the above overview, the following points must be considered:

- the need for a daily increase in energy production capacities,
- a strong inertia in the modification of the relative partition of various types of energy sources (Figure 1.10),
- the important role of carbonaceous fossil fuels (circa 75%), and
- the overall reserves (in all possible forms) of these fuels.

In this context, there is no doubt that the role of carbonaceous fossil fuels will be as important in future decades as it is now, even if (as is probable) it is necessary to implement considerable modifications in the production and use of carbonaceous products.

1.4.3 Some Immediate Consequences: Costs of Energy and Greenhouse Effect

Geopolitical facets of energy are not here considered. These factors are influenced by the Middle East's leading position regarding oil (with South America and Africa following close behind) and the leading position of Russia, the Middle East, and some other countries concerning gas. Regardless of these factors, as there is no doubt that the easiest way to satisfy the increase in the world energy demand is to produce more carbonaceous fuels, the dramatic rise in the price per barrel of oil in recent years is not surprising.

Despite the difficulties this situation creates, including the instability of prices, in terms of sustainable development these facts are not necessarily regrettable. Reserves of oil and gas are limited (albeit not negligible), therefore following the principle of economy, to significantly increase the price of hydrocarbons is an effective way to reduce demand, and to promote processes and techniques which enable reductions in the consumption.

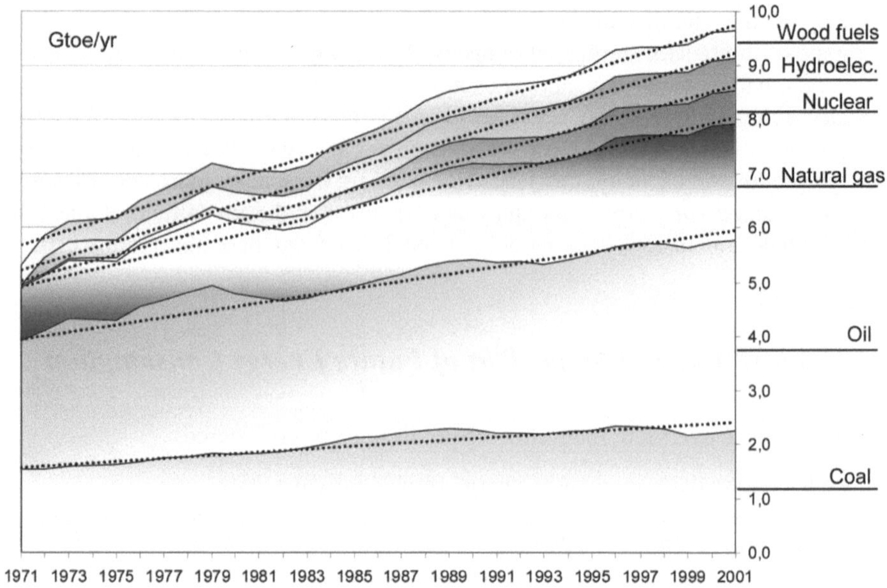

Fig. 1.10 Evolution of the world's main explicit consumptions.

At the same time, because anthropogenic effects have largely contributed to the spectacular increase in greenhouse gases (CO_2 production, CH_4 emissions, etc.), it is not possible to disregard the impact of these energy supplies on the environment.

The international scientific community working in this field (see, for instance, all the reports of the Intergovernmental Panel on Climate Change, IPCC) agrees that from the beginning of the last century there has been an increase both in the average surface temperature of the ESOA system and in the average concentration of carbon dioxide in the atmosphere.

Since the beginning of the industrial era, the average temperature of the Earth's surface has increased by about 0.6°C, with, depending on the various scenarios, a future additional increase of between +1.5°C and +6°C in the next hundred years. In the same period, the concentration of CO_2, a major contributor to the greenhouse effect, increased from 280 to 370 ppm in the year 2000, with possible increases of between 550 and 1,000 ppm up until 2100.

Cautionary attitudes therefore strongly advocate that we (i) resort as little as possible to carbonaceous products – which is consistent with the principle of economy because of limited reserves, at least for oil and gas – and (ii) develop all techniques available to reduce the emissions of carbon in the atmosphere – or at least promote methods for the capture and sequestration of carbon dioxide.

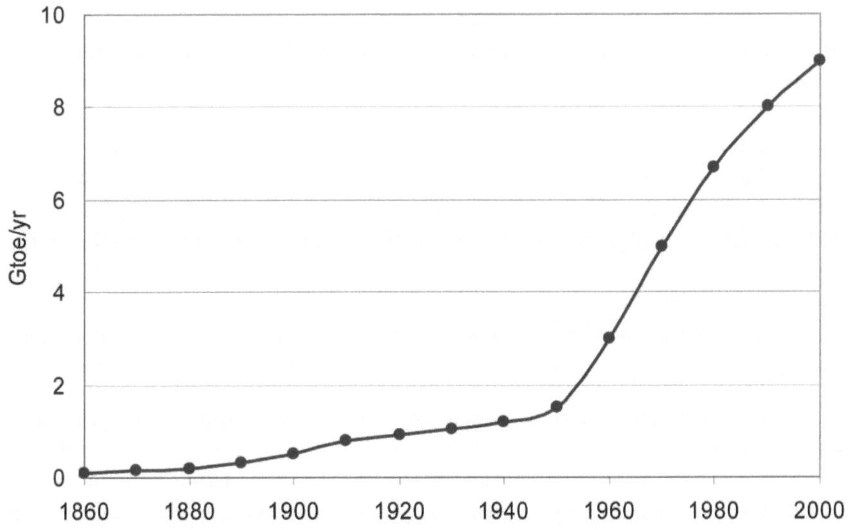

Fig. 1.11 World primary energy use (Gtoe/yr).

1.4.4 Energy for All: A System Which Can Be Controlled?

Before the oil crises of the 1970s, world energy consumption was growing faster than the Earth's population. The individual average consumption grew as technical progress spread globally.

Due to adjustments in the price of oil, followed by parallel evolutions in the price of all forms of energy, a phase out has been observed in certain important sector activities. For example, between 1973 and 1985:

- In Japan – in industry the energy cost of a tonne of steel dropped from 0.43 toe to 0.34, i.e., a reduction of about 20%,
- In the US – in transport the fuel consumption of cars, expressed in toe/yr, decreased from 2.25 to 1.69, i.e., 25% (due to changes in habits and the reduction in vehicle mass),
- in France – in housing, the consumption per family was reduced from 1.96 toe/yr to 1.58, i.e., a 20% reduction.

Since 1975 world energy use has grown at the same rate as that of the population (Figure 1.11). But there are huge differences between various countries and communities.

A simple way of measuring the necessity and urgency of substantial modifications in our behaviour and habits can be derived from observations based on Figures 1.8 and 1.9:

- If everybody on our planet were to live an "energetic way of life" analogous to that of North America (Figure 1.9) the energy consumption of the world would multiply by more than 5. That means that with the same relative percentage of hydrocarbons in the energy sources as now, there would be an increase of anthropogenic emissions of greenhouse gases by a factor of 5, resulting in a huge modification in all the cycles between the various components of the ESOA system (Figure 1.5).

- If the objective is the European standard, then the world increase would correspond to a multiplying factor of about 2.5, which is still excessive.
- A less ambitious objective of raising every person to the level of the present average energy individual consumption of about 1.5 toe/(inh.yr) would imply an increase in the production of energy of about 30%.
- All these observations are being made without taking into account the near-future increases in the world population, which will range between 20 and 50%.

We do not discuss here the important work which has been done in formulating and reaching international agreements to stabilise and even reduce the anthropogenic production of greenhouse gases. It is common knowledge that, in spite of the difficulties, some international protocols have already been agreed and, in some situations and/or countries, implemented (see, for example [15]).

So we urgently need to increasingly control our demands for energy. The time has passed when we could talk about solar energy *versus* hydrocarbons, hydrocarbons *versus* nuclear energy, fossil fuels *versus* renewable facilities. We need now to use every possibility and on every scale; individual, local, regional, national, and global. Now is the time for us to put everything to account.

A variety of possible actions can be considered in relation to the following three dimensions – which are involved in any project, design, decision or realisation: (i) energy savings, (ii) systemic approaches and innovations, and (iii) technical advances [16]. The development of a completely new concept cannot be excluded, even though inertial factors are prevalent throughout the field of energy because of the size of investments and devices. In this section the above three dimensions are discussed along with some examples, followed by some comments on the capture and sequestration of carbon dioxide, illustrated by a typical example, the case of forest areas.

1.4.5 Energy Savings

The first dimension to highlight is energy saving, including all possible methods. It is often said, particularly in developed countries, that the “first source of energy is energy saving”. This elementary concept should be a key feature of any energy policy, and at all decision levels.

This approach (often stimulated by price increases) should be promoted throughout all structures of society, including in the family and at all different levels in schools and educational systems. Public campaigns and various projects aimed at increasing awareness of energy issues among the broad public and various professional communities, explaining the main challenges and what can be done to save energy and reduce the greenhouse effect, should be widely promoted [17].

As for practical recommendations, the International Energy Agency has pointed out [18] that there is a considerable lack of information in households and small businesses. They often do not care much about electricity costs, and when they do, they do not have accurate information on how to make savings.

Permanent increases in fuel prices and increasing worries about greenhouse effects will undoubtedly influence modes of transportation. Where several options exist, for instance in land transport in Europe, there is no doubt that there will be an evolution in the relative roles of trains *versus* trucks, or railways *versus* roads and/or highways.

1.4.6 Systemic Approaches and Innovations

In the 1970s, when people began to wake up to the necessity of saving energy, there were many initiatives regarding main energy vectors and sources, including many different types of devices and processes:

- There was a tendency, especially in Europe where energy supply dependence from abroad was high, to promote low-exergy heat sources (geothermal sources for instance). Numerous systemic approaches were promoted for heating houses, with the aim of enhancing the value of every calorie, even at a relatively low temperature level.
- The above technique was extensively implemented combined with a number of improvements in insulation techniques and materials, particularly in residential and tertiary buildings. The term "positive energy buildings" refers to highly insulated buildings with partial recovery of heat from air exhaust and local production of energy, including passive solar supply.
- In industrial areas and on large residential sites, combined heat and power generation was widely implemented.
- One of the numerous examples where the necessity of reducing the consumption and costs of energy was promoted is a new, combined cogeneration plant developed in Dunkerque (France) by Gaz de France and Arcelor. This project, called DK6, first started producing electricity in 2005. It is a high efficiency 790 MWe power-station which combines two gas turbines (165 MWe each), two boilers with post-combustion and two steam turbines (230 MWe each). This plant is fuelled by natural gas and gases produced at the adjoining iron factories of Arcelor.

Despite the numerous initiatives, according to the International Energy Agency [18], "current levels of energy R&D investment are unlikely to be adequate given the magnitude of the climate challenge". Energy R&D investments in IEA member countries peaked in 1980 and declined substantially thereafter (about half being in nuclear fission). Investment in renewables has been lower in recent years than in the period from 1977 to 1986. This dramatic decrease in energy investment and urgent needs to combat the climate change trends stress the necessity for greater and sustained commitment to technology demonstration, to further development of underlying basic sciences, and to promotion of the fastest market uptake of new technologies.

1.4.7 Technical Advances

Of the various attempts which have been made in this area over the past few decades, many have resulted in real technical advances. Those presented below are mentioned primarily because of their key importance in relatively broad areas of application, while not necessarily being the most significant:

- Improvement in materials used in rotating machines has already been noted in relation to the development of wind energy. Because of this progress, the relative mass of the rotating parts in gas turbines has been reduced and/or the rotation speeds increased. This has resulted in a more than 10% increase in efficiency in the last 15 years.
- With a world average of more than 1 car per 10 inhabitants, and with more than 1/5 of oil resources devoted to transport by car, the industry has, not surprisingly, made tremendous efforts to reduce the nominal consumption of new cars. All the "down-sizing" efforts made on car engines are directed towards this goal.
- Efforts towards the same objective mean that the ongoing development of electricity-powered vehicles, which are more efficient than the current models, will be preceded by hybrid types of motorisation with combined thermal engines and electric motors.

1.4.8 The Capture and Sequestration of Carbon Dioxide

The relative amount of carbon fossil fuel in overall world energy consumption will probably not decrease in the short and middle term because of the high increase in energy demand and the low flexibility of nuclear and renewable energies. Therefore, in parallel with improvements in the production of energy, all methods of carbon sequestration must be investigated, in particular biological sequestration.

Among the many possible technologies of carbon dioxide sequestration, either concentrated or diffuse, ocean sequestration is today still questionable because little is known about its long-term effects on the ocean. Two other approaches are usually considered: the first concerns concentrated carbon dioxide storage (underground storage, etc.), suitable for concentrated sources (power plants, industries, etc.) and the second concerns diffuse storage (biomass), which appears as a way of storing diffuse emissions (residential heating, transports, and so on).

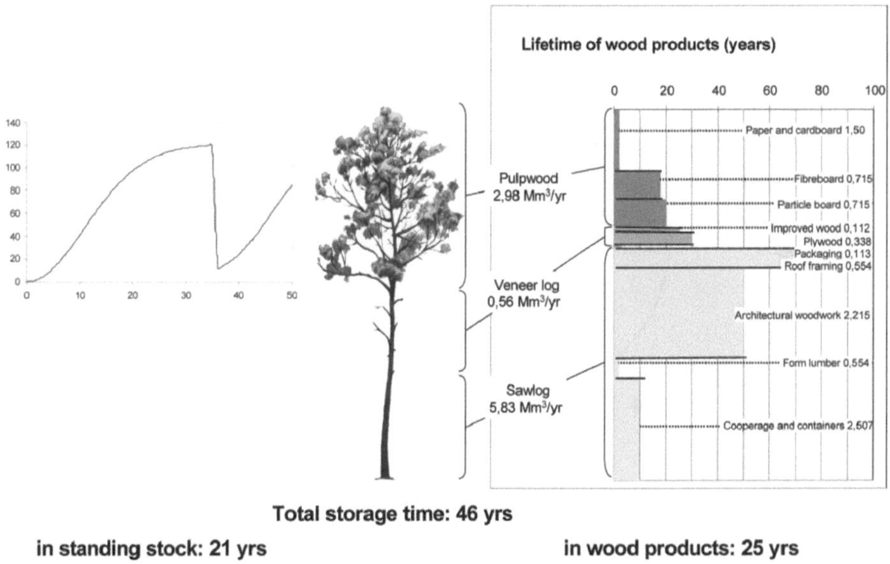

Fig. 1.12 Average storage time of a carbon atom (maritime pine forest of the Landes d'Aquitaine, France, 13,600 km^2 [22].

1.4.9 An Example of an Integrated Approach: The Case of Forest Areas

We examine an example of an integrated approach in a system which partially satisfies our goals: the production of energy and the reduction of greenhouse effects [19–21]. Forest ecosystems can play an important role in storing diffuse carbon, as well as in producing wood energy and other materials. These two aspects are often studied separately.

Vegetal biomass forms a significant stock of carbon on a global scale (about 500 Gt C [13] for an overall terrestrial biomass of 2,000 Gt C). Important fluxes of carbon are exchanged between vegetal biomass and the atmosphere (about 110 Gt C/yr, Figure 1.5) through implicit consumption of solar energy. Fluxes exchanged between the atmosphere and oceans are of a similar magnitude. Compared to the fluxes exchanged between those two components, the anthropogenic CO_2 emission flux is much lower, but unlike the other exchanges mentioned, it is not compensated by an inverse equivalent flux.

In order to expose the capability of forests and various products to store significant amounts of carbon, several analyses have been developed. We consider two studies:

- The first one consisted of a real case study [22] of the actual average time of storage for one C atom. The case studied is the one of the largest forests in Europe. Starting from concrete data concerning the growth of trees (maritime pines, *pinus*

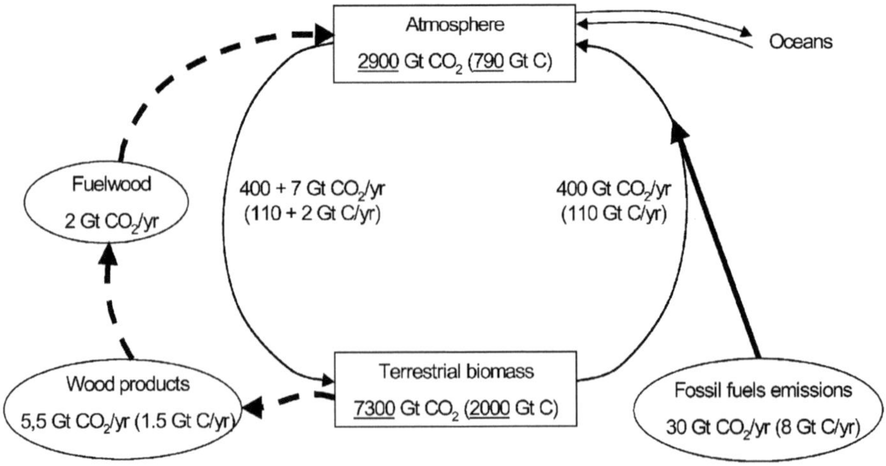

Fig. 1.13 Carbon cycle in three components of the ESOA system, for 2005 (the whole ESOA system includes the oceans, see Figure 1.5).

pinaster, which are usually cut at the age of 35 years, left part of Figure 1.12) and the average lifetimes of all the derived products of the forest (right part of Figure 1.12), an average total storage time of 46 years was evaluated. Although it is not very high, this value calls for some attention.

- The second study was an evaluation of the potential of forest areas all over the world (about 1/3 of emerging lands is covered by forests) in terms of annual carbon storage, compared to the present anthropogenic production of carbon dioxide. Significant performances can be achieved by modifying existing forest management and increasing the potential by planting new forests [14].

Wood, like any other product from the forest, therefore, has a double function: (i) to capture carbon and store it and (ii) to work as a substitute for fossil fuel energy, either directly or after being used as a material in various human activities [23–25]. Therefore forest ecosystems are particularly well suited to catch and store the carbon diffused in the atmosphere as well to stabilise the use of fossil fuels through extended uses of wood and other products.

Figure 1.13 summarises the potential of using forest resources in the attempt to control more efficiently the carbon cycles between the components of the "Earth Surface, Oceans and Atmosphere" system. This is possible, and only partially, provided materials from forests are used in the most efficient way – as fuels, as well as materials [26] – and also only if any substitution of fossil fuel by forest product automatically involves a reduction of fossil fuel consumptions.

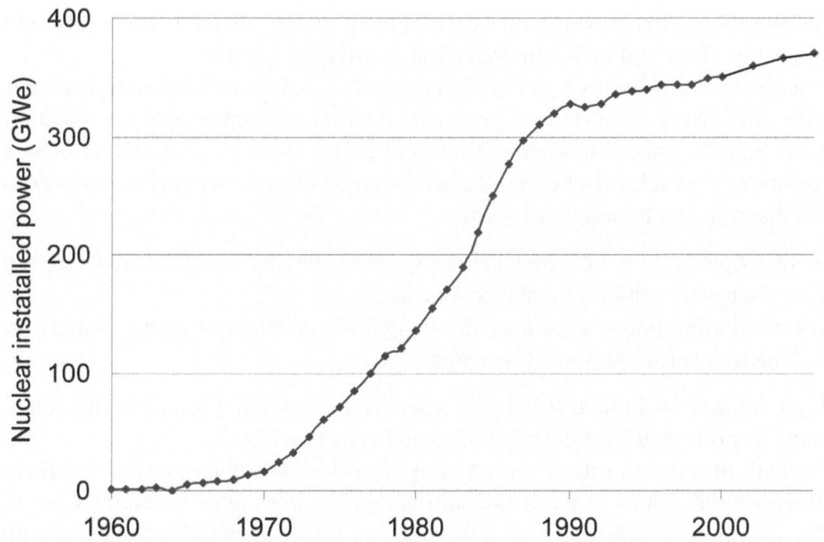

Fig. 1.14 Evolution of nuclear electricity capacity since its origin.

1.5 Closing Remarks

Before concluding this chapter, the following additional points should be noted:

- The examination of energy data is a particularly challenging task due to the diversity of the data sources. Because the data are gathered from very different geographical areas, there is significant risk of deficiencies and inaccuracies. This applies also to the data presented in this chapter.
- When working in the field of energy, one has to bear in mind that we are dealing with complex systems of very large sizes and every major undertaking involves risky decisions and requires substantial means in investment and major human engagement. As a consequence, the development of related technologies is relatively slow. Rigorous safety regulations add to these constraints, compounding the urgent problem of solving our energy needs. This applies to the development of civil nuclear energy (Figure 1.14) and will also apply to fission as a means to produce electricity, even if its technological and industrial development is eventually successful.
- Many sources and vectors of energy have been omitted from this presentation. This is due not so much to the fact that a comprehensive and exhaustive description of so wide an area is not possible in the present chapter, but rather because the attempts at developing and promoting those technologies have been unsuccessful. Sea energy – i.e., how to use the temperature differences in the various layers of seas and oceans depending on their depths – is one example. Other examples, such as hydrogen as a vector of energy, have also been omitted because,

despite their current worldwide development and/or progress, their substantial impact on short and mid-term evolution is still questionable.

- Finally, because energy is often linked to other major elements in human activities, the energy solution to a practical situation is often strongly connected to other equally important commodities and issues. Such is the problem of water resources – which is linked to climate change issues – and projects concerning the disposal and treatment of waste.

One other significant point which has neither been detailed nor discussed here is the human, energy and environmental cost of wars.

As concluding points regarding the worldwide problem of energy supply, two complementary principles must be emphasised:

- First, because of limited fossil fuel reserves, the rules and principle of economy must be promoted, particularly for oil and gas resources.
- Second, in order to satisfy the pressing need for anthropogenic carbon dioxide emission reductions, a set of precautionary principles must be established with the intention of reducing CO_2 emissions as much as possible, and promoting techniques for its capture and storage.

1.6 Conclusion

Despite significant inertia effects in the endeavours to redistribute relative participation between energy sources and vectors, the following elements of any energy planning programme are essential [12]: on the local and operational scale a variety of activities must be promoted – energy savings, the development of all forms of energy production, the development of systems for carbon capture and storage. The latter is particularly important in relation to the impact of energy production industry on the behaviour of the ESOA system. Clearly, all of these points must be included in a broader "vision of the world", aiming for a more balanced, homogeneous geographical distribution of resources.

Acknowledgements

This work emerged from the research programme realised by TREFLE Laboratory ("TRansferts Ecoulements Fluides Energétique") in cooperation with several research institutions: Bordeaux 1 University, the Ecole Nationale Supérieure d'Arts et Métiers (Centre de Bordeaux-Talence), the Ecole Nationale Supérieure de Physique et Chimie de Bordeaux and the CNRS (Centre National de la Recherche Scientifique – UMR 8508). Some of the results presented, especially those on forests and wood uses, are from the doctoral thesis of Anne Prieur [22] in the laboratory TREFLE, with the financial support of Gaz de France and ANRT (Association Nationale de le Recherche Technique). Anne Prieur is now working for the IFP (Institut Français du Pétrole).

References

1. Bonnet, J.F. and Combarnous, M., Du rayonnement solaire à la production de biomasse végétale terrestre: Une vision schématique des flux d'énergie, *Revue Générale de Thermique* **35**, 1996, 527–542.
2. Bonnet, J.F. and Combarnous, M., Conversion du rayonnement solaire dans la production de biomasse végétale et animale, *Entropie* **233**, 2001, 3–11.
3. Lutz, W., Sanderson, W. and Scherbov, S., The end of the world population growth, *Nature* **412**, 2001, 543–545.
4. Léridon, H., Certitudes et incertitudes de la prévision démographique, *C.R. Geoscience* **335**, 2003, 627–635.
5. Web Site FAO, http://www.fao.org.
6. BP, BP Statistical Review of World Energy, June 2007 (www.bp.com).
7. World Energy Council, *WEC 2004 Survey of Energy Resources*, Elsevier, London, 2004, 446 pp.
8. Lanier, D., Heavy oil – A major energy source for the 21st century, in *Proceedings of 7th Unitar Heavy Crude – Tar Sands International Confernce*, Beijing, 27–30 October 1998, Paper No. 1998.039.
9. Alazard, N. and Montadert, L., Ressources pétrolières pour le XXIème siècle: Quel avenir?, *Revue de l'IFP* **48**, 1993, 69–82.
10. Alpern, B. and Lemos de Sousa, M.J., Documented international enquiry on solid sedimentary fossil fuels; Coal: Definitions, classifications, reserves-resources, and energy potentiel, *Coal Geology* **50**, 2002, 3–41.
11. McDonald, G.J., The future of methane as an energy source, *Annu. Rev. Energy* **15**, 1990, 53–83.
12. Combarnous, M. and Prieur, A., Les consommations d'énergie dans le monde: Une méthode robuste de prévision de leur évolution à moyen terme; premières conséquences, *C.R. Geoscience* **335**, 2003, 943–950.
13. Batjes, N.H., Management options for reducing CO_2 concentrations in the atmosphere by increasing carbon sequestration in the soil, Report 410-200-031, International Soil Reference and Information Centre, Wageningen, The Netherlands, 1999, 114 pp.
14. Prieur, A., Bonnet, J.F. and Combarnous, M., Les surfaces boisées à l'échelle de la planète: Usages conjoints pour la séquestration du carbone et la production d'énergie, *C.R. Geoscience* **336**, 2004, 1323–1335.
15. Dechamps, P. and Pilavachi, P.A., Research and development actions to reduce CO_2 emissions within the European Union, *Oil & Gas Science and Technology – Revue de l'IFP* **59**(3), 2004, 323–330.
16. Martin, J.M., Energy technologies: Systemic aspects, technological trajectories and institutional frameworks, *Technological Forecasting and Social Change* **53**, 1996, 81–95.
17. Jankovici, J.M., *L'avenir climatique – Quel temps ferons-nous?*, Science Ouverte, Seuil, Paris, 2002.
18. Philibert, C., Technology innovation, development and diffusion, OECD and IEA Information Paper, 2003.
19. Cooper, C.F., Carbon storage in managed forests, *Canadian Journal of Forest Research* **13**, 1983, 155–166.
20. Wigley, T.M.L. and Schimel, D.S., *The Carbon Cycle*, Cambridge University Press, 2000.
21. Nandey, D.N., Carbon sequestration in agroforestry systems, *Climate Policy* **2**, 2002, 367–377.
22. Prieur, A., Les ressources forestières: Produits du bois, usages énergétiques, capture et stockage du carbone, Ph.D. Thesis, University Bordeaux 1, 2004, 183 pp.

23. Prieur, A. and Rauch, E., Carbon sequestration in forest ecosystems and gas industry: Sustainable development and clean energy production from fossil fuel, in *Proceedings 22nd World Gas Conference*, Tokyo, 1–5 June 2003.
24. Prieur, A. and Combarnous, M., Forêts et cycles de vie: Aspects énergétiques, produits, carbone stocké, in *Proceedings 12ème Congrès Forestier Mondial*, Montréal, 21–28 September 2003.
25. Richardson, J., Bjorheden, R., Hakkila, P., Lowe, A.T. and Smith, C.T., *Bioenergy from Sustainable Forestry: Guiding Principles and Practice*, Kluwer Academic Publishers, 2002, 344 pp.
26. Bernard, C., Rabot-Querci, M.L., Rogaume, Y., Rogaume, C. and Zoulalian, A., Optimiser la combustion pour un développement durable du bois-énergie, *Oil & Gas Science and Technology – Revue de l'IFP* **61**(2), 2006, 203–211.
27. WEC, *2004 Survey of World Energy Resources*, Elsevier, London, 2004, 446 pp.

Chapter 2
Sustainability Concept for Energy, Water and Environment Systems

Naim Hamdia Afgan

Abstract. Sustainability has become an unavoidable issue in all major planning and undertakings that involve future use of energy, water and any other natural resources. In order to ensure sustainable growth of our society, we must satisfy the sustainability criteria and meet the constraints imposed by the finiteness of all natural resources and the dynamics of their natural renewal. But how do we quantify and measure sustainability and how do we ensure that the sustainability requirements are fulfilled? We consider the sustainability concept for energy, water and environmental systems and its interaction with current major global trends: globalisation, democratisation and decentralisation. In the assessment of global energy and water resources we consider the current resource consumption and possible scenarios for meeting future demands. We then move to the definition of the sustainability concept. In order to introduce a measure of sustainability, we focus on possible definitions of respective criteria with specific application to energy system design. It is argued that multi-criteria sustainability measurements of options for an energy system must be based on four sets of indicators: technological efficiency, economic, environmental and social indicators.

Key words: Sustainability, energy, water and environment system, resources, limits, sustainability measures.

2.1 Introduction

Our civilisation has developed under economic, social and ecological limitations. The industrial revolution brought with it the recognition that in order to improve the quality of life there must be coordination of development among different commodities. This has resulted in economic and social benefits based on the natural capital available for technological development.

K. Hanjalić et al., Sustainable Energy Technologies: Options and Prospects, 25–49.

Throughout the history of human society, the cyclic development of human structures has affected changes in the pattern of social structures. Critical states at specific periods of time created new complexities in human society. The industrial revolution added a new complexity by introducing new commodities. As we near the end of the industrial revolution, it has become evident that complexity indicators such as population, economics, material resources, social structure and religious devotion have reached another critical state.

A number of scholars have highlighted the unique state of our current civilisation. Attention has been focused on indicators related to material resources and the environment. In our lifetime there have been many attempts to emphasise different aspects of the use of material resources. Some of those are based on ethical and religious principles regarding our responsibilities to humanity and to the divine. Warning signs have been given that we are nearing certain limits after which the changes will be irreparable. The first and second energy crises have shown our vulnerability. Recent claims have been made that the concentration of CO_2 in the Earth atmosphere is reaching a limit that may trigger irreversible changes in the environment with catastrophic consequences for life on our planet. Sustainable development encompasses economic, social, and ecological aspects of conservation and change. The essence of sustainability is the need to preserve the natural cycles of renewal, i.e., the balance between consumption and natural resource recovery. Or, as postulated by the World Commission on Environment and Development (WCED), "to satisfy our needs without compromising the ability of future generations to meet their own needs". This definition is based on the ethical imperative of equity within and between generations. Moreover, apart from meeting the basic needs of all, sustainable development implies sustaining the natural life-support systems on Earth, and extending to all the opportunity to satisfy the aspirations for a better life. Hence, sustainable development is more precisely defined as "a process of change in which the exploitation of resources, the direction of investments, the orientation of technological development, and institutional change are all in harmony and enhance both current and future potential to meet human needs and aspirations".

The first issue is concerned with the objectives of sustainable development; that is, "what should be sustained" and "what kind of development is preferred". These are normative questions that involve value judgements about society's objectives with respect to social, economic, and ecological system goals. These value judgements are conveniently expressed in terms of a social welfare function, which allows an evaluation of trade-offs among the different system goals.

The second issue deals with the positive aspect of sustainable development; that is, the feasibility problem of "what can be sustained" and "what kind of system we can devise". It requires one to understand how the different systems interact and evolve, and how they could be managed. Formally, this can be represented in a dynamic model by a set of differential equations and additional constraints. The entire set of feasible combinations of social, economic and ecological states describes the temporal transformation space of the economy in the broadest sense.

Complexity is the property that describes the state of a complex system [1, 2]. It is a multi-criteria indicator that comprises all the individual characteristics of the system. A complex system is an entity that characterises a structure with a large number of interacting elements. Elements have different structures. Elements in biology are structured to perform specific functions. A typical example is the DNA molecule, comprising a large number of elements interacting among themselves. In information theory the structure of elements is described as an internet network with a large number of nodes for information exchange. In the energy system we can describe a complex system as a system that produces, transports and utilises different energy sources. The complexity of these systems is the internal property of the system expressed as the wholeness property. This implies that the complexity describes the essential characteristic of the system.

If the complexity is described in thermodynamic terms, it represents the internal parameter of the system expressed by agglomerated indicators describing the specific property of the system. If we take into consideration only the material system, we can take the entropy of the system as the macroscopic property of the system. These can be applied to chemically bounded molecules. Prigogine [3, 4] has determined the characteristic property of these systems as entropy generation. This means that every interaction between elements accompanied by mass, momentum and energy exchange is ultimately connected and contributes to the entropy generation in the system. It should be taken into consideration that the entropy generation is defined per unit mass of the system and represents a specific property of the system. So the entropy generation represents the complexity property of the system.

If we take into consideration a non-material system where complex properties include entities which are not defined per unit mass of the system, we have to introduce a notion which represents the wholeness of the system. A good example of this type of complex system is the Internet system. Large numbers of nodes are connected in a large net which transfers information among the nodes. If we assume that transfer of information contributes to the increase of the informativity of the system, it follows that the increase of informativity is equivalent to the increase of the complexity of the system. In this respect the informativity is equivalent to complexity.

In order to comprehend the full complexity of sustainability and to develop a rational sustainability concept, we need to draw on up-to-date scientific knowledge from different fields. For that purpose, it is imperative that we not only inform, but that we alarm, the scientific community at large, that we focus their attention on sustainability and urge their contribution of relevant knowledge to synthesising a comprehensive notion of sustainable science that will aim at better understanding of the future of our planet [5].

2.2 Sustainability and Global Processes

In the complexity definition of the sustainability concept there are three clusters of indicators which are used to describe the state of the global system: the resource, environmental and social clusters. There are also three processes that are immanent to the development of our planet: globalisation, democratisation and decentralisation [6]. Even though there are many cases in which these processes are in conflict, the full scale effects of these processes are of fundamental importance for the understanding of issues such as the behaviour of complex systems. In order to relate these processes to sustainability and to account for their effects in the estimate of the sustainability of a system, it is necessary to introduce criteria for their quantification and measure.

Globalisation. Recently it has become evident that economic forces drive the transfer of capital, material, resources and manpower throughought the world without obstacles imposed by local, state and regional boundaries. The contemporary revival of interest in the field of international political economy has coincided with an unprecedented restructuring of the world economy, referred to as globalisation. The forces of change associated with globalisation have been felt through states and societies to such an extent that it has become the focus of much research across the social sciences. Yet, despite recognising that globalisation represents a critical issue in the development of political economy, the interdisciplinary approach combined with economic theories pursued so far has not provided a foundation that is adequate for consolidation of the field. Interdisciplinary insight into the global economy needs knowledge of the structure of the system as well as the parameters that describe the state of the system. The intensity of globalisation is assessed by the quantities that are used as indicators reflecting the state of the system under consideration. This implies that the process of economic reform will be measured by changes in respective indicators over time. The globalisation process is taking place within the system, so that the system parameters are supposed to be the measuring parameters of the intensity of processes in the system. In this respect the intensive parameters of the system are to be used in the determination of the state of the system. In the terminology of Thermodynamics, the intensive parameters are specific quantities of the respective extensive parameters. In this case, it could be understood as specific capital, specific material, specific resources and specific manpower. In engineering practice, in order to become operational, indicators have to be measured as the state parameters of the system. Since we are interested in measuring the change of the state of the system, it is necessary to introduce as indicators of the globalisation process the respective change of the intensive parameters of the system. So, as indicators for the globalisation process, the following parameters can be adopted: rate of change of specific capital, rate of change of specific material, rate of change of specific resources and rate of change of specific manpower.

Democratisation. Democracy is the principle of equity of rights, opportunity and treatment [7]. The process leading to the establishment of social organisation based on the principles of democracy is democratisation. So, the democratisation process can be defined within different boundaries including the local, regional and global environment. The intensity of the democratisation process is dependent on the number of attributes reflecting the ethnic, religious, cultural and educational

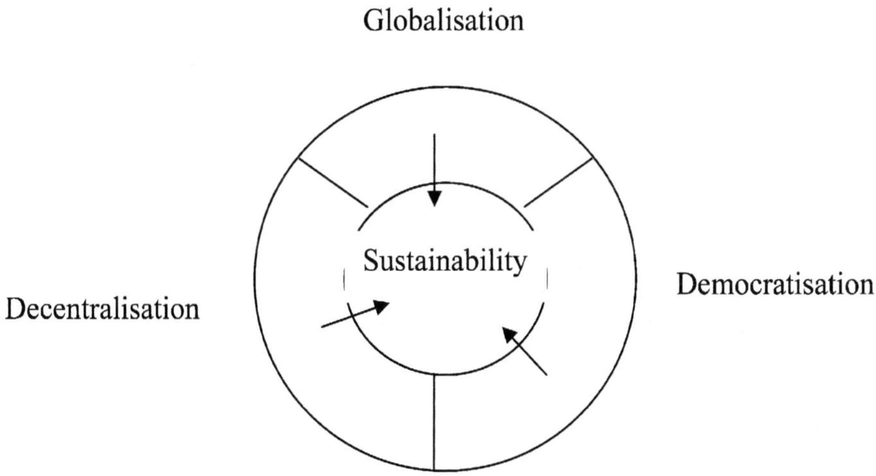

Fig. 2.1 Schematic presentation of interaction: Sustainability – Globalisation – Democratisation – Decentralisation.

environment. For every structured social system the respective indicators reflecting different aspects of the democratisation process can define the intensity of that process. Individual parameters defining specific characteristics of democratisation can be used to measure the intensity of that process. Among those are: equity of rights, job opportunity and treatment. Each of the parameters can be defined as the specific value of the internal parameter of the system under consideration. As the internal parameters, the indicators of the democratisation process can be defined as: the specific number of citizens participating in the voting system, the specific number of job opportunities in the system etc. Since the respective indicators cluster also defines the democratisation process, it is of interest to make the assessment with reference to the effect of the social parameters defined by the democratisation of the observed system. Again, we have to form a respective aggregation function which will describe the state of the system. In this respect the sustainability assessment can be used as the decision-making paradigm for the system assessment.

Decentralisation. It has long been considered that large energy and water systems are economically better justified than small systems. The economic indicators and constraints prevailing at the time were indisputably in favour of large systems, acting as the driving force in the decision-making process. With a new wave of miniaturisation, it has become evident that in a complex system assessment priority may be given to smaller systems [8]. This trend applies to all segments of life and human activities, ranging from technology to societal organisation. For example, the new trend in governance systems is to encourage regional autonomies in state organisation and to transfer more and more decision rights to local and regional governments. The same can be said of the development of energy, water and environmental systems. Smaller cogeneration units have become an attractive solution in many areas, leading to better economic, environmental and social values in the use of available resources. The modern micro gas turbine in conjunction with cogeneration has proved to be justified by the complex assessment method. Of course, this should be proved by applying a network system to investigate options and pathways for an accelerated transition towards sustainable energy technologies and systems. However, in a number of cases it has been shown that the appropriate selection of the criteria and respective indicators with corresponding decision-making procedures will give priority to decentralised systems.

The aggregation function for the decentralisation assessment should include all parameters that are of importance to the assessment of the system. Adoption of this procedure, which will lead to the formation of clusters of indicators represented by the respective sustainability indicator, will enable us to define an appropriate sustainability index which represents the quality of the system.

In addition to the three global processes discussed above, there are other factors and processes that also play an important role in quantifying sustainability. Recall that there are three essential life support systems, namely, energy, water and environment. Each of these systems comprises a large number of subsystems defined with different indicators reflecting economic, environmental and social criteria. In everyday life there is a need for decision-making actions governed by a number of criteria. Fundamental advances in our ability to address such issues as the behaviour of complex systems will be required. Besides those criteria leading to the decision for large systems, there will be more and more criteria which will make it possible to apply knowledge about different aspects of the system, including its interaction with other systems, interaction with the environment, and its *reflection* to the social environment. In this respect, there will be a need to have an education system which will accommodate basic knowledge with operational sustainability assessment.

2.3 Limits

Energy, water and environment are essential commodities, necessary for human life on our planet. These three commodities have been fundamental resources in the economic, social and cultural development of our civilisation. In the early days of human history it was believed that there were abundant reserves of these commodities. With the industrial revolution the use of resources drove economic and social development. With the increase of population and a respective increase of the standard of living, natural resources have become a scarcity in some regions. With the further increase in demand it has become evident that the scarcity of natural resources may become a global problem and affect human life on our planet. The Club of Rome was among the first to draw world attention to the potential limits of our planet's natural capital. The energy crises in 1972 and 1978 focused the attention of our community at large on the limits of energy resources [9] and various institutions launched programmes to investigate the scarcity of natural resources on a global

Table 2.1 Assessed energy resources (WEC Members Commission 2003).

	Total 10^9 toe	CPE %	North America %	Latin America %	West Europe %	Africa %	Asia-Pacific %	Middle East %
Coal	606	5.5	27.9	2.1	28.4	27.0	0.4	8.7
Oil	148	8.9	4.9	12.8	5.5	7.6	13.1	2.0
Gas	147	7.6	4.7	3.6	9.9	31.5	41.6	0.8

scale. It has become clear that modern society has to adopt a new philosophy, based on limited natural resources.

2.3.1 Energy

In 1886 Boltzmann [10], one of the Fathers of modern physical chemistry, wrote that the struggle for life is not a struggle for basic elements or energy, but a struggle for the availability of energy transferred from the hot Sun to the cold Earth. In fact, life on Earth requires a continuous flux of solar energy to support the energy capturing by photosynthesis [11]. The Sun is an enormous machine that produces energy by nuclear fusion and offers planet Earth the possibility of receiving large quantities of available energy (exergy). Every year the Sun sends 5.6×10^{24} joules of energy to the Earth and produces 2×10^{11} tonnes of organic material through photosynthesis. This is equivalent to 3×10^{21} joules/year. In the billions of years since the creation of planet Earth this process has led to the accumulation of enormous energy in the form of different hydrocarbons. Mankind's energy resources rely heavily on the chemical energy stored in fossil fuel. Table 2.1 shows assessed energy resources [12, 13].

Energy and matter constitute the Earth's natural capital, which is essential for human activities such as industry, amenities and services. Our natural capital, as the inhabitants of the planet Earth, may be classified as:

- Solar capital (providing 99% of the energy used on the Earth).
- Earth capital (life support resources and processes including human resources).

Many suggest that this natural capital is being rapidly depleted. Many also suggest that contemporary economic theory does not appreciate the significance of natural capital in techno-economic production.

All natural resources are theoretically renewable, but over widely different time scales. If the time period for renewal is small, they are said to be renewable. If the renewal period is longer but takes place within the time frame of our lives, they are said to be potentially renewable. The renewal of some natural resources is only possible through geological processes which take place on such a long time scale that for all practical purposes they are regarded as non-renewable. Our use of natural material resources involves no loss of matter as such. All Earth matter remains with the Earth, but in a form which cannot be easily used. When the quality or useful

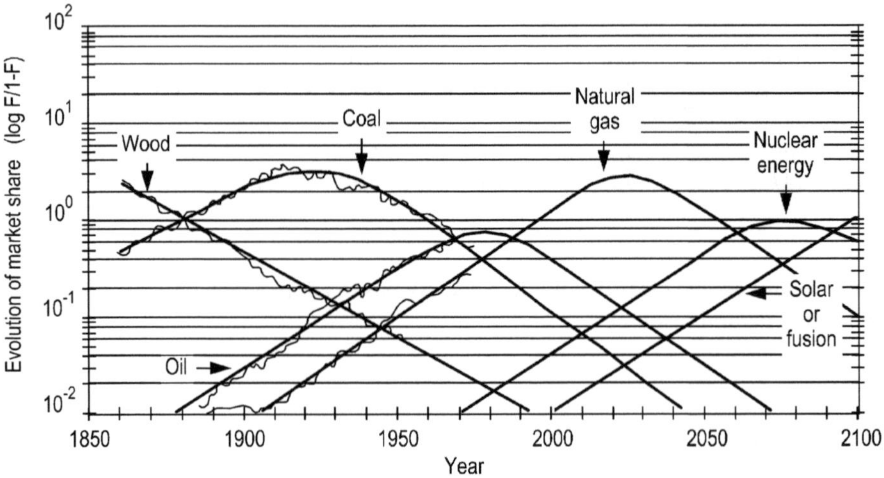

Fig. 2.2 Market penetration of primary energy sources.

part of a given amount of energy is degraded due to use we say that its entropy is increased.

Because of the abundant energy resources in the early days of the industrial revolution, the development strategy of our civilisation was based on the assumption of unlimited energy resources, as well as the assumption that no other limitations existed which could affect the development of human welfare. It has been recognised that the pattern of energy resource use has been highly dependent on the development of technology. It is therefore instructive to observe the change in the consumption of different resources through the history of energy consumption. Worldwide use of primary energy sources since 1850 is shown in Figure 2.2 [14, 15].

In Figure 2.2 "F" denotes the market fraction of each primary energy source at a given time. Two factors affect the energy consumption pattern. The first is related to technological development, and the second to the availability of the respective energy resources. Obviously, these patterns of the use of energy source are developed under constraints pertaining to the total level of energy consumption and reflect the existing social structure both in numbers and diversity [16–18]. World energy consumption is shown in Figure 2.3.

Looking at the present consumption pattern of energy sources, we see that oil supplies about 40% of total energy, coal around 30%, natural gas 20% and nuclear energy 6.5%. This means that fossil fuels currently supply 90% of our total energy. In the past several decades our civilisation has witnessed changes which have brought our long-term prospects into question. Non-recyclable fossil fuel is an exhaustible natural resource that will one day no longer be available. As they are our main source of energy it is of common interest to learn how long these fossil fuel resources will be available. This question has attracted the attention of a number of distinguished authorities attempting to forecast the energy future of our planet.

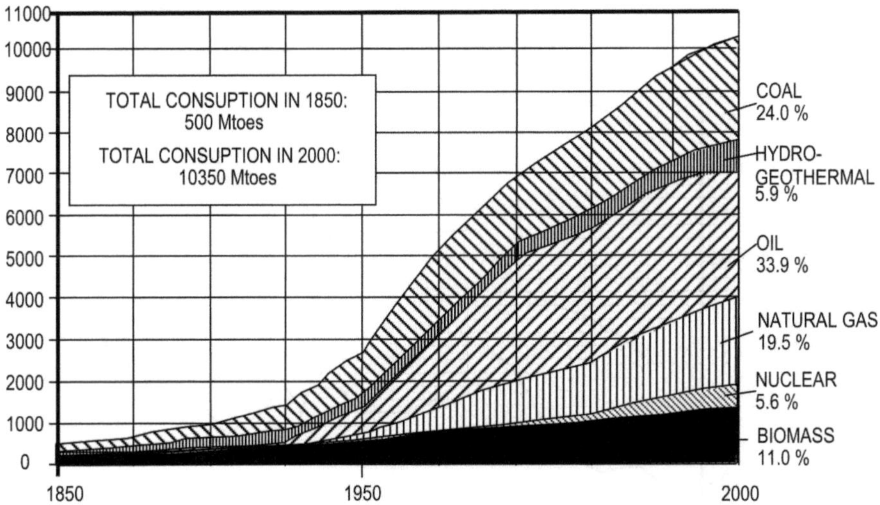

Fig. 2.3 World energy consumption.

The Report of the Club of Rome "Limits to Growth", published in 1972, was among the first which pointed to the finite nature of fossil fuels. After the first and second energy crises the community at large became aware of the possibility of the physical exhaustion of fossil fuels. It is known that the estimates of the exploitable reserves depend on the involved exploitation cost. For oil, based on the exploitation costs of under $20 per barrel, it was estimated that the proven reserves have over the past twenty years levelled off at 2.2 trillion (10^{12}) barrels, Over the last 150 years we have already consumed one-third of that amount, or about 700 billion barrels, which leaves only the remaining 1.5 trillion barrels. Compared with the present consumption, that means that oil is available for the next 40 years only.

Figure 2.4 shows the estimated residual life forecast for the three main fossil fuels over the last half of the century. According to the present forecast, coal is available for the next 250 years and gas for the next 50 years. It is also evident that while fuel consumption is increasing, new technologies for the discovery of new resources are becoming available and the continuous increase in price is shifting the limit of what is considered as exploitable, leading to an extension of the time period for exhaustion of the available energy sources.

It is known that energy consumption is dependent on two main parameters; namely, the amount of energy consumed per capita and the growth of population. It has been proved that there is a strong correlation between the Gross Domestic Product and energy consumption per capita. Figure 2.5 shows the economic growth and energy consumption for a number of countries in 1991.

Whatever the accuracy of our prediction methods and models, it is clear that uncertainty in our calculation may affect the time scale but not the essential under-

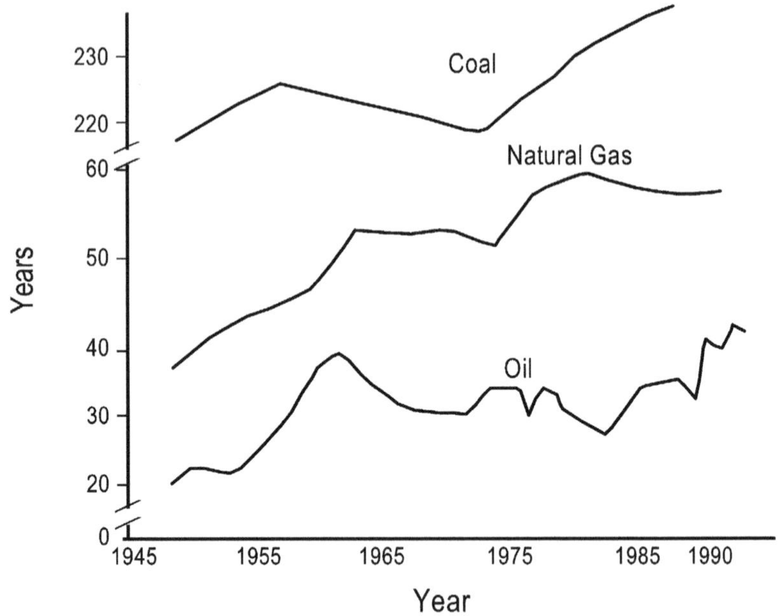

Fig. 2.4 Residual life forecast of energy resources.

standing that there is a real depletion of energy resources and that human action is required before adverse effects become irreversible [20].

Figure 2.4 shows also that each of these energy resources became significantly scarcer during the 1970s. The situation reversed during the 1980s as new reserves were discovered. Nevertheless, the alarming realisation of the depletion of energy resources and the coincidental energy crisis in the 1970s left a lasting imprint and far reaching implications on future economic growth, proving that scarcity of resources and economic growth are closely interrelated. To be sure, it has not been proven that the short term scarcity fluctuation of energy resources has substantial implications on long term economic growth. However, the need for active involvement in the allocation of scarce, non-renewable energy resources and the potential negative effects of uncontrolled consumption on economic growth has become evident.

2.3.2 *Environment*

The use of primary energy resources is a major source of CO_2 emissions [21–24]. Because fossil fuels have been shown to be economical, in recent years more than 88% of primary energy in the world has been generated from fossil fuels. However, the exhaust gases from combusted fuels have accumulated to such an extent that the global environment is being seriously damaged. The accumulated amount of CO_2

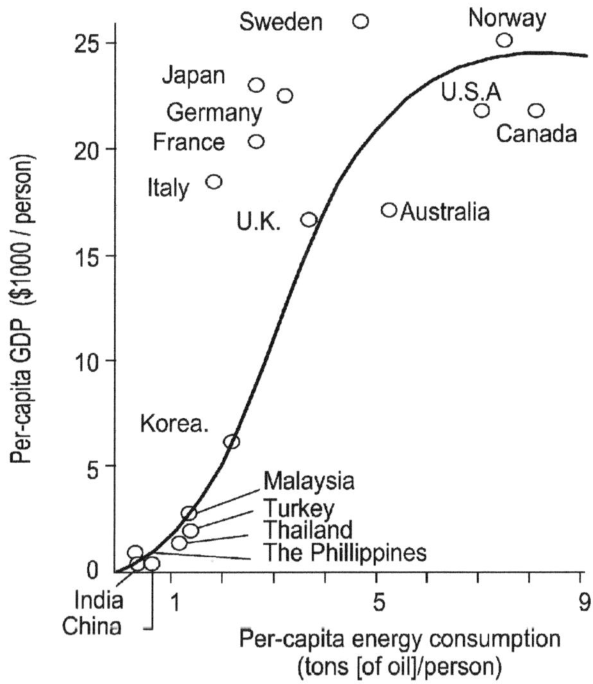

Fig. 2.5 Correlation between income per capita and energy consumption levels per capita of selected industrialised and developing countries (Source: Herman Daly, Steady-state Economics, Washington, Island Press, 1991).

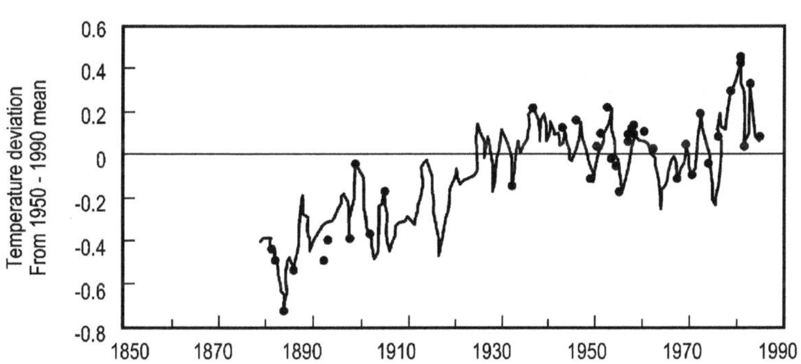

Fig. 2.6 Global warming trend 1900–1990.

in the atmosphere is estimated at about 2.75×10^{12} t. The global warming trend from 1900 to 1997 is shown in Figure 2.6 [25]. The future trend of carbon dioxide concentration in the atmosphere can be seen in Figure 2.8.

Obviously, further increases of CO_2 emissions will lead to disastrous effects on the environment. Emissions of SO_2, NOx and suspended particulate matter will further contribute to this.

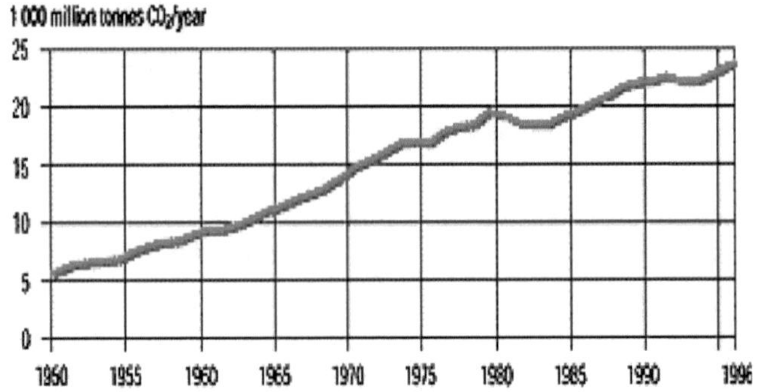

Fig. 2.7 Cumulative CO_2 production.

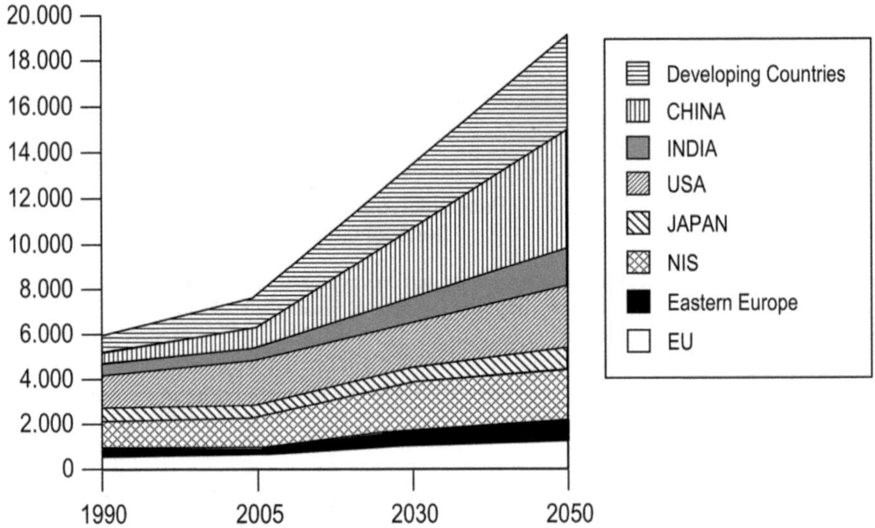

Fig. 2.8 Forecast for CO_2 emission.

On the world scale, coal will continue to be a major source of fuel for electric power generation. Many developing countries, such as China and India, will continue to use inexpensive, abundant, indigenous coal to meet growing domestic energy needs. As the economy of developing countries continues to expand, the use of coal worldwide is greatly increasing. The major long-term environmental concerns regarding the use of coal have changed from acid rain to greenhouse gas emissions – primarily carbon dioxide from combustion. Coal is expected to continue to dominate China's energy picture in the future. The share of coal in primary energy consumption is forecast to be no less than 70% during the period 1995–2010. In 1993 China produced a total of 1114 billion (10^9) tonnes of coal, in 2000 1.5 trillion (10^{12}), and in 2010 it will be 2 trillion. Since China is the third largest energy consumer in the

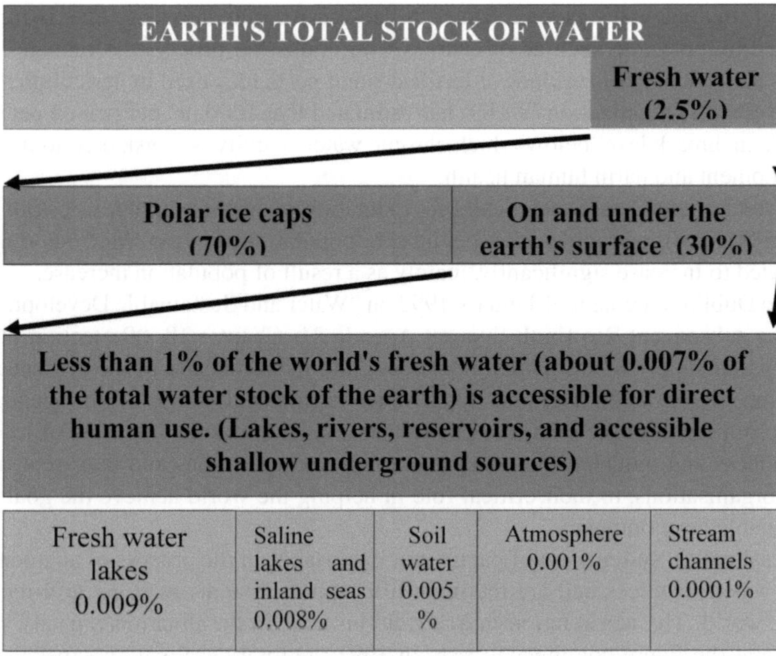

Fig. 2.9 Global stock of water.

world after the USA and Russia, its contribution to the global accumulation of CO_2 will be substantial unless adequate mitigation strategies are adopted. The example of China is instructive in the assessment of the future needs of developing countries and their accelerated economic development.

2.3.3 *Water*

The sustainability of desalination systems, an essential component of human-made or built capital, is discussed in this section with respect to its important contribution to life support systems. Figure 2.9 shows the distribution of the global stock of water [26, 27].

Of the total global water stock, 97.5% is saline and only 2.5% is freshwater. Approximately 70% of this global freshwater stock is locked in polar icecaps and a major part of the remaining 30% lies in remote underground aquifers. In effect, only a miniscule fraction of the freshwater available in rivers, lakes and reservoirs (less than 1% of total freshwater, or 0.007% of the total global water stock) is readily accessible for direct human use. Furthermore, the spatial and temporal distribution of the freshwater stocks and flows is hugely uneven. Hydrologists estimate the average annual flow of all the world's rivers to be about 41,000 km^3/yr. Less than a third of this potential resource can be harnessed for human needs. This is further reduced by pollution such as discharges from industrial processes, drainage from mines and leaching of residues of fertilisers and pesticides used in agriculture. The World Health Organization (WHO) has estimated that 1000 m^3 per person per year is the benchmark level below which chronic water scarcity is considered to impede development and harm human health.

Several countries are technically in a situation of water scarcity, i.e., with less than 1000 m^3 of renewable water per head of population per year. Water shortage is predicted to increase significantly, mainly as a result of population increase.

The Dublin Statement of January 1992 on "Water and Sustainable Development" and the subsequent Rio Earth Summit Agenda 21, Chapter 18, "Protection of the quality and supply of freshwater resources", are most relevant to the present context since desalination augments fresh water resources. Chapter 30 of Agenda 21 is also important in the context of desalination as it draws the attention of leaders of business and industry, including international corporations and their representative organisations, to their critical role in helping the world achieve the goals for sustainable development.

Desalination systems are of paramount importance in the process of augmenting fresh water resources and are the main life support systems in many arid regions of the world. The world has seen a 22-fold increase in desalination capacity since 1972 and the figure continues to rise. In 1997 the total desalination capacity was 22,730,000 m^3 of fresh water per day. That represents a doubling in global capacity over 10 years and a 22-fold increase over 25 years. Yet desalinated seawater represents only about one thousandth of the fresh water used worldwide. Desalinated water costs several times more than the water supplied by conventional means. The countries in the Arabian Gulf Region heavily subsidise the costs to render it affordable. In some of these countries, water is subsidised so heavily that users make little effort to curb their use. Water consumption would be greatly reduced if the price were closer to the true cost of production.

2.4 Sustainability Definitions

Over the past few years "sustainability" has become a popular buzzword in the discussion of the use of resources and environmental policy. Before any further discussion of the subject, it is necessary to define and properly assess the term we are going to use. So, what is sustainability? Among the definitions most often adopted are the following:

a. World Commission on Environment and Development (Brundtland Commission) [28] "Development that meets the needs of the present without compromising the ability of future generations to meet their own needs".
b. Agenda 21, Chapter 35 [29]
   "Development requires taking long-term perspectives, integrating local and re-

gional effects of global change into the development process, and using the best scientific and traditional knowledge available".

c. Council of Academies of Engineering and Technological Sciences [30]
"It means the balancing of economic, social, environmental and technological consideration, as well as the incorporation of a set of ethical values".
d. Earth Chapter [31]
"The protection of the environment is essential for human well-being and the enjoyment of fundamental rights, and as such requires the exercise of corresponding fundamental duties".
e. Thomas Jefferson, September 6 1889 [32]
"Then I say the Earth belongs to each generation during its course, fully and in its right no generation can contract debts greater than may be paid during the course of its existence".

All five definitions emphasise a specific aspect of sustainability. Definitions (a) and (e) imply that each generation must bequeath enough natural capital for future generations to satisfy their needs. Even if there is some ambiguity in these definitions, it is implied that we should leave our descendants the wherewithal to survive and meet their own needs. There is no specification for the form or amount of these resources as it is difficult to anticipate future scenarios.

Definitions (b) and (c) incorporate the political requirement at global, regional and local levels to stimulate the United Nations, governments and local authorities to plan development programmes in accordance with scientific and technological knowledge. Note the inclusion in definition (c) of the ethical aspect of actions to be taken to meet sustainable development. Definition (d) reflects also religious beliefs which assume responsibilities and duties toward nature and Earth.

2.5 Sustainability Concept

Sustainable development encompasses economic, social, and ecological perspectives of conservation and change. The WCED definition of sustainability (a) is based on the ethical imperative of equity within and between generations. Moreover, apart from meeting the basic needs of all, sustainable development implies sustaining the natural life-support systems on Earth, and extending to all the opportunity to satisfy their aspirations for a better life. Hence, sustainable development is more precisely defined as "a process of change in which the exploitation of resources, the direction of investments, the orientation of technological development and institutional change are all in harmony and enhance both current and future potential to meet human needs and aspirations" [33, 34].

This definition involves an important transformation and extension of the ecologically based concept of physical sustainability to the social and economic context of development. Thus, the terms of sustainability cannot be defined exclusively from an environmental point of view or on the basis of attitudes. Rather, the challenge is to define consistent and operational terms of sustainability from the perspective of an integrated social, ecological, and economic system. This gives rise to two fundamental issues that need to be clearly distinguished before integrating normative and positive issues in an overall framework.

The first issue is concerned with the objectives of sustainable development; that is, "what should be sustained" and "what kinds of development are preferred". These are normative questions that involve value judgements about society's objectives with respect to social, economic, and ecological system goals. These value judgements are usefully expressed in terms of a social welfare function, which allows an evaluation of trade-offs among the different system goals.

The second issue deals with the positive aspect of sustainable development; that is, the feasibility problem of "what can be sustained" and "what kind of system we can devise". It requires an understanding of how the different systems interact and evolve, and how they could be managed. Formally, this can be represented in a dynamic model by a set of differential equations and additional constraints. The entire set of feasible combinations of social, economic and ecological states describes the inter-temporal transformation space of the economy in the broadest sense [35–37].

2.6 Sustainability Measurement

Measuring sustainability is a major issue in planning and realisation of sustainable development. The development of a tool that reliably measures sustainability is a prerequisite for identifying non-sustainable processes, informing designers of the quality of products and monitoring impacts on the social environment. The multiplicity of indicators and measuring tools being developed in this fast-growing field shows the importance of the conceptual and methodological work in this area. The development and selection of indicators require parameters related to the reliability, appropriateness, practicality and limitations of measurement [38–42].

In order to cope with the complexity of sustainability-related issues for different systems, the indicators have to reflect the wholeness of the system as well as the interaction of its subsystems. Consequently, indicators have to measure the intensity of the interactions between elements of the systems and between the system and its environment. From this point of view, there is a need for indicator sets related to the interaction processes that allow an assessment of the complex relationship of every system and its environment.

2.6.1 Characteristics of effective indicators

Indicators can be useful as proxies or substitutes for measuring conditions that are so complex that there is no direct measurement. For instance, it is hard to measure the "quality of life in my town" because there are many different things that make up quality of life and people may have different opinions about which conditions

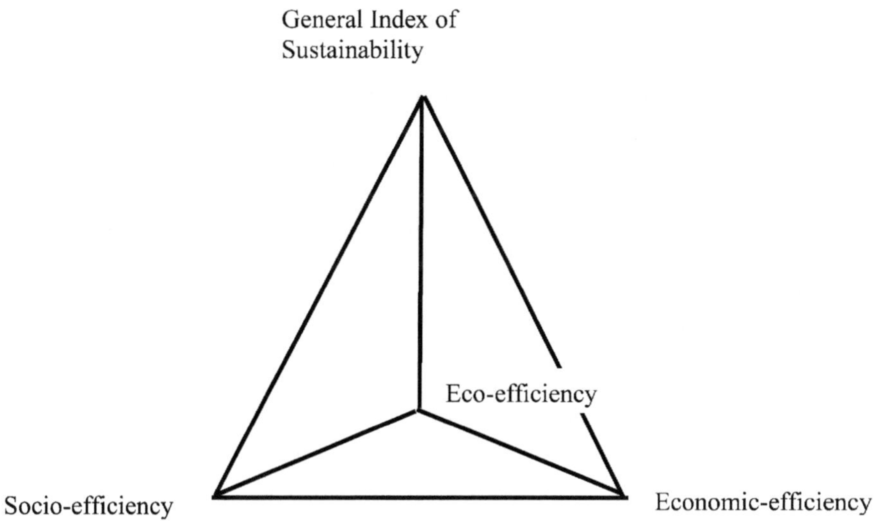

Fig. 2.10 General index of sustainability.

count most. A simple substitute indicator is "The number of people moving into the town compared to the number moving out".

Sustainability can be presented in the form of a triangle pyramid, where every corner on the base represents one of the efficiencies to be included in the assessment of any system. The fourth corner represents the Sustainability Index value (Figure 2.10). The Sustainability Index is obtained when a balance is found between the issues of all three efficiencies reflecting imposed constraints. In order to obtain the Sustainability Index for the option under consideration the weighting coefficient for the efficiencies has to be determined. The decision-making theory is used to calculate weighting coefficients. Non-numerical constraints are generated to represent constraints between the criteria.

The interactions between the three aspects of sustainability emphasise that sustainable development is not a static concept which can be easily translated and quantified. It is a dynamic concept, the result of a process of social learning involving many actors. For instance, in order to know which system is more sustainable it is necessary to formulate shared visions about the value of non-economic elements like biodiversity or cultural heritage. Because these visions and the underlying eco and social values change over time, it is imperative to consider all three aspects of sustainability as a continuous process, including the process of social learning and the environmental global change.

For the assessment of the system , attention will be focused on the three efficiency definitions.

2.6.2 Economic Efficiency

The traditional method for assessment of systems is based on the econometric justification of the use of capital needed for unit production. This method has been the basis of the decision-making process in the selection of systems; it has been a driving force for development of economic welfare in industrial societies. One of the basic assumptions of this method was that of the abundance of resources. With the development of the notion that scarcity of resources imposes limits on the use of resources it has been realised that, in addition to the limits of resources, there are other limits which play an important role in the decision-making process. Indicators for economic efficiency assessment are: investment cost including material cost, fuel cost, thermal efficiency and operation and maintenance cost. These indicators are a result of the optimisation procedure adopted for the respective optimisation function and respective design parameter of the system.

2.6.3 Ecological Efficiency

Following recognition of the adverse effect of combustion products on the environment, new indicators have been introduced in the decision-making procedure for system selection. The Kyoto Protocol has imposed local, regional and global limits of emission of CO_2, which should be incorporated in the design, operation and selection of new energy systems. This has led to the development and introduction of indicators for the ecological aspects of any energy system.

Indicators for ecological efficiency assessment include the concentration of combustion product species which are considered to have an adverse effect on the local, regional and global environment. Ecological efficiency is evaluated by the monitoring and assessment of those indicators which affect the quality of the environment.

2.6.4 Social Efficiency

The social aspect of any human endeavour is of paramount importance in the selection of possible options. It has become evident that the social aspect of any engineering system is an important part of its total quality. Criteria which assess the social aspect of a system are therefore as important as the economic and environment criteria. To formulate social criteria it is necessary to create a system of indicators of sustainable development which provide a reference for the respective type of system, and which may be used in the numerical evaluation of the system.

Indicators for social efficiency assessment are: job opportunity, diversity of qualifications, community benefits and consequences for local safety. The job opportunity indicator takes into consideration the number of jobs created by the respective system.

2.7 Sustainability Index Definition

The decision-making process comprises several steps that need to be followed in order to obtain a mathematical tool for the assessment of the rating among the options under consideration [43–46]. The first step in the preparation of data for the multi-criteria sustainability assessment is arithmetisation of the data. This step consists of the formation of particular membership functions $q_1(x_1),\dots,q_m(x_m)$. For every indicator x_i we have

1. to fix two values $\min(i)$, $\max(i)$;
2. to indicate if the function $q_i(x_i)$ is decreasing or increasing when argument x_i is increasing;
3. to choose the value of the exponent λ for the increasing and decreasing functions $q_i(x_i)$ in the formula

$$q_i(x_i) = \begin{cases} 0 & \text{if} \quad x_i \leq \min(i) \\ \left(\dfrac{x_i - \min(i)}{\max(i) - \min(i)} \right)^{\lambda} & \text{if} \quad \min(i) < x_i \leq \max(i) \\ 1 & \text{if} \quad x_i > \max(i) \end{cases}$$

The functions $q_1(x_1),\dots,q_m(x_m)$ indicate the formation process is being completed with a matrix $(q_i^{(j)})$, $i = 1,\dots,m$, $j = 1,\dots,k$, where element $q_i^{(j)}$ is a value of i-th particular criterion for j-th option. In this analysis it is assumed that the functions $q_1(x_1),\dots,q_m(x_m)$ are linear. In our case when there are four indicators q_1, q_2, q_3 and q_4, the membership functions are adapted as the decreasing functions. There is no constraint as regards to increasing or decreasing functions. They are defined in accordance with the indicators values.

The general index method comprises the formation of an aggregative function with the weighted arithmetic mean as the synthesising function defined as

$$Q(q;w) = \sum_{i=1}^{m} w_i q_i,$$

where w_i presents the weight-coefficients elements of vector $\mathbf{w}$ and q_i presents the indicators of specific criteria.

In order to define the weight-coefficient vector, the randomisation of uncertainty is introduced. Randomisation produces stochastics with realisations from corresponding sets of functions and a random weight-vector. It is assumed that the measurement of the weight coefficients is accurate to within a step $h = 1/n$, with n being a positive integer. In this case the infinite set of all possible vectors may be approximated by the finite set $W(m,n)$ of all possible weight vectors with discrete components.

Table 2.2 Energy system indicators.

	Efficiency %	Instalation USD/kW	Elect. cost c/kWh	CO_2 $kgCO_2/kWh$	Area km^2/kW
Coal	43	1000	5.4	0.82	0.04
Solar Thermal	15	3500	17	0.10	0.08
Geothermal	8	2500	8	0.06	0.03
Biomass	1	2500	14	1.18	5.20
Nuclear	33	2300	4	0.025	0.01
PV Solar	10	4500	75	0.1	0.12
Wind	28	1100	7	0.02	0.79
Ocean	3	10000	25	0.02	0.28
Hydro	80	2000	8	0.04	0.13
Gas	38	650	4	0.38	0.04

For non-numeric, inexact and incomplete information $I = OI \cup II$, where OI is ordinal information and II is incomplete information, is used for the reduction of the set $W(m,n)$ of all possible vectors **w** to obtain the discrete components set $W(I;n,m)$, defining a number of constraints reflecting non-numeric information about mutual relations among the criteria under consideration.

2.7.1 Multi-Criteria Assessment of Energy Systems

As the non-numerical information we will impose a condition that should define the mutual relation of the individual criteria. This will give us the possibility to introduce a qualitative measure of the relations among the criteria.

As an example for the multi-criteria assessment of energy system we will take a number of energy system options with respective values for the five indicators including economic, environment and social indices [46]. Options of energy systems are defined with respective indicators (see Table 2.2).

The group of cases is designed to give priority to a single indicator with other indicators having the same values. Each case will represent a different set of the priority of criteria as they are used in the definition General Index of Sustainability. Among the cases which are designed with the preference of single options are:

CASE 1 Efficiency > Investment = Elect. Cost > CO_2 = Area
In this case, priority is given to the efficiency criteria of the energy system. As shown, the efficiency of systems with different basic principles is not a very realistic indicator to use for comparison of the system. This suggests that in the evaluation of efficiency criteria it would be better to use the relative value of the efficiency for each system. For example, for the heat conversion system the Carnot efficiency should be used as the absolute efficiency.

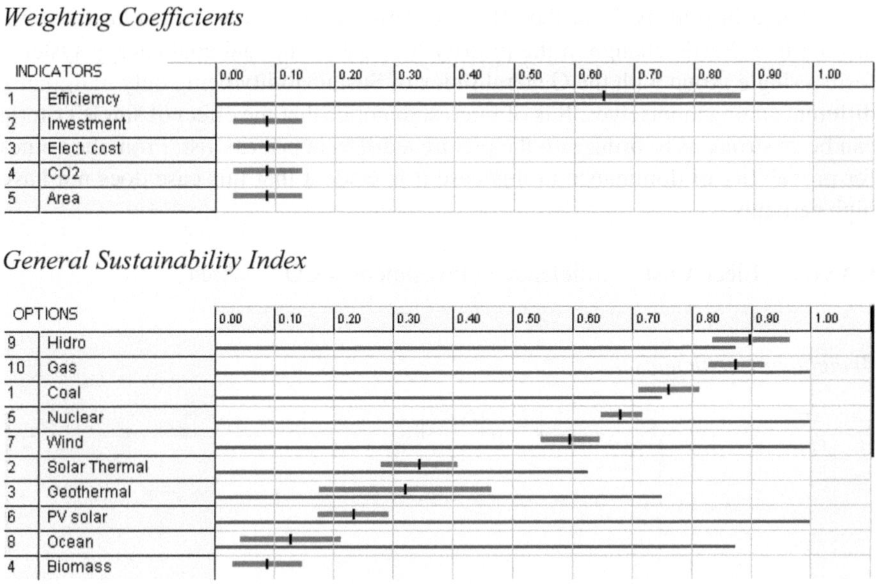

Fig. 2.11 Case 1: Sustainability index and weighting coefficients.

CASE 2 Investment > Efficiency = Elect. Cost > CO_2 = Area

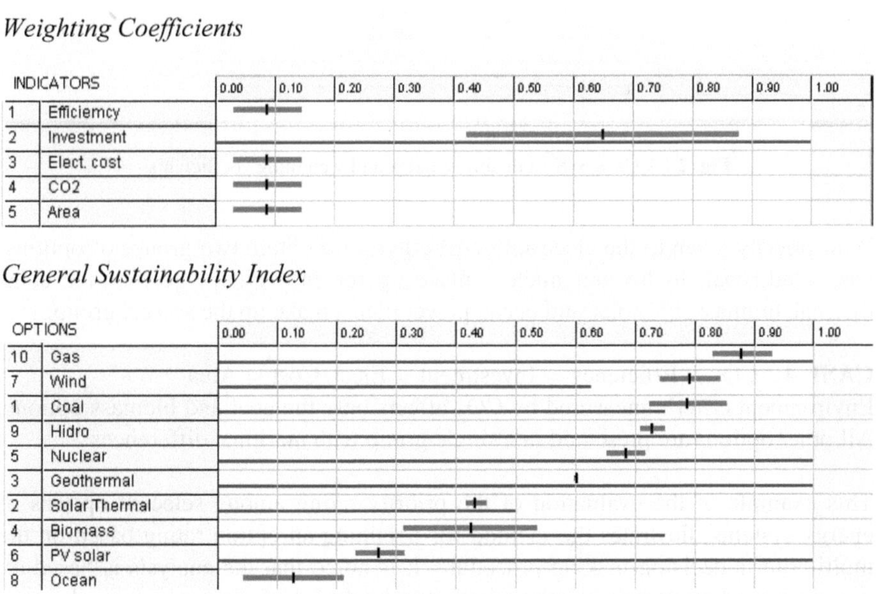

Fig. 2.12 Case 2: Sustainability index and weighting coefficients.

The change in priority from the efficiency criteria to the investment cost criteria has led to a drastic change in the priority list. Gas, wind and coal energy systems form a single group with the General Index of Sustainability being only marginally different among themselves. It is of interest to notice that the effect of single criteria can be so strong as to bring into the picture a different priority list. From the values for probability of dominance in this case it is evident that this case does not have high certainty.

CASE 3 Elect. Cost > Efficiency = Investment = CO_2 = Area

Weighting Coefficients

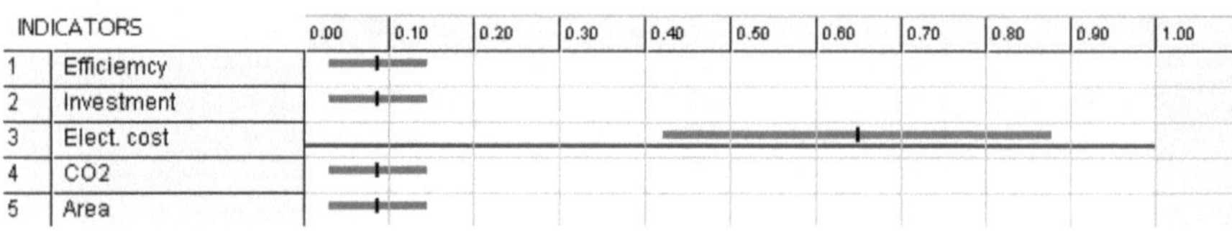

General Sustainability Index

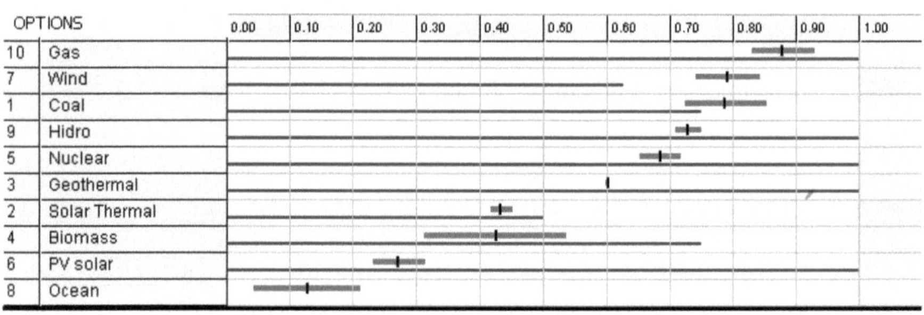

Fig. 2.13 Case 3: Sustainability index and weighting coefficients.

With priority given to the electricity cost criteria we obtain two groups of options: gas, wind, coal, hydro and nuclear make up the first group; geothermal, solar thermal, biomass, PV solar and ocean power plants make up the second group.

CASE 4 CO_2 > Efficiency = Investment = Elect. Cost = Area
Environment criteria measured by CO_2 affects only the coal and biomass options. All other options are presented in a single group with marginal differences.

This example of the evaluation of the priority rating among selected options of energy systems illustrates the method for obtaining an option rating based on the multi-criteria decision-making procedure. It is noted that the analysis is based on non-numerical information as the criteria for the design of cases which result in the respective rating among the options.

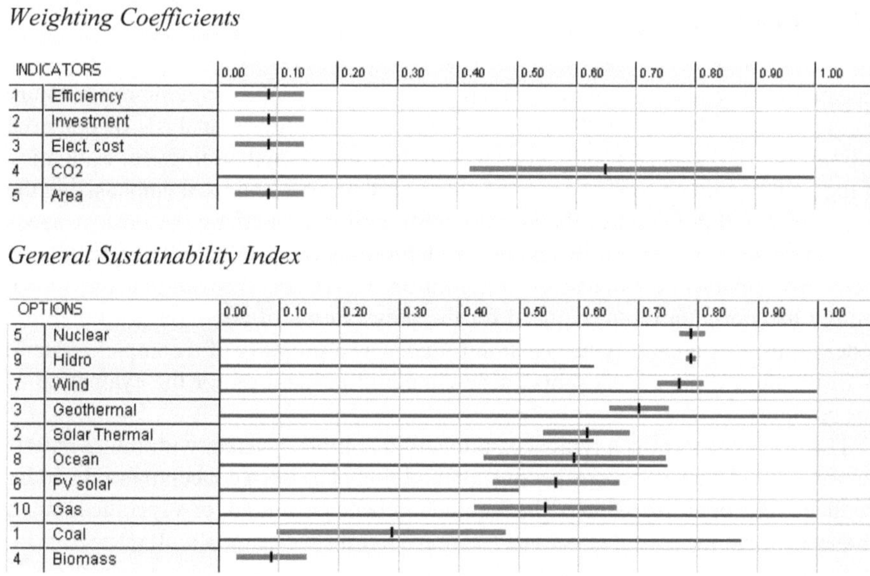

Fig. 2.14 Case 4: Sustainability index and weighting coefficients.

The selection of the group of cases illustrates an evaluation of the options with constraints, with the possibility of having a predetermined relation between indicators. Even though this analysis is based on a limited number of considered cases, it can be noticed that the priority of the rating list is the result of the respective relation among the criteria under consideration. In the group of cases, it can be noticed that the option which is the first on the rating list is closely related to the respective indicator priority and its value. If priority is given to a single criterion, with the other criteria having a respective value of indicators for each individual option, it may affect the rating list of the options. If the efficiency criteria are given priority there are changes in the rating list compared to the single criteria rating, and priority is obtained for hydro, coal, and gas options. The same can be noticed if priority is given to the other indicators. For the case of the Investment Indicator priority, gas, wind and coal power plants are the first on the rating list of the options under consideration. For the case of the Electricity Cost Indicator priority, the first places on the priority list are the gas and nuclear options. If the CO_2 Production Indicator and Area Indicator are given priority, the gas and nuclear power plants are rated in first place. A special case is designed with criteria that introduce a relation among individual Indicators. Besides the changes in the first place on the rating list, it can be noticed that there are changes in the rating among options. Options with renewable energy power plants have gained a higher place on the rating list in comparison with the case with an equal weighting factor for all indicators.

2.8 Conclusions

Sustainability assessment of power plants is immanent to the development of future energy strategy. It implies the need to verify a multi-criteria analysis of potential options. In this respect it is of interest to introduce a new methodology for evaluation of the power plant options under consideration. This article aims at emphasising the importance and potential of the sustainability notion in meeting the present needs of modern society. Starting from current definitions of sustainability and showing their close relation to the historical definition, it has been proven that our global society has borne in mind the need for the preservation of commodities for future generations. Sustainability has been introduced as a property of a complex system. With its multi-dimensional scope, a new method is required for the evaluation of this complex system.

The energy system is a typical example of a complex system with multi-criteria assessment. The need for a respective methodology has been conceptualised and introduced. The demonstration of the method has been presented for a specific number of energy systems, taking into consideration a number of criteria with respective indicators.

References

1. Kates, R.W. et al., Sustainability science, *Science* **292**, 2001, 641–642.
2. Ravetz, J., *Scientific Knowledge and Its Social Problems*, Transaction, New Jersey, 1996.
3. Nicolis, G. and Prigogine, I., *Self-Organisation in Non-equilibrium Systems: From Dissipative Structure to Order Through Fluctuation*, John Wiley & Sons, 1977.
4. Prigogine, I. and Stengers, I., *La Novelle Alliance. Métamorphose de la Science*, Gallimard, Paris, 1979
5. van Kroonenberg, H.H., Energy for sustainable development: Post Rio challenges and Dutch response, *Resource, Conservation, Recycling* **12**, 1994.
6. Managing Globalisation, *Foreign Policy*, March, 2004.
7. Parr, S. and Morrissey, D., A democratic 21st century?, *The ACP-EU Courier* **194**, 2002, 2–5.
8. Decentralisation, http://www.fact/index.com
9. Meadows, D., Meadows, H., Randers, D.L. and Behrens, J., *The Limits of Growth*, Universe Book, New York, 1972.
10. Botzmann, *Vorlesung über Gas Theorie*, Vol. 1, Leipzig, 1896.
11. Ohta, T., *Energy Technology*, Elsevier Science, 1994.
12. Marchetti C., Check on the Earth – Capacity for Man, *Energy*, **4**, 1979, 1107–1117.
13. Master, C.D., World resources of crude oil, natural gas, bitumen's and shale oil, Topic 25, World Petroleum Congress Publ., Houston, 1987.
14. Marchetti, C., The 'historical instant' of fossil fuel, symptom of sick world, *International Journal of Hydrogen Energy* **16**, 1991, 563–575.
15. Marchetti, C., Society as learning system, *Techno. Forecast. Soc. Changes* **18**, 1980, 267–282.
16. Arnold, M.St.J., Kersall, G.J. and Nelson D.M., Clean efficient electric generation for the next century: British coal topping cycle, in *Combustion Technology for a Clean Environment*, M.G. Carvalho, W.A. Fiveland, F.C. Lockwood and Ch. Papadopoulos (eds.), Gordon and Breach, 1995.
17. Mazzuracchio, P., Raggi, A. and Barbiri, G., New method for assessment of the global quality of energy system, *Applied Energy* **53**, 1996, 315–324.

18. Noel, D., A recommendation of effect of energy scarcity on economic growth, *Energy* **2**, 1995, 1–12.
19. Farinelli, U., Alternative energy sources for the Third World: Perspective, barriers, opportunity, Pontifical Academy of Science, Plenary Session, October 25–29, 1994.
20. Keatiny, M., *Agenda for Change*, Center for Our Common Future, 1993.
21. Mackey, R.M. and Probert, S.D., National policy for achieving thrift, environmental protection, improved quality of life and sustainability, *Applied Energy* **51**, 1995, 243–367.
22. Price, T. and Probert S.D., An energy and environmental strategy for the Rhymney Value, South Wales, *Applied Energy* **51**, 1995, 139–195.
23. Mackey, R.M. and Probert, S.D, NAFTA countries energy and environmental interdependence, *Applied Energy* **52**, 1995, 1–33.
24. Mackey, R.M. and Probert, S.D, Energy and environmental policies of the developed and developing countries within the evolving Oceania and South-East Asian Trading Block, *Applied Energy* **51**, 1995, 369–406.
25. Hought, R.A. and Woodwell G.M., Global climatic change, *Scientific American*, April, 1989.
26. Al Gobaisi, D., Sustainability of desalination systems, in *EURO Course on Sustainability Assessment of Desalination Plants*, Vilamore, 2000.
27. Al Gobaisi, D., Sustainable use of our planetary natural capital for life support on the Earth, in *IEEE Systems, Man and Cybernetics Conference*, Tunisia, 1998.
28. *Report of The United Nation Conference on Environment and Development*, Vol. 1, Chapter 7, June, 1992.
29. Agenda 21, Chapter 35, *Science for Sustainable Development*, United Nations Conference on Environment and Development, 1992.
30. *Declaration of the Council of Academies of Engineering and Technological Sciences.*
31. Thc Earth Chapter: *A Contribution toward Its Realization*, Franciscan Centre of Environment Studies, Roma, 1995.
32. Jenkinson, C.S., *The Quality of Thomas Jefferson's Soul*, White House Library, 1978.
33. Annan, K.A., *WE, The Peoples of United Nations in the 21st Century*, United Nations, New York, 2000.
34. Kates, R.W. et al, Sustainability science, *Science* **292**, 2001, 641–642.
35. National Research Council, Board on Sustainable Development, *Our Common Journey: Transition toward Sustainability*, National Academic Press, DC, 1999.
36. Watson, R. et al., *Protecting Our Planet, Securing Our Future*, United Nations Environmental Programme, Nairobi, 1998.
37. Hammond, G.F., Energy, environment and sustainable development: A UK perspective, *Trans. ICHemE Part B* **78**, July 2000.
38. Binswangen, M., Technological progress and sustainability development: What about the rebound, *Ecological Economics* **36**, 2001, 119–132.
39. Pemberton, M. and Ulph, D., Measuring income and measuring sustainability, *Scandinavian Journal of Economics* **10**(1), 2001, 25–40.
40. *Indicators of Sustainable Engineering*, Physical Sciences Research Council, December 1996.
41. D'Angelo, E., Perrella, G. and Bianco R., *Energy Efficiency Indicators of Italy*, ENEA Centro Ricerche Casaccia, Roma, RT/ERG/96/3.
42. Cafier, G. and Conte, G., *Rome as a Sustainable City*, Agency for a Sustainable Mediterranean Development, 1995.
43. Afgan, N.H. and Carvalho, M.G., *Sustainability Assessment Method for Energy Systems*, Kluwer Academic Publisher, New York, 2000.
44. Afgan, N.H., Al Gobaisi, D., Carvalho, M.G. and Cumo, M., Energy sustainable development, *Renewable and Sustainable Energy Reviews* **2**, 1998, 235–286.
45. Afgan, N.H., Carvalho, M.G. and Hovanov, A.N., Energy system assessment with sustainability indicators, *Energy Policy* **28**, 2000, 603–612.
46. Afgan, N. and Carvalho, M.G., Multi-criteria assessment of new and renewable energy power plants, *International Journal ENERGY* **27**, 2002, 739–755.
47. Afgan, H.N. and Carvalho, G.M., *Quality, Sustainability, Indicators of Energy Systems*, Begell House Publisher, New York, 2006.

Chapter 3
Efficient Production and Use of Energy: Novel Energy Rationing Technologies in Russia

Sergey V. Alekseenko

Abstract. Starting with an overview of the current state of the Russian energy situation and power industry, we consider future prospects and developments in energy production and consumption in Russia. Energy efficiency, safety, economic efficiency, and ecological safety are considered to be priorities in the further development of energy conversion utilities and in the power industry. To this purpose, the significance of the role of energy- and resource-saving techniques are considered, followed by an overview of various novel and promising technologies.

Key words: Energy, fuel, coal, energy-saving, renewable energy sources.

3.1 Russian Power Engineering: Problems and Solutions

The economic development of any country and the standard of living of its population essentially depend on the condition of the country's energy sector and its power industry. Figure 3.1 illustrates the correlation between the gross domestic product (GDP) and energy consumption per capita.

It is believed that the higher the energy consumption of a country, the higher its standard of living. It is also assumed that after reaching a certain critical level of GDP, about $18,000 per capita, a society feels prosperous and comfortable, and further growth in GDP does not have any radical impact on the life of the society.

In the lower part of the chart we see countries with low energy consumption and a low standard of living, such as China and India. Unfortunately, Russia is somewhere in the lower half of the chart, despite the fact that it shows a high level of energy consumption. The EU countries, Japan, the USA, and Canada are above the critical line. It is interesting that we can also distinguish two groups of countries with a high standard of living. The same high standard can be achieved with a different structure of energy consumption. There are several factors that influence this: different eco-

K. Hanjalić et al., Sustainable Energy Technologies: Options and Prospects, 51–73.

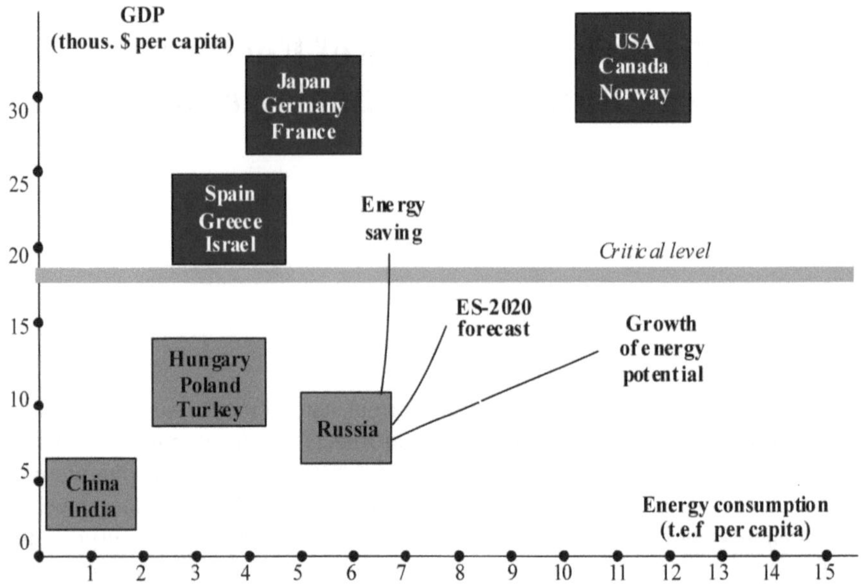

Fig. 3.1 Correlation between the Gross Domestic Product (GDP) per capita and specific energy consumption. (ES–2020 is "The Energy Strategy for Russia for the period up to 2020".)

nomic structures, climatic conditions, and different approaches to energy efficiency and energy saving.

Since the ultimate goal of the power industry is to provide a sufficient energy supply to a country's economy, we see that this level is achieved not only by the total amount of energy produced, but also through energy-saving measures. There is a close relationship between energy production, consumption, and energy saving.

A high energy intensity of GDP (the ratio of GDP and energy consumption) is certainly one of the most severe problems for the Russian economy. In 2000, the energy intensity of GDP for Russia was 3.2 times higher than that of the EU countries, 2.2 times higher than that of the USA, and 3.6 times higher than that of Japan. Greater differences can be revealed while analysing different industries. The share of energy cost in manufacturing in Russia is about 18% (compared to 3–10% in the Soviet period). But if we consider the chemical and petrochemical industry, this parameter increases up to 40–45% (and even 70% for some factories). This results in non-competitiveness of domestic products in both world and local markets and increases the share of imports. As a consequence of these high costs, enterprises leave the market. Another dangerous consequence is a decrease in investments – why invest in a plant with no profit return?

What is the reason for the high energy intensity of the Russian economy? There are several natural and inescapable reasons – harsh climate, huge distances and difficult access to energy deposits. But there are other factors as well. The Russian economy has a high proportion of energy-consuming industries with a low level of high-tech. These technologies are often energy-wasteful both in terms of production and consumption. There is also a lack of economic motivation for the introduction of energy-saving technologies. This is partially because of low domestic fuel prices (especially gas), a poor monitoring and book-keeping of energy consumption, inefficient operation modes and sometimes even a total lack of energy-consumption control facilities. Here are just a few examples. Flaws in the design and operation of boilers and heating plants result in a 30% increase in fuel costs. The efficiency of heat power plants (HPP) with steam turbines is about 35% (and only 25% for small-size plants), in contrast to the 50–60% efficiency of the combined-cycle plants (CCP) operating in many countries. As for fuel prices, a simple analysis reveals two negative contributing factors. First, the low domestic price of fuel: the price of crude oil is 72% of the export price, for coal it is 57%, and for gas 20%. Second, there is a huge difference in prices for different kinds of fuel – the energy produced from gas is half of the cost of that produced from coal (16.6 $/toe *versus* 30.3 $/toe for year 2001). It should actually be the reverse, since gas is a valuable raw material for other industries as well. This obstructs the development of coal technologies and creates an abnormal proportion of gas and coal in the domestic power industry. In electricity production in Russia, the share of gas is 42.6%, and only 20.1% for coal. In the EU countries, these parameters are 17.5% and 27.4% respectively, and for the US 19% and 50%. The share of gas fuel in the Russian power industry overall is even higher, namely 60–64%.

To make an analysis of the energy supply for Siberia, we must incorporate even more complex conditions. Siberia is the coldest zone in the world, where 60% of fuel is consumed for heat supply. Moreover, 62% of the heat in West Siberia is produced by 28,000 boiler plants with very low energy efficiency and extremely poor environmental parameters.

Guidelines for development of the power industry and the efficient use of energy resources are formulated in "The Energy Strategy for Russia for the Period up to 2020" [1], published in the year 2000. The strategy is based on two possible scenarios of economic development – optimistic and pessimistic (the latter could be considered also as a "moderate" scenario). The first scenario assumes a 5% annual growth of GDP (i.e., with a GDP 2.6 times higher in 2020 than in 2000), and the second scenario is based on a 3.5% annual growth (with GDP doubling over 20 years). The revised version of the Strategy (dated 28.08.2003) is based on higher growth rates, i.e., 6% and 4.2% annual growth, (predicting an increase in GDP by 3.3 times and 2.3 times, respectively) for the same period. The objective of rapid economic growth is to approach the high standards of living of developed countries. Quantitatively this level can be described in terms of GDP per capita. Here Russia is behind EU countries by a factor of 3 ($7,500 *versus* $21,900).

Today Russia is among the leading countries in economic growth. According to the estimates of UN experts, in the year 2004 the rate of GDP growth in Russia was 6.7%. In comparison, the world average was 4.9%; economic growth in the US was 4.4%, in the EU 2.4% and China showed 9.1% growth. However, this high index for Russia is a temporary parameter, caused by favourable conditions. For the near future, some experts predict a growth of 3–6%. However it is difficult to forecast with any degree of certainty. Analyses by International Energy Agency experts (among others) give a more reliable figure – about 3% (i.e., by 2020 GDP will increase by 1.9 times). This means that the gap in economic parameters and standards of living between Russia and industrialised countries will still remain high. One of the obstacles to the fast development of the Russian economy is high energy intensity of GDP combined with the unlikelihood of swiftly reducing this parameter. In the framework of the Energy Strategy, to maintain the projected rates of economic growth would require a drastic reduction of energy intensity for the economy (by 2.2 times in an optimistic scenario and by 1.8 times in a pessimistic scenario). According to the estimates of the Melentyev Institute of Energy Systems, this index cannot exceed 1.7, and the International Energy Agency gives an even lower index – 1.4. This forecast means that by the year 2020 the large difference between the energy intensity of the Russian economy and the economy of highly developed countries will remain almost the same as today.

According to the Energy Strategy, the decrease in GDP energy intensity should be carried out through two approaches: 2/3 through economic restructuring (higher input of high-tech or low-energy-intensity production), and 1/3 through technical and organisational energy saving. Here we come to an important conclusion about the crucial role of energy saving in the national economy. Instead of remaining a "desired objective", energy saving is now an absolute necessity for Russia to achieve high economic growth. A recent analysis of the Russian power industry [2] by European experts is titled "Energy strategy of Russia – key role of energy saving". It additionally informs the reader that Europe is interested in the reduction of GDP energy intensity in Russia because the supply of Russian gas to Europe depends on the difference between gas production and domestic consumption. The latter factor depends directly on the efficiency of energy consumption in the Russian economy.

The experts' estimates demonstrate that Russia has a huge potential for energy saving – more than 40% of its total energy consumption (400–500 million toe per year). One third of this potential is concentrated in the fuel/energy conversion complex, another third lies in the manufacturing and building industries, and one fourth should be pursued in the municipal utility sector. It should be emphasised that the realisation of this potential is costly. According to the Energy Strategy, up to 20% of potential energy saving can be pursued at a cost of about 15 $/toe, which is comparable to the fuel price. The most expensive part of the programme will require more than 60 $/toe (this is 15% of the potential energy saving). But the major part of the potential energy-saving measures will demand costs from 15 to 60 $/toe. Thus, this programme requires considerable, but targeted investments.

However, the indicators presented in the "Energy Strategy – 2020" published in 2004) already need serious correction, since the situation changes rapidly both in Russia and worldwide. Even now we face the problem of doubling the energy production in Russia by the year of 2030 – from 1 to 2 billion kWh. This means that Russia has to bring onto the energy market about 10 million kW of new power production facilities every year. The investment in the Russian power industry should be kept at the level of above $10 billion in the coming years and at $35–40 billion at the end of the stated term. The problem must be solved not only through construction of new facilities, but also through modernisation of existing energy equipment, which is physically too old (50–70% of the life time) and has gone beyond obsolescence. The government should formulate an energy policy with clear priorities: energy efficiency, energy safety, economic efficiency, and environmental safety.

Regarding specific methods for achieving higher efficiency in energy production, the trends of the Russian power industry are congruent with the global ones, though the degree of development and the significance attached to specific approaches are different for different nations. These global trends are as follows:

- *A higher degree of efficiency* in the conventional fossil-fuel power plants.
- *The use of combined methods* (*cogeneration*), e.g., combined production of electric and thermal energy.
- *Integration* of autonomous power sources into the national power supply grid.
- *A comprehensive stage use of energy*, i.e., fuel consumed not only for heat and electricity production, but also for chemical products, building materials, etc.
- *A reduction of the high proportion of heavy oil and gas in the power industry* compensated by a higher share of coal and *nuclear energy* and, to a smaller extent, by *renewable energy sources*.
- *The use of heat pumps* for energy saving and the utilisation of the potential of non-conventional energy sources.
- The development of *new types of energy sources*; this primarily concerns hydrogen energy and fuel cells.

Several specific approaches and technologies developed in Russia are discussed in the following sections (see also [3, 4]). The last section deals with the problems of higher efficiency in energy consumption. In this section, we will also discuss the Russian programmes in power engineering. Besides the above mentioned "Energy Strategy for Russia for the Period up to 2020" (adopted in 2000), there is another important detailed document on the state's R&D programme, namely "Ecologically Clean Power Engineering" (1989). This programme comprises four main areas: "Safe Nuclear Plants", "Ecologically Clean Solid Fuel Heat Plants", "Nonconventional Energy Sources", and "Fuel for the Future" [5]. In 1991, five more documents were added to this programme: "Ecologically Clean Methods for the Immobilization of Nuclear Waste", "Underground Nuclear Power Plants", "The Non-Conventional Concept of NPP (Nuclear Power Plant) with Natural Safety", "Hydrogen Energy and Technology", "Electricity and Heat Production from Coal Slurry". Within this programme, many projects, selected through various competitions, were implemented. These projects focused on various topics and ideas and inventions, which could alleviate many emerging problems of the Russian power industry, providing solutions of high level technologies within strict environmental standards. We will mention some of these projects later. Unfortunately, in the years of *Perestroyka* insufficient financial means obstructed most of these valuable projects. Nevertheless, the challenge of power production problems is so great that the Russian government decided to develop a new version of the "GOELRO" plan (i.e., reminiscent of the first industrialisation plan for electric power systems adopted in the USSR in the years

1920–1935), which successfully provided a plentiful power supply for the emerging Soviet Russia.

3.2 Higher Efficiency for Gas-Fueled Thermal Power Plants

It was mentioned above that in Russia the share of gas fuel in the production of electric energy is 43%, which is twice that of the average worldwide level. This demonstrates the urgent need to achieve higher efficiency in gas-driven heat power plants. The maximum efficiency of modern steam-turbine plants (fueled by gas) is about 38–40%. But in Russia this indicator is significantly lower due to the use of outdated equipment. At the same time the efficiency of a combined-cycle plant (CCP) can reach up to 60%, that is, 50% higher. A typical CCP is comprised of a gas-turbine unit (GTU) and a steam-turbine unit (STU). The combustion products from the GTU are fed to the utilisation boiler, which now serves as heat source for the Rankine steam-turbine cycle, running in parallel with the GTU cycle. If we assume that the typical level of efficiency of an STU is 38%, and of a GTU 35%, the total efficiency of electric energy production will be 60%. Thus, CCP should be considered as the mainstream for the modernisation and development of the power industry in our country. Currently, however, only one powerful CCP operates in St-Petersburg, with a capacity of 450 MW.

GTU with steam injection into the combustion chamber – the so-called CCP-STIG (Steam Injected Gas Turbine) is a promising version of CCP. Here there is no steam turbine. This idea was put forward by S. Khristianovich from the Institute of Theoretical and Applied Mechanics (Novosibirsk, Russia). Later on, in 1980 the process was patented in the USA as a "Cheng" cycle, which is used today in several power plants with a power capacity from 2 to 100 MW and efficiency up to 43%. With an increase in the inlet parameters the efficiency can reach 55%.

Several projects for power engines with a capacity of 180 and 300 MW are currently being considered in Russia. An aviation engine is taken as the base for the development of an industrial prototype of CCP-STIG with the capacity of 100 MW. A pilot installation of 60 MW was planned to be launched in Moscow in 2006.

A rather simple method for enhancing the efficiency of small-size plants (50–500 kW) was put forward in the project [6]. The key point here is additional steam superheating up to 800–1000°C before the turbine cycle, aimed at achieving greater efficiency and power. For this purpose, a hydrogen-oxygen mixture is burnt in a special chamber and the water steam produced is mixed with the main steam. The expected efficiency (of electric energy) is 50–55%.

Fig. 3.2 Power setup in Novosibirsk Scientific Center on the basis of 6 MW steam back-pressure turbine (Turbokon Company).

3.3 Cogeneration

Cogeneration, i.e., combined production of electricity and heat, is a generally accepted technology used for the enhancement of energy production efficiency. It has become traditional in the Russian power industry. This concept implies that vapour condensation heat, generated in condensers in the condensation thermal power plants (CPP), which only produce electricity, and is usually dissipated into the environment, could be efficiently utilised for various purposes. In addition, electrical energy could be generated in boiler plants (primarily in steam-driven ones) where heat is produced for heating and hot water supply. This could be accomplished, for instance, by replacing a steam expander ("pressure reducer") with a backpressure turbine [7]. This would provide basic essential economy. Fuel consumption for the production of 1 kWh of electric energy equals 140–150 g.e.f. as compared with 335–345 g.e.f. at RAO UES (Russian Power Grid Joint Stock Company). The power of turbines for boiler plants ranges from 0.2 to 6.0 MW. One of the first pilot turbines of 6 MW is currently being installed in the boiler plant in Academgorodok, Novosibirsk (as a part of the "Turbokon" company project of the city of Kaluga, Figure 3.2). Natural gas-fueled boiler plants could be retrofitted with a gas turbine unit, which is inefficient when used just in a gas-cycle regime. Thus, a boiler plant can be easily transformed into a mini cogeneration plant.

Note that production of electric energy is possible not only using steam and gas turbines, but also by employing so-called hydrosteam turbines [7]. These are driven by hot water which boils up in the expansion nozzles fastened on a Segner wheel, and escapes out of these nozzles in the form of a two-phase mixture. The principle of operation is based solely on the force of the reaction. Such turbines can be used in boiler plants, in geothermic and other systems with sufficiently heated hot water (over 85°C). Pilot plants with a capacity of 10 kW are already in operation, while 300 kW turbine units are in development.

The total energy potential for the use of turbines in steam boiler plants is estimated to be equal to 25,000 MW, which is 12% of all power generating units of RAO UES. Since the cost of electric energy production in this case is twice as low as that of RAO UES, the fuel economy will amount to saving 30 million toe, whereas CO_2 emissions will reduce down to 150 million tonnes. Note also that the cost of 1 kW of installed capacity of a steam turbine of low and moderate power is minimal, $150–300. In comparison, the cost of a steam turbine for large thermal power plants is $1200/kW.

3.4 Coal Technologies

As previously mentioned, an increase in the proportion of coal in the power industry is required in Russia. It is especially important for Siberia, where coal is in abundance and is the main power-generating fuel. In general we can affirm that the development of coal-driven "ecologically clean" power engineering is a priority now and for the future. The use of coal of different grades is a feature in Russia. Thus, on the one hand, it is necessary to develop different methods of combustion (or utilisation) depending on the coal type, and on the other hand it is necessary to increase the grade of applied power-generating coal. Let us consider possible technologies, both those ready for application, and those in development. Regarding the development of boilers, two interrelated problems present the key challenges: an increase in economic yield and a decrease in harmful emissions.

One obvious method for higher thermal efficiency in a traditional steam-turbine cycle is to increase the inlet pressure and temperature of steam, i.e., transition to supercritical steam. In the framework of the "TERMI" international programme, European countries plan (up to 2015) to build coal-fueled plants with a power of 400–1000 MW, steam temperature of 600–700°C, pressure 30 MPa, and efficiency up to 55%. In Russia, similar research work is being carried out based on the operational experience of a plant with a performance of 720 tonnes of vapour per hour, pressure of 30 MPa and temperature of 650°C – in the Kashira State Regional Power Plant (in operation since 1963), and in the boiler plant of the Russian Thermal Engineering Institute (in operation since 1949).

The main principles, used by modern technologies for "ecologically clean" thermal plants, are as follows: staged combustion, coal gasification with subsequent utilisation in combined-cycle plants (CCP), fluidised bed combustion (FBC) and combustion in a circulating fluidised bed (CFB), combustion in melted slag, special methods of flue gas cleaning, etc. The above-mentioned Russian programme "Ecologically Clean Solid Fuel Heat Plants" (1989) consists of nine basic projects for four types of coal: Kansk-Achinsk brown coal, Kuznetsk and Ekibastuz bituminous coals, Donetsk anthracite culm of lower quality. We list these projects, aimed at developing and improving some of the above-mentioned technologies.

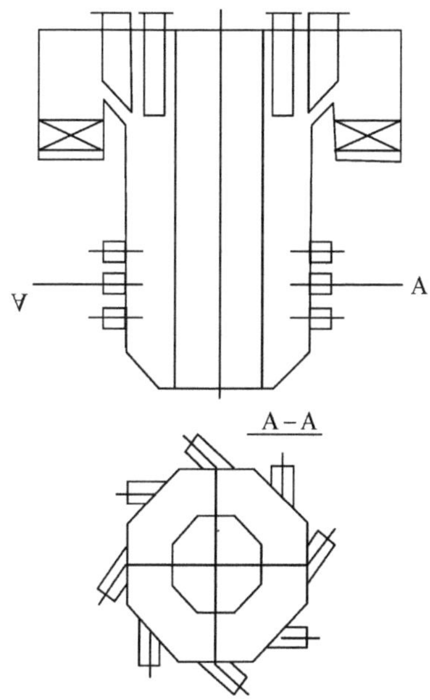

Fig. 3.3 Diagram of an annular furnace.

Kansk-Achinsk coal. Berezovskaya state district power plant-2
Version 1: 6.4 GW, units of 800 MW. Steam boilers with staged combustion in a vortex chamber with liquid slag-tap removal.
Version 2: 5.0 GW with units of CCP-600. Combustion and gasification in melted slag.

Ekibastuz coal. State district power plant in Yuzhno-Uralsk
Version 1: Condensation power plant of 4 GW, units of 500 MW. Boilers with tangentially fired furnace.
Version 2: Condensation power plant of 4 GW, blocks of 500 MW. Boilers with air-plume primary furnaces for preliminary coal gasification and fuel combustion in a circulating fluidised bed.

Kuznetsk bituminous coal. Petrovskaya thermal power plant
Version 1: 6 steam-turbine plants of 180 MW. Combustion in CFB.
Version 2: 2 CCP of 300 MW. Pressure cycle gasification in a fluidised bed and high-pressure steam generators.

Anthracite culm coal. State district power plant in Rostov and Nesvetay, 2.4 GW, units of 300 MW
Version 1: Combustion in CFB.
Version 2: Combustion in primary furnaces with slag-tap removal and regenerative rotating filters with fine filling.
Version 3: Combustion in melted slag.

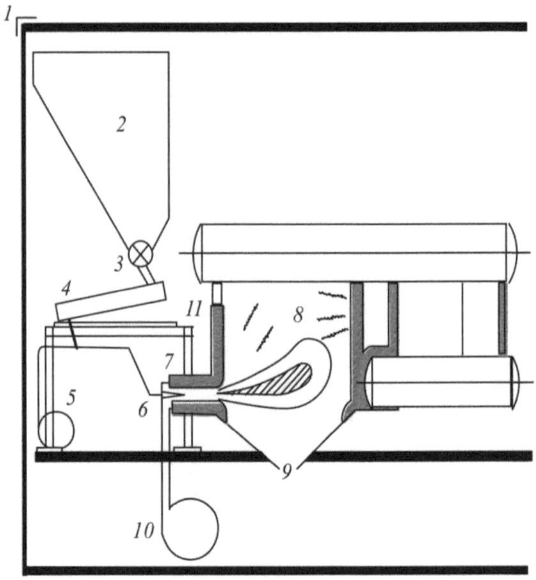

Fig. 3.4 Diagram for ultra-fine pulverised coal combustion. 1 – belt conveyor; 2 – hopper for crushed coal; 3 – coal feeder; 4 – mill for production of ultrafine pulverised coal; 5 – high-pressure fan; 6 – coal-dust burner; 7 – furnace extension; 8 – boiler furnace; 9 – cold funnel; 10 – blow fan; 11 – mill shelf.

These projects are all currently under development. During the last 20 years only several large boilers were put into operation in Russia. Among them, we would like to note the pilot boiler of 820 MW with an annular furnace, produced by Sibtekhenergo, installed in Irkutsk (Figure 3.3). Two screens and good aerodynamics increase heat release and decrease the boiler height by 30–40%, which is important for very large boiler units.

Let us consider some other new technologies currently explored in Russia. As in many other countries, heavy oil (or gas) is used for ignition and combustion stabilisation of low-grade coal in a pulverised coal boiler. Heavy oil is an expensive fuel and difficult to use in cold climates with the low ambient temperatures typical in Russia. To avoid the use of heavy oil, the plasma method for ignition and lighting of pulverised coal boilers has been used. This method was developed by the Center of Plasma Technologies (Gusinoozersk) and the Institute of Thermophysics. The main element for this technology is an arc plasmatron. The plasma system has been patented in many countries, and is now installed in over 70 boilers of different sizes, ranging from 75 to 670 tonnes of steam production per hour.

One of the modern directions in power engineering is the use of ultra-fine pulverised coal, with the average size of coal particles of 5–40 μm [8]. Its reaction properties are close to those of heavy oil. Therefore, one of its applications is substitution of heavy oil in the igniting and lighting systems of pulverised-coal boilers, similar to the plasma system. Another method is direct combustion of ultra-fine

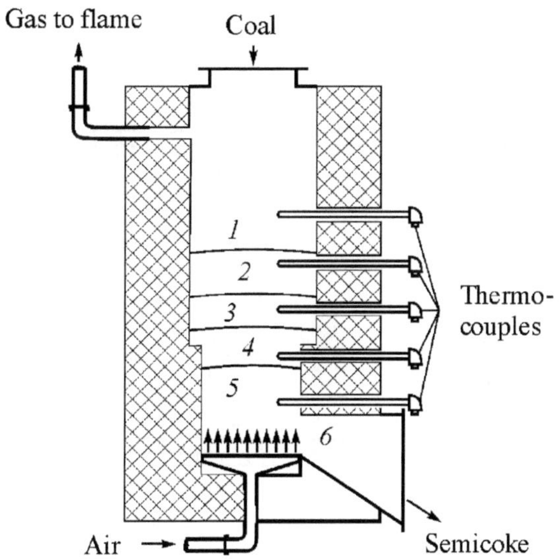

Fig. 3.5 Diagram of fuel-bed coal gasification according to "Carbonica" technology. 1 – activation zone; 2 – carbon oxidation zone; 3 – zone for pyrolysis and volatile combustion; 4 – zone for pyrolysis and volatile combustion; 5 – wet coal; 6 – gas distributing device.

coal in the gas-and-oil fired boilers of moderate power. A combustion stand with the power of 1 MW was developed at the Institute of Thermophysics, where different types of coal were combusted. A gas and heavy-oil fired boiler DE-6 of 4 MW power, produced by the Biisk boiler plant, has been assembled in Novosibirsk for the combustion of micronised coal (Figure 3.4). Special mills are used for coal micronising, in particular a vibration-centrifugal mill with productivity from 100 kg/h to 5 t/h, developed at the Siberian Branch of RAS.

Another known, but not yet implemented method deals with the possibility of direct combustion of micronised coal (5–10 μm) in a gas turbine, included in the gas-steam cycle (CCP). This project is being performed in the framework of a special programme of the Siberian Branch of RAS.

As for gasification, we should note that even in 1958 in the USSR, there were 2500 gas generators with a productivity of 15 million tonnes of coal per year. Currently there are several interesting developments. The "Carbonica" technology has been brought up to the industrial level in Krasnoyarsk [9]. This is the process of layer gasification of volatile coal components at air blast under the mode of so-called "reverse heat wave" (Figure 3.5). The main purpose of this technology is the production of a high-grade coal adsorbent. This technology can be easily adapted for the purposes of power generation.

Much attention is paid to plasma gasification technologies. New technologies, using both ordinary arc plasmatrons and new-type plasmatrons with liquid electrodes, were proposed at the Institute of Thermophysics (Figure 3.6). The new-type plasmatrons use melted metal as an electrode, so there is no problem of electrode wear-

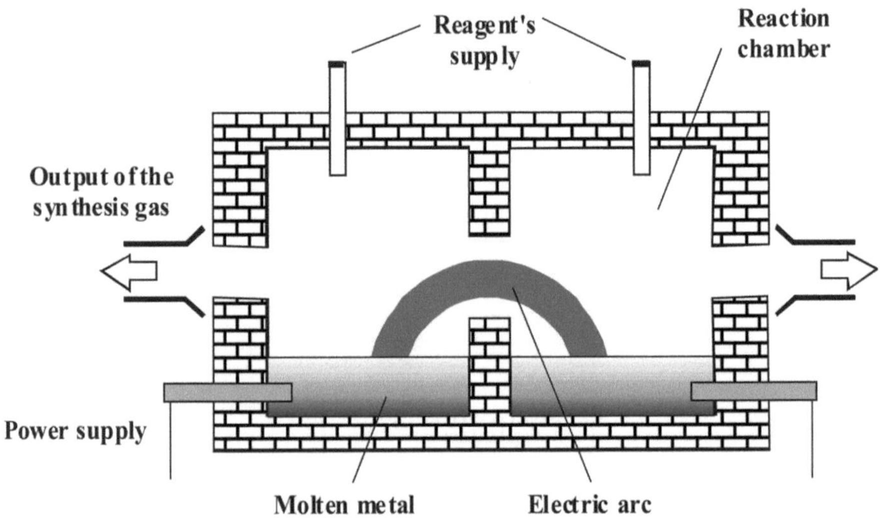

Fig. 3.6 Plasmatron with molten-metal electrodes.

ing [10]. This plasmatron has been aimed in particular at the treatment of wastes from coal industry. A pilot equipment of 0.5 MW power is now available. We should note that there are no limitations to power.

An original approach [11] has been proposed at the Institute of Thermophysics for the application of water under the supercritical state (temperature above 374°C, pressure above 22 MPa). Under this state water can unrestrictedly dissolve organic substances and oxygen. Various applications are possible: treatment of coal, heavy residue, biological silt, sewage runoff with production of liquid hydrocarbon fuel (LHCF) or complete combustion of organic substances, aimed at the generation of high-enthalpy working media, to be used in a combined-cycle plant. Such a project, completed recently in Novosibirsk, utilises sewage runoff with a 25% share of organic substances. A tube reactor with a volume of 39 litres will produce 70 kg/h of LHCF by consuming 234 kg/h of sewage runoff. A bundle of 100 tubes with the area of 2.5 m^2 will produce 7 t/h of LHCF.

3.5 Renewable Energy Sources

The potential of renewable energy sources in Russia is huge. Considering the technical potential, i.e., the potential that can be utilised by currently available technologies, the estimated capability for Russia is 4.6 billion toe. This is five times larger than the total energy consumption of the country, or, by considering different types of renewable energy sources (RES), the potential of each is: biomass –

Table 3.1 Existing power of RES plants – worldwide data (2000).

Sources	Electric energy, GW	Heat, GW
Small rivers	70	–
Biomass	30	200
Wind	31	–
Geothermal energy	8	17
Photovoltaic	0.94	–
Solar HPP	0.4	–
Solar collectors	–	17
RES contribution to electric energy production = 1.6%		

Table 3.2 Role of RES.

Share of RES in the total energy consumption, %			
	2000	2010	2020
Russia	1.2	1.9	4.3
EU	4	12	4.3
Share of RES in the total electric power production, %			
	2000	2010	2020
Russia	0.5	1.0	1.5–2.0
EU	2.9	12	
Denmark	12.0		

53 million toe, solar energy – 2300, wind energy – 2000, geothermal energy – 180, low-potential heat – 115, energy of small waterways – 125.

The current contribution of RES in the energy balance is obvious from Tables 3.1 and 3.2. Table 3.1 demonstrates the installed capacity of RES plants and the second table shows the role of RES in energy consumption and electric power generation. This table evidences that biomass provides the maximum contribution to heat production, whereas biomass, small rivers, and wind plants contribute to electric energy production. But the total contribution of RES to worldwide electricity production is very low – just 1.6%. Table 3.2 illustrates the role of RES in the energy balance of Russia and EU countries.

We see that today (and in the near future up to 2020) the role of RES in the energy balance of Russia is extremely low – 1–2% of the total electric power production. This is close to the forecast of the Russian Energy Strategy. At the same time the EU countries have more impressive plans concerning RES, projecting much higher percentages than Russia. According to recent data, by the year of 2012, 48 countries (including 14 developing countries) will produce from 5 to 30% of their electric energy from RES. Economic data evidence a large increase of investment in the development of RES (about 30 billion dollars) in 2004. This is 20–25% of total investments in the power industry.

What can motivate business to use the renewable energy sources in Russia, and why its share is so small? The motivation is the same as for energy saving. Firstly, there are limited deposits of organic fuel. According to official statistics, there is enough stock of easily recoverable gas in Russia for 80 years, and of oil for 20 years. Another motive is based on energy safety issues. The next point involves environmental concerns. We know that the greatest pollution of the environment comes from organic fuel consumed by the conventional power industry. But the Kyoto Protocol has declared the objective to keep CO_2 emissions for 2008–2012 at the 1990 levels – which means a need for a drastic reduction in combustion of organic fuel through conventional methods. However this is not an actual problem for Russia today because of the recent decrease in industrial production.

It appears that the specificity of Russian geography could be the main motivation for further RES development, since in many regions in Russia (especially in Siberia) access to centralised systems of energy supply is difficult. By some estimates, up to 50–70% of the territory of Russia (with a population of 20 million) is not provided for by centralised systems of energy supply. For these places, therefore, the renewable energy sources would be the only sources of energy that could provide a comfortable living environment for the population. Unfortunately, Russian government bodies do not pay sufficient attention to the development of non-conventional power engineering and RES-based plants. In general, only initiative-level projects are carried out. The energy-saving programme of the Siberian Branch of RAS for 2005 dealt with 42 financed projects, 13 of which were directly linked to non-conventional power engineering, including features such as heat pumps, processing of biomass, deep processing of fuel, gasification, solar voltaic, etc. All profiled institutes of SB RAS are involved in these projects, though we still feel the lack of a targeted policy for the development of non-conventional power engineering.

Let us consider some of the main varieties of renewable energy sources and their development in Russia. We will start with heat pumps – the key elements in utilisation of low-potential heat. However, there is a more general concept of a thermal transformer, which can operate in different modes: as a heat pump, cooling machine, and a machine for combined production of heat and cold. The main feature of a heat pump is that the produced heat is greater than the heat taken from the high-potential heat source. The point is that heat is produced not only from the main energy source (gas, coal, steam, electric energy), but is also contributed to by the thermal energy taken from the low-potential source (this is a source with a lower temperature – geothermal springs, liquid waste effluents from industry and households, the warmth of air, soil, river). With the use of commercial machines, the fuel saving may be 20–70%. The temperature of the low-potential source may range broadly (from +80° to −17°C).

The energy-saving policy in many industrialised countries is based on the deployment of heat pumps. For example, 22% of Swedish houses (350,000) are heated by heat pumps. Nowadays the world runs 40 million heat pumps, although Russia uses only 140 heat pumps.

It is estimated that by the year 2020 the impact of heat pumps on the total heat supply in developed countries will amount to 75%. Russian business is indifferent to

Fig. 3.7 Absorption lithium bromide heat pumps and refrigerating machines of 2 MW.

heat pumps. The key developers and producers of heat pumps in Russia are situated in Novosibirsk. The research and development basis is provided by the Institute of Thermophysics. "Teplosibmash" company is the leading manufacturer of adsorption machines (Figure 3.7). Up to now, this company has produced six heat pumps and seven cooling machines with an overall capacity of 23 MW [12]. "Energia" and "SKB IPI" are the companies that manufacture vapour-compressor heat pumps and cooling machines, with a capacity of up to 5 MW. These are the only companies that produce heat pumps in Russia.

Insufficient attention is paid to the Stirling engine, which is a versatile thermal machine. This is an external combustion engine with a simple design and which is independent on the type of fuel. The working medium is a gas with low molecular weight like hydrogen or helium – thus making the engine itself environmentally friendly. It is now attracting the attention of developers because of its use in autonomous energy supply systems. Yet its usage is still limited. In Austria there is a good example of the use of such an engine coupled with RES at a wood-fueled demo plant, with the power of 35 MW (elec.) and 20% efficiency. Although this is quite moderate efficiency, it should be recalled that the efficiency of a steam-power cycle in the same conditions is two times lower. An original design of such an engine with liquid pistons is currently being developed at the Institute of Thermophysics.

Solar energy has the highest potential among renewable sources. Solar power engineering is developed in three basic directions: solar water-heating devices, solar electric power stations, and photovoltaic transformers. A solar water-heating setup uses a planar or curved (parabolic or circular) solar collector for heating water, air or other heat carriers. Here the main characteristic is the heating surface area. The total (worldwide) surface area of solar collectors is 50–60 million m^2, which is equivalent to 5–7 million toe per year. In Russia there is as yet no significant application of solar energy. However, even under the Siberian conditions we could expect a positive output. A project producing solar collectors for individual housing, the ECODOME project, is currently underway. Different types of solar collectors including underground heat accumulators will be assembled at the construction site of the ECO-village near Academgorodok.

A regular steam-power cycle is used for Solar Electric Power Stations: this requires a solar energy concentrator. The USA has now seven such stations with a total capacity of 354 MW. However, this type of station is not being considered as attractive for Russia.

As for photovoltaic transducers (PVT), today there is a great boom in this technology. In 2000, the total power of all PVT in the world was 260 MW. Japan, with 80 MW, was the leader. The contribution of Russia is very low, just 0.5 MW. The efficiency of PVT can be as much as 24% for monocrystalline devices, 17% for polycrystalline, and 11% for amorphous devices. The key material is silicon. Unfortunately, this is the most expensive method of electricity production. Photovoltaic panels cost up to \$4,000 per kW, and a whole plant with panels would cost around \$10,000 per kW. The cost of the electric energy produced is also high: 0.15–0.40 \$/kWh. There are several prospective directions in the photovoltaic field: PVT with solar energy concentrators, PVT based on GaAs–GaAl, and thin-film solar cells.

In our opinion, the thin-film solar cells could be of particular interest as they show good results and are relatively cheap because of the low mass of photosensitive material and inexpensive technologies. The jet plasma chemical deposition of silicon films, developed at the Institute of Thermophysics, is a good example of such a technology, based on the following principles [13]. A supersonic jet of monosilane flowing into argon is irradiated by an electron beam and the excited jet forms a silicon deposition on the substrate (a thin film of amorphous or polycrystalline silicon). The rate of silicon deposition is very high: 500 times higher than for diffusion-based methods. This technique also demonstrates high energy efficiency. The combination of these factors should lower the cost of the product and make it available for mass production. According to our current estimate, the investment in a solar cell plant operating on this technology will be by one order of magnitude lower than that for foreign analogues. Moreover, the production time, as well as estimated investments in a plant for production of solar cells of the same power, are expected to be also less by an order of magnitude. Figure 3.8 shows a photograph of the pilot plant built at the Institute of Thermophysics for silicon film production with the jet plasma chemical method. We plan to launch solar silicon production in Novosibirsk on an industrial scale.

Germany, USA and Denmark have a large interest in the development of wind energy projects. In 2002, the total (worldwide) power of wind plants was 31.1 GW, with a projection to increase at a higher rate in the near future. However, the drawbacks of wind energy are the high level of noise and the large area of land required for wind farms. In Russia, wind energy applications are not common and they are usually based on imported machinery. However, the Institute of Theoretical and Applied Mechanics (Novosibirsk) has developed an original design for a wind-energy plant with rotating cylinders. This design has advantages at low wind velocities (2–6 m/s). It is planned that the industrial manufacture of this design will be carried out in Omsk (Russia).

Fig. 3.8 Pilot plant for silicon film deposition at Institute of Thermophysics.

Geothermal energy, as another RES, could have good prospects in Russia, as Russia has a huge potential of geothermal energy, with West Siberia being the richest source in the country. It is believed that geothermal sources with a temperature higher than 100°C are suitable for the production of electric energy in geo-power plants (GeoPP). If the temperature is slightly below 100°C, the hot water can be used for heat supply, whereas lower temperatures require a heat pump. The institutes of the Siberian Branch of RAS have gained extensive experience in geothermal energy. In 1970, Paratunka GeoPP (a project of the Institute of Thermophysics) was built on the Kamchatka peninsula. The power plant is equipped with a Freon turbine with a capacity of 815 kW (operating with hot water of 80°C). Currently the total power of worldwide geothermal heat supply is 17 GW, and the capacity of GeoPP is 10 GW. At present, Kamchatka (Russia) has Verkhnemutnovskaya GeoPP (12 MW) and the first unit of Mutnovskaya GeoPP (50 MW) in operation, with ultimate plans for 220 MW. Note that the territory of the Novosibirsk region has geothermal sources of warm water (39°C), and in the Tomsk region these sources have a temperature of 85°C. We have also experience in building heat pumps for agricultural farms. One method for the production of electric energy from hot springs is based on the use of the hydro-steam turbine, mentioned above. This technology was developed independently by the "Energia" and "Turbokon" companies.

Solid municipal wastes (SMW) can be considered a kind of secondary renewable energy source. Municipal and industrial waste treatment is a big problem for modern society. However the point is that the SMW can be used for the production of thermal and electric power. At present the USA disposes of an unprecedented amount of SMW – 250 million tonnes per year. About 10% of this waste is used for production of heat/electricity at incineration plants. The US had 125 waste incineration plants in 1993. Japan runs 1800 incineration facilities, burning 72% of municipal waste. Several countries have adopted national programmes for the processing of waste and by-production of thermal and electric energy. Russia produces 60 million tonnes of SMW, but has only five incineration plants. Only two of them are equipped with imported modern technical facilities. However, several large modern incinerations plants are in the offing: the Moscow authorities plan by the year 2010 to incinerate at least half of the total amount of the municipal waste by production of energy. The city of Novosibirsk, which produces 0.5 million tonnes of solid waste annually, is also considering such an option. There are, of course, some problems associated with waste-processing energy-production techniques, and these have been discussed extensively in the literature.

There are also other methods for obtaining energy from SMW, which may be more suitable for some regions and communities than incineration. One of these results in the production of biogas due to anaerobic fermentation at disposal tips. Biogas is a binary mix of methane and carbon dioxide. It can be cleaned and used for combustion. Today Russia has only two demo setups. Another method for waste treatment is based on thermal plasma (at high temperatures), where all organic components are released and no hazardous emissions (dioxines and furans) are produced. We have also developed methods for pyrolysis, gasification, and combustion of solid wastes with output of syngas and heat in facilities with an electric-arc plasmatron, and a plasmatron with liquid-metal electrodes (Figure 3.6) [10].

In our opinion, this is a relatively costly and complicated technology, more suited for disposal of hazardous wastes or for special purposes. The mass processing of municipal waste is better performed at the District Complex Heat Station [14]. This facility is a joint project of experts from the Institute of Thermophysics, "Tekhenergokhimprom", and VNIPIET (Figure 3.9). The project has been developed using a rotating drum kiln with further combustion of combustible gases in a vortex afterburner. The project includes deep cleaning of flue gases to meet the EU standards on hazardous emissions. The distinguishing feature of this project is the opportunity to produce thermal and electric energy as well as construction materials. The capacity of this station is 40,000 tonnes of SMW per year. This covers the needs of a town/district with a population of 100,000 and produces 100,000 Gcal of heat as a by-product. The investment required is 8–10 million dollars, which is several times lower than similar foreign projects. Currently a number of projects have been developed for two sites in the Soviet district of Novosibirsk – the Gusinobrodsk scrap-heap of Novosibirsk, the city of Berdsk, and in the Sakhalin region.

3.6 New Types of Energy Sources

Hydrogen energy based on the use of hydrogen as fuel is the most promising direction for power engineering of the future. The caloric efficiency of hydrogen (120 MJ/kg) is 2.5 times higher than that of natural gas, 3 times higher than petrol,

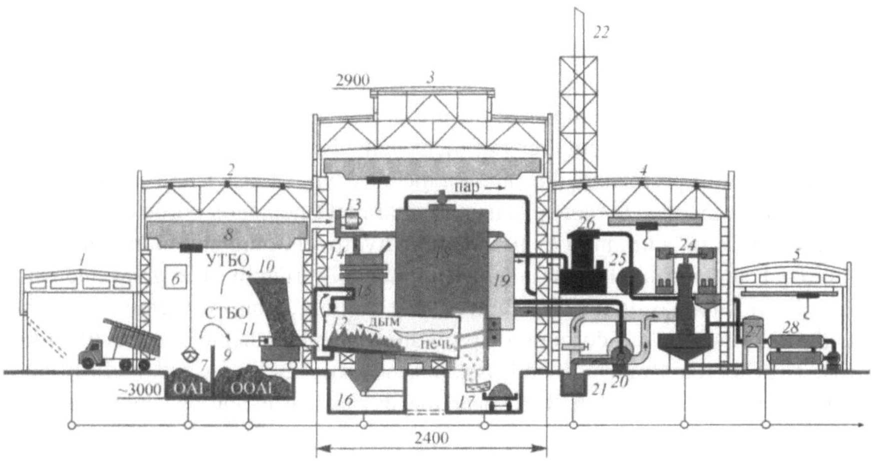

Fig. 3.9 Schematics of a District Complex Heat Station with the input of 40,000 tonnes of solid municipal waste (SMW) per year 1 – SMW discharge room; 2 – bin storage; 3 – SMW incineration section; 4 – flue gas cleaning section; 5 – heat processing section; 6 – window for bin storage operator; 7 – receiving bin of fresh SMW; 8 – charging crane; 9 – bin for "averaged" SMW; 10 – charging funnel for SMW; 11 – feeder to the furnace; 12 – rotating drum furnace; 13 – blow fan of combustion system; 14 – first flue from furnace; 15 – after-burner; 16 – ash collector pit; 17 – slag discharge pit; 18 – steam boiler-utiliser; 19 – gas heater; 20 – smoke exhaust; 21 – exit flue; 22 – smoke stack; 23 – jet-foam fly-ash collector; 24 – alkaline absorber; 25 – mist absorber; 26 - condenser; 27 – water conditioning apparatus; 28 – heat pumps and boilers of thermal section.

and 6 times higher than coal. It is supposed that hydrogen will be used in fuel cells (FC), which in fact are electrochemical generators for direct conversion of chemical energy into electric energy. The high-temperature FC also gives useful heat which could be utilised. Unlike an electrochemical cell, here we have the consumable materials – fuel and an oxidiser. The most popular method is using hydrogen as fuel and oxygen as an oxidiser. Water is the only output product of this chemical reaction; so the fuel cell is an energy source with zero emission. As for the energy conversion, fuel cells are highly efficient because their maximal efficiency (coefficient of transformation of chemical energy into electric) can go up to 50–70%. However, sufficiently high reaction rates require the use of a catalyst – i.e., platinum group metals. Although the idea of fuel cells was proposed 150 years ago, fuel cells are still far from widespread commercialisation because they, and the energy they produce, remain expensive. There are also engineering problems – which could be solved at a new level of technical development. Today we are witnessing a new wave of interest in fuel cells. Many companies and research organisations are considering different schemes and work on applications for fuel cells. Power engineering, space technology, transport, and microelectronics are the main focuses for applications. However, large-scale commercialisation of FC is expected within the next 20–30 years. One of the technical problems is that the electrode (cathode) and membranes require a very extensive surface. Here success very much depends on the development of nanotechnologies, which will make it possible to produce nanotubes, nanocones, fullerenes (with sizes of only several nanometers). These nanostructures might become the basis for new and efficient fuel cell components. Today there are several examples of the use of fuel cells in power plants, though as yet their total capacity is only tens of megawatts. Note, that though hydrogen fuel cells are the most promising, they are not the only type of FC. There are several developments of small-size fuel cells using liquid fuel (methanol, boron compounds), as well as FCs on aluminum as fuel. Unlike hydrogen, pure aluminum and boron compounds are absolutely safe and environmentally friendly. The great interest in fuel cells in Russia is driven by a joint project of the Russian Academy of Sciences and the "Norilsk Nickel" corporation providing annual financing of up to 40 million USD. The Siberian Branch of the Russian Academy of Sciences and the Institute of Thermophysics is conducting R&D in all the above areas including nanotechnology.

3.7 Methods of Efficient Use of Energy

The above described novel and efficient energy technologies and measures belong to the energy production sector, where the proportion of total energy-saving potential is equal to one third of the whole Russian economy. This amounts to 150 million tonnes of equivalent fuel per year. The greatest potential is aimed at the consumer, first in industry (1/3), and in the housing and communal services (1/4). Because of the great variety of different subjects we will not provide a detailed analysis here. We will mention only common considerations and approaches.

A qualified energy audit of a certain subject – enterprise, building or technical complex – should be carried out before executing energy-saving measures. This is necessary for an evaluation of the energy-saving potential and for the determination of efficient measures both in the short term (quick payback) and the long term. The availability of control instruments for the measurement of energy consumption, as well as the existence of a control system (for instance, to maintain a given indoor temperature dependent on external air temperature, time of day and season) is of great importance for a higher standard of performance. The latter measure can save up to 30% of the expended energy. An overwhelming amount of heat loss in buildings takes place through the walls, doors and ventilation system. Therefore, it is very important to achieve and maintain the new standards of thermal insulation, and to develop and use new materials and heat insulators during the construction of new buildings. Double and triple glazing should be applied to windows, as well as heat-saving optically transparent glass coatings – which still require further research and development. Ventilation causes a great problem. Huge heat losses (up to 60%) are caused by influx-and-extract ventilation systems, which are usually unavoidable in most chemical industries. Therefore the development of ventilation systems based on heat and moisture recuperation is of particular importance. The same applies to air conditioning systems. The problem of energy efficiency for individual accommodations (cottages, village houses) is a separate issue. The problem here

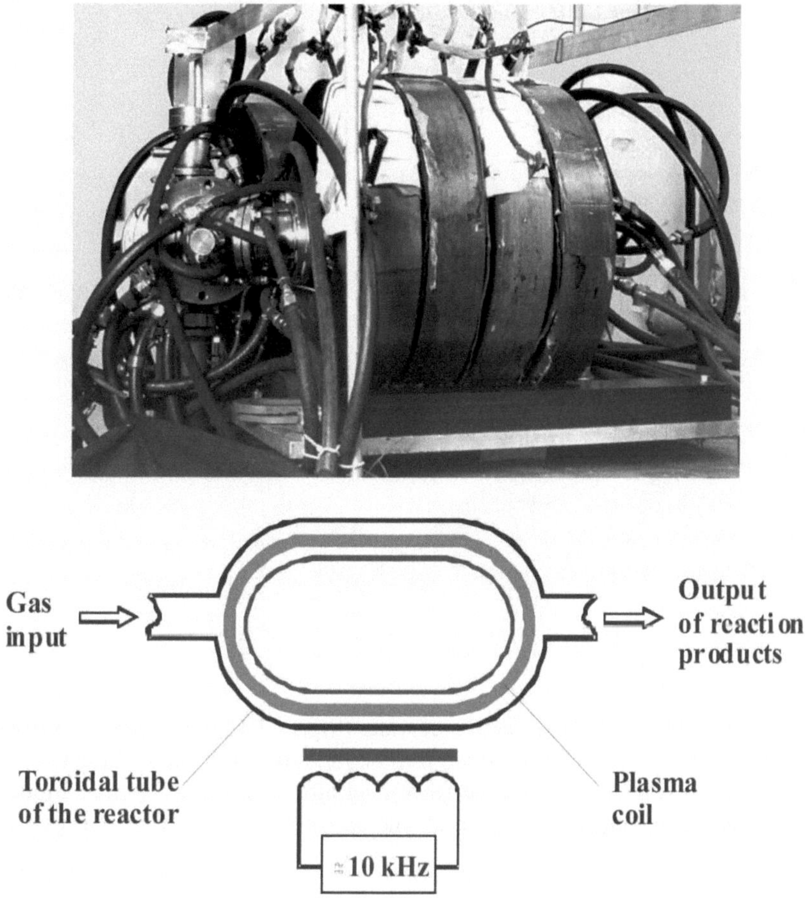

Fig. 3.10 Transformer plasmatron.

concerns energy registration (which for isolated small objects is economically inexpedient), autonomous power suppliers, and enhanced heat losses compared with tenement-houses. However in this area it is possible to produce solutions based on non-traditional methods, which can sometimes result in unexpected and significant gains.

Quite a large amount of energy is allocated for lighting systems, which spend up to 20% of the total electric energy produced worldwide. A focused search for new energy-saving and resource-saving light sources with a high light output, up to 100–150 Lm/W, and a lifetime of up to 10 years is a priority. Light-emitting diodes and gas-discharge lamps including inductive electrode-free lamps [15] are currently among the most practical developments. The latter type of lamp is based on the principle of a transformer plasmatron (Figure 3.10). Since there are no electrodes, the durability of such a lamp may extend to 10 years. Induction mercury lamps used

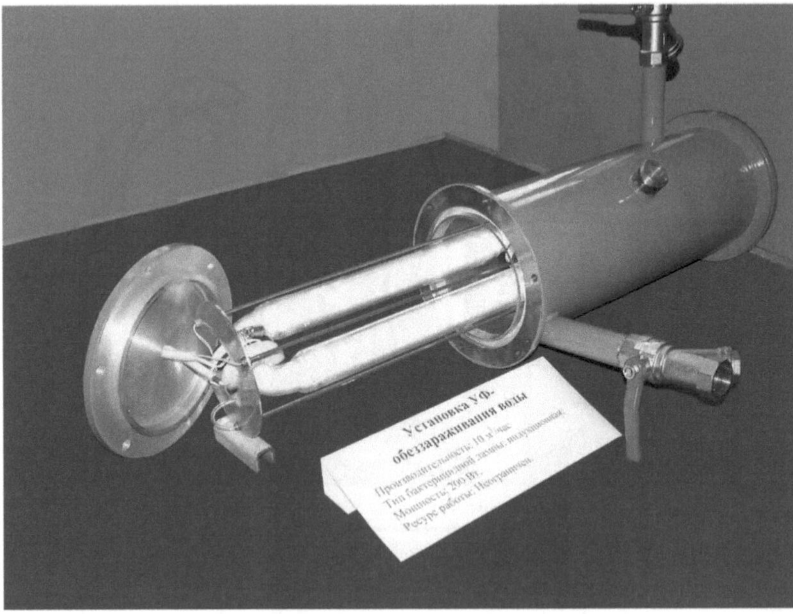

Fig. 3.11 Unit for ultraviolet disinfection of water.

in devices for ultra-violet disinfection of water (Figure 3.11) are of great interest and potential. Sodium lamps are considered to be the most efficient for street lighting. Note that there is a Council on Energy Saving affiliated to the Siberian Branch of RAS, which coordinates research in all the above mentioned basic energy-saving problems [3].

3.8 Conclusions

While analysing the state of the Russian energy situation and its power industry, taking into consideration global trends and strict demands in terms of ecology, economy, efficiency, and safety, the following conclusions were made.

The future stable development of the Russian power industry is linked with the necessity of a drastic increase in investments aimed at the modernisation of out-of-date energy equipment and installations (estimated to exceed 50% of their planned life time), and the introduction of new "ecologically clean" technologies. The priorities are as follows: an increase in the efficiency of plants operating on natural gas; a significant increase in the share of coal for power engineering based on the integration of new technologies; and a subsequent growth in nuclear power engineering. Specific attention should be paid to energy and resource saving as the most important component of the energy supply in Russia. As for renewable energy sources and other technologies, e.g., hydrogen power engineering, they should undoubtedly be developed. However, their near-term contribution to national energy production will be insignificant, despite their potential importance for particular regions of Russia.

References

1. *Power Engineering in Russia. Development Strategy*, Scientific Ground of Energy Policy, Russian Ministry of Energy, Moscow, 2003.
2. Laponsh, B., Filimon, T. and Zhgenti, V., Russian-European dialog on Russian power engineering and energy strategy – Determinant role of energy saving, *Teploenergetika* **7**, 2004, 64–73.
3. Alekseenko, S.V., Energy saving solutions, in *Proceedings of a Scientific Seminar of the Presidium of Siberian Branch of RAS "Russian Power Engineering Development Problems"*, Novosibirsk, February 24, 2005, pp. 145–159.
4. Alekseenko, S.V., Power engineering, non conventional power engineering and energy saving resources, in *Proceedings of a Scientific Seminar of the Presidium of Siberian Branch of RAS "The Problems of Non Conventional Power Engineering"*, Novosibirsk, December 13, 2005, pp. 12–31.
5. Salomatov, V.V., *Nature-Saving Technologies at Thermal and Nuclear Power Plants*, Novosibirsk State Technical University, Novosibirsk, 2006.
6. Leontiev, A.I., Fedorov, V.A. and Milman, O.O., Power system on a combined fuel, Patent of RF for utility model No. 30848 of 10 October 2002.
7. Favorsky, O.N., Leontiev, A.I., Fedorov, V.A. and Milman, O.O., Efficient technologies for production of electric and heat energy using the organic fuel, *Thermal Engineering* **9**, 2003, 19–21.
8. Alekseenko, S.V., Burdukov, A.P., Chernova, G.V. and Churashev, V.N., Power efficient and environmentally responsible technologies for reconstruction and modernization of coal power engineering, *Izvestia RAN. Energetika* **2**, 2003, 52–63.
9. Islamov, S.R., Stepanov, S.G., Morozov, A.B. and Groo, A.A., Production of carbonaceous reducing agents from D grade coals based on "Carbonica" technology, in *Collection: Modern Technologies for Development of Mineral Resources*, Krasnoyarsk State University of Non-Ferrous Metals and Gold, Krasnoyarsk, 2005, Vol. 3, pp. 289–294.
10. Predtechensky, M.R., Kuropyatnik, I.N. and Tukhto, O.M., Novel plasma arc reactor with molted metal electrodes for coal gasification technology, in *Proceedings of ISPC 15*, Orleans, France, July 9–13, 2001, pp. 1187–1190.
11. Vostrikov, A.A. and Psarov, S.A., The explosive oxidation of hydrocarbons in a supercritical water, *Technical Physics Letters* **28**(9), 2002, 776–778.
12. Baranenko, A.V., Timofeevsky, L.S., Dolotov, A.G. and Popov, A.V., Absorption heat converters, St. Petersburg State University of Low-Temperature and Food Technologies, St. Petersburg, 2005.
13. Sharafutdinov, R.G., Skrypnikov, A.V. and Polisan, A.A., The deposition method for the films of hydrogenised silicon, Patent of RF No.2100477 of 27 December 1997.
14. Alekseenko, S.V., Basin, A.S., Gordienko, A.A. and Malykhin, V.A., Development of demonstration unit of the combined district heat plant, *Problemy Energosberezhenia* **1**, 2003, 10–12.
15. Ulanov, I.M., Nonelectrode discharge lamp, Patent of RF No. 2156008 of 10 September 2000.

Chapter 4
Perspectives on Wind Energy

Gijs van Kuik, Bart Ummels and Ralph Hendriks

Abstract. Wind energy has conquered many startup problems and has attained in a new, more mature phase. The annual market volume increases by 20–30% per year, making significant contributions to the national electricity supply in some countries (DK, DE, ES). Wind turbines have increased in size, achieved compliance with severe grid connection requirements and have cautiously gone offshore. The cost of wind-generated electricity approaches the cost of fossil-fuel-generated electricity. With this success, new challenges arise: the integration of wind energy into the electricity supply system, the step from wind turbines into (offshore) wind power plants, the associated extreme reliability requirements and the prospect of a continued cost decrease. The chapter presents the state of the art, and discusses perspectives on these issues.

Key words: Wind energy, wind turbines, grid integration, generation cost, research.

4.1 Introduction

Modern wind energy is a relatively young (about 30 years) but rapidly expanding power source. Over the past decade, global installed capacity has increased from 2.5 Gigawatts (GW) in 1992 to almost 75 GW at the end of 2006, at an annual growth rate of near 30%, see Figure 4.1. Seventy percent of this capacity has been installed in Europe. Penetration levels in the electricity sector (percentage of wind energy in the national electricity supply) have reached 20% in Denmark, about 6% in Germany and 8% in Spain [1]. If positive policy support continues to develop, the European Wind Energy Association (EWEA) has projected that wind power will achieve an installed capacity of 75,000 MW in the EU-15 by 2010. This would represent an overall contribution to electricity supply of 5.5%. By 2020, this figure is expected to increase to more than 12% [2]. The present status and the perspective for 2020 and beyond show that wind energy has evolved from a "small and nice"

K. Hanjalić et al., Sustainable Energy Technologies: Options and Prospects, 75–97.

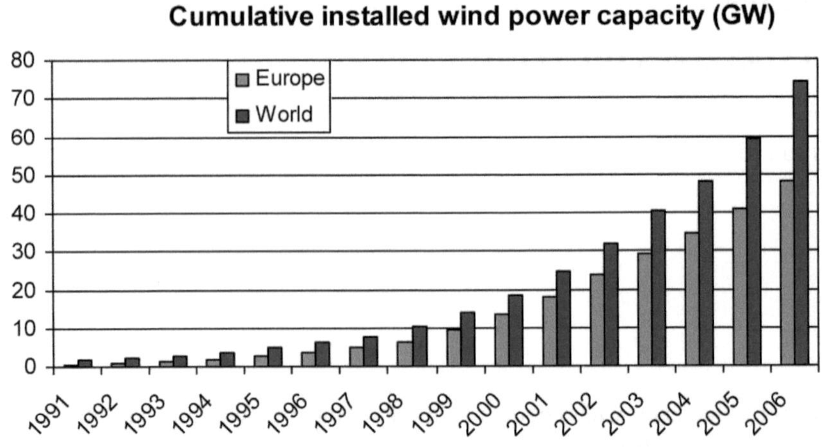

Fig. 4.1 Cumulative development of installed wind power capacity, in GW.

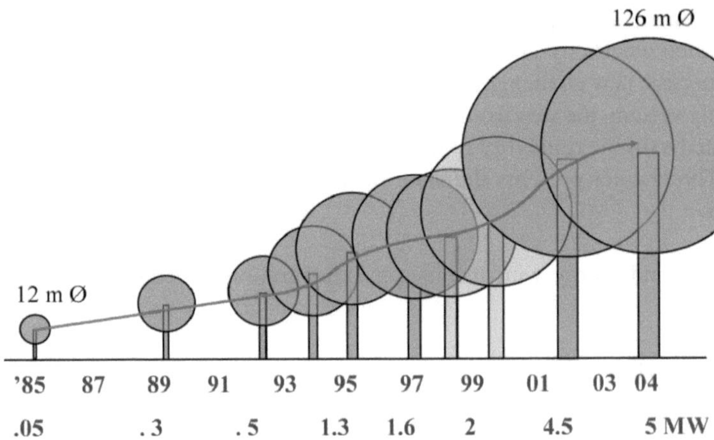

Fig. 4.2 The development of wind turbine size.

alternative energy source driven by environmentally concerned people, to one of the fastest growing energy sources nowadays, together with which several countries will satisfy a large portion of their Kyoto obligations to reduce greenhouse gasses. It is not only the market volume that shows that wind energy has overcome its initial growing pains. The wind turbine industry is led by large multinational companies like Vestas, Siemens, General Electric and Mitsubishi, who will construct within a few years the first wind power plants with the size of conventional power plants (500–1000 MW). The turbines themselves have increased in size too. Turbines with an installed generator capacity of 5 to 6 MW and a diameter of 110–120 m have now been installed, see Figure 4.2.

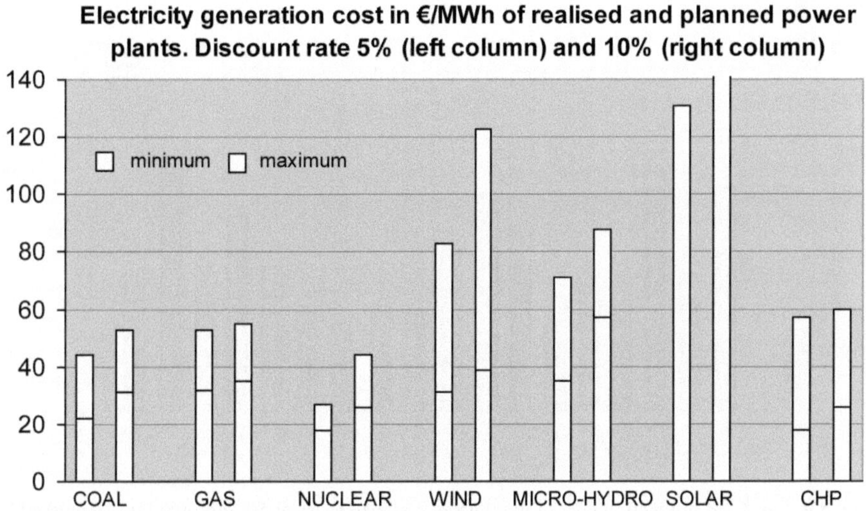

Fig. 4.3 Comparison of electricity generation costs, according to the IEA-NEA [3]. Levelised cost calculation, investment recovery time 40 y, for wind 20 y. Based on 2003 US$, exchange rates €/$ 1.14.

So far, the development of wind energy has been successful. However, further penetration of wind in the electricity supply system depends on two major challenges: a continued cost reduction that makes state incentives eventually superfluous, and efficient measures to integrate wind energy production in the electricity supply system. Both topics will be discussed below.

4.2 Market and Cost Perspectives

Wind energy is nowadays one of the sustainable energy sources that is closest to current electricity market price levels. Figure 4.3 shows wind energy generation costs compared with other generation costs from a recent comparative study by the Nuclear and International Energy Agency [3]. All generation methods are compared on an equal basis: levelised production cost, based on discount rates of 5 and 10%, and an investment recovery time of 40 years for fossil and nuclear power plants, 20 years for wind. The costs for integration of each source into the electricity grid are not taken into account (see the section on Grid Integration), nor the environmental damage or benefits. The calculations for wind energy are based on realised and planned projects in several countries, amounting to 1000 MW of installed power in total, of which 740 MW is offshore. Figure 4.4 shows the results per project. Unlike Figure 4.3, the cost is expressed in US-2003 $/MWh. The average for all countries, weighted with the installed capacity, is 65 $/MWh, or 57 €/MWh (based on an exchange rate of 1.14 €/$ used in the report).

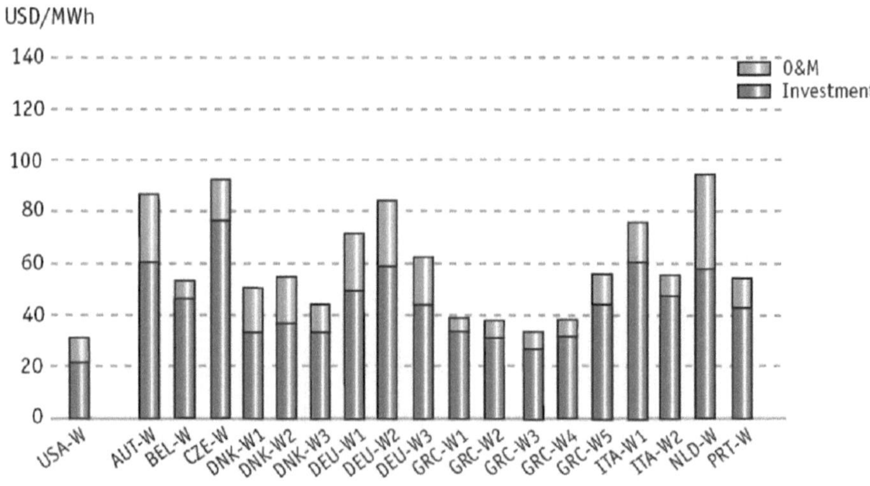

Fig. 4.4 Levelised production cost of wind generated electricity at 5% discount rate, expressed in US $/MWh [3].

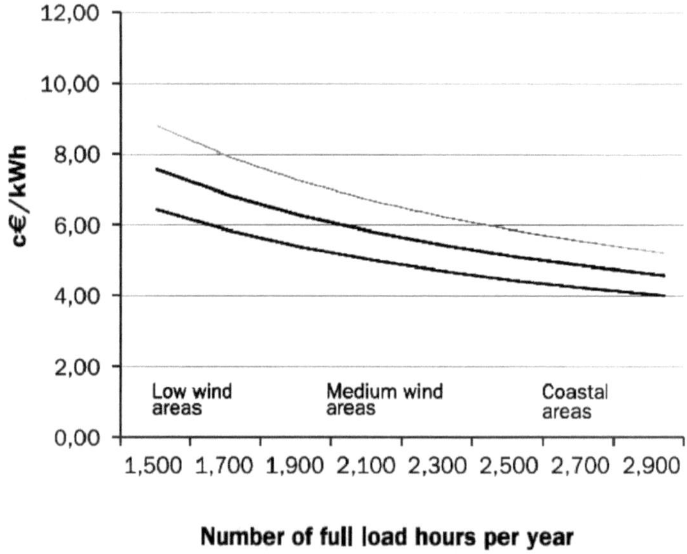

Fig. 4.5 The costs of wind power as a function of wind speed (or number of full load hours) and discount rate. The lowest line holds for 5% p.a., the middle for 7.5% and the highest for 10% discount rate [4].

The results are quite sensitive to the costs of the large offshore projects. The offshore contribution exists of four projects: two realised in Denmark (DNK-W1 and W2) and two in preparation (Germany, DEU-W1, and the Netherlands, NDL-W). If the latter two are omitted since these costs are mere estimates, the average level

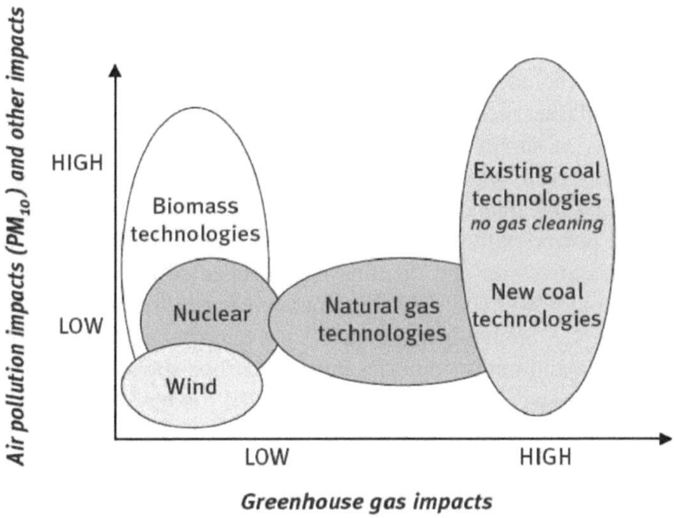

Fig. 4.6 Comparison of energy sources with regard to air pollution and greenhouse gas emission. Source: ExternE study [6].

becomes 57 $/MWh, or 50 €/MWh. This agrees well with the costs of wind power as calculated by the European Wind energy Association [4]. Figure 4.5 shows again levelised costs, now showing the influence of the wind speed and discount rate,[1] calculated as levelised production costs. Modern wind turbines at windy sites reach a generation cost level of 0.04 to 0.05 €/kWh, which comes close to the generation cost of fossil-fuelled power plants. It should be noted that usually these cost levels for wind projects are not reflected in real life. The reason for this is that the financing conditions of real projects are much more difficult than assumed in the levelised cost calculations. Often depreciation periods of less than 20 years are required. Since financial institutions consider wind projects to have a high risk profile, cost of financing can also be high. Usually the wind energy project developers are smaller, with less financial means, than those in the fossil and nuclear power industry. The difference between the levelised production costs as presented here, and the resulting generation costs based on total project cost can be very large. In the Netherlands project costs can be twice as much as the levelised costs presented here [5]. As a consequence, wind energy still needs some financial incentives to meet current market prices, except for big projects in very windy places developed by large project developers.

Many governments provide these incentives because of the sustainable origin of a wind-produced kilowatt-hour (kWh). Figure 4.6 shows results from the ExternE study [6], where wind energy appears to have the lowest contribution to greenhouse gas emissions and air pollution of all energy sources under consideration. According

[1] The input for this calculation is: investment costs in the range of 1,100 €/kW, O&M costs averaging 1.2 EURcents/kWh over a lifetime of 20 years.

to this study, wind-generated electricity has a benefit of 0.03–0.04 €/kWh compared to coal-generated electricity when the additional costs for neutralising the environmental impact are included (the so-called external costs). Other arguments in favour of cost and market incentives are the industry's job creation potential, and the (political) security of supply. Fast depletion of the European natural gas reserves, the Kyoto obligations and high oil prices have increased the sense of urgency. These arguments validate the financial incentives by governments to support the development of wind power. Although governments change opinion from time to time, switching the incentives on and off, this is not reflected in the history of worldwide installed wind power (Figure 4.1, cumulative) or in the assessments of future prospects (Figure 4.7, annual installation). An inventory of the wind power developments in Europe and many countries outside Europe shows a continued growth in annual installed power [1]. The forecast until 2011 is based on actual plans, the prediction is an estimate. The contribution to the world's electricity consumption by wind energy was 0.8% by the end of 2006 and is expected to rise to 4% in 2015, since the annual growth rate of wind-generated electricity (approximately 30%) is much higher than the global increase in electricity demand (3–3.5%). For Europe, being the leading continent in wind energy, these numbers are higher. At the end of 2005, 2.8% of the European electricity demand was covered by wind energy. According to the scenarios of the European Wind energy Association (EWEA), this will rise to 5.5% in 2010, 13% in 2020 and 23% in 2030 [2]. The scenarios of EWEA are known to be conservative: all previous scenarios have been surpassed by reality. The same holds for the scenarios of the European Commission. Between 1996 and 2003, the Commission's estimate of how much wind power would be built in 2010 was exceeded by a factor of nine.

The driving force behind these optimistic expectations is continued cost reduction. The progress ratio (PR) is a parameter that expresses the rate at which unit costs decline each time the cumulative market size doubles. For example, a PR of 0.8 implies that after one cumulative doubling of installed capacity, unit costs are only 80% of the original costs, i.e., a 20% cost decrease. For the onshore wind power development so far, the PR is evaluated to be 0.82% [7]. The cumulative market doubles approximately every three years (see Figure 4.1). Based on a less optimistic PR of 0.9, it is expected that new onshore wind power plants in coastal regions will cross the break-even point with coal-generated power by 2010. For offshore wind power, there is not yet sufficient statistical data. A detailed study of possible cost reduction measures applied to the cost and technology level of 2003 shows a possible reduction to 60–75% in 2020, depending on the chosen offshore market scenario and several other parameters like distance to the shore [7].

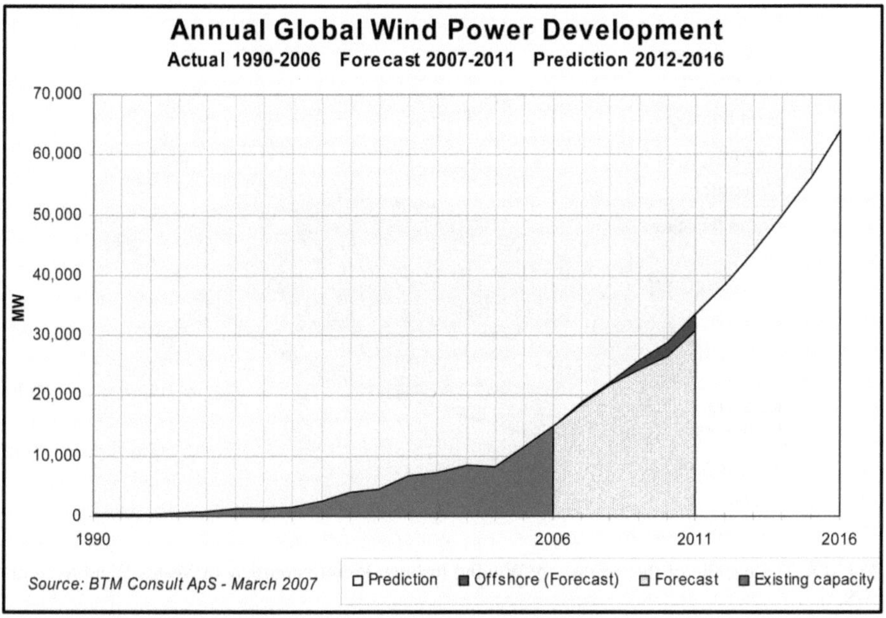

Fig. 4.7 Realisation and forecasted development of annual installation volume [1].

4.3 Technology Development

4.3.1 The State of the Art

The size of the turbines has increased more than tenfold in 20 years. Figure 4.2 shows the development of the size. Modern wind turbines are the largest rotating machines on earth: the rotor area is larger than a soccer field. The development towards “as large as possible” is driven by several aspects. Most of the installed wind power is in Europe where good sites with high wind speeds become scarce. The windy areas are either densely populated (North-West Europe) or natural areas (Spain). Large turbines are a solution as long as the infrastructure (roads, cranes, etc.) is available. Another solution is to go offshore. All countries with Baltic- or North-Sea shores are developing offshore wind energy. For the time being, the larger the better is certainly true here, since many costs depend little or not at all on size (e.g., project preparation, transport and assembly, electrical infrastructure). Furthermore, large turbines are often preferred from the viewpoint of spatial planning: the rotational speed of a turbine is inversely proportional to the diameter. A few large turbines rotating slowly present a better view than a series of smaller turbines with a higher rotational speed.

Wind turbine technology is in good health: the availability of turbines is ∼98%, which means that during 2% of the time they cannot produce electricity due to main-

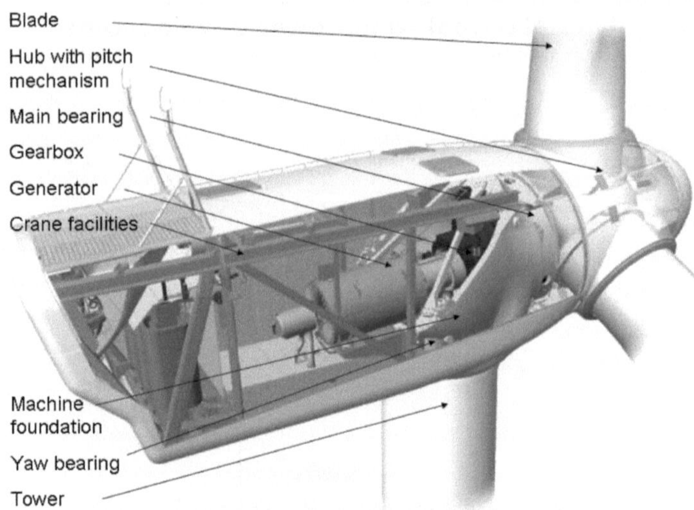

Fig. 4.8 The nacelle of the Vestas 3 MW wind turbine. Picture courtesy of Vestas Wind Systems A/S.

tenance or failures. The technology of wind turbines has evolved from robust and simple (the successful Danish concept) to turbines with a maximum controllability that produce the number of kilowatt-hours specified by the electric utilities. Figure 4.8, taken from the website of the manufacturer, shows an example of a modern commercial wind turbine, the Vestas 3 MW machine. Unlike for the smaller Vestas' turbine blades made of glass fiber reinforced material, the load carrying part is made of carbon fibers because of the much better strength and stiffness properties. The non-load-carrying part is still made of cheaper glass fibers. The position (pitch angle) of each blade can be adjusted individually for power control and for safety reasons. In case of an emergency stop, each blade is turned to a brake position driven by a co-rotating emergency power supply and control computer. The rotor load is taken by one bearing in the front of the nacelle, with the gearbox mounted directly at this bearing. The rotor loads and moments are guided by the machine foundation directly to the tower. The turbine is designed for offshore application, so crane facilities are built-in to exchange components when necessary.

This Vestas turbine and similar products of other manufacturers are the result of a 30-year process of evolution of design and technology. Alternative rotor concepts (only one or two blades) have been built but without commercial success. The two viable power control concepts, constant rotor speed with passive load control due to flow separation, and variable rotor speed with active load control by blade pitching, were both commercially successful but with the constant rotor speed concept (the Danish concept) as the clear winner in terms of market volume. The control of constant speed turbines is easy and cheap, but results in high loads, and therefore in heavy turbines. Moreover, the power quality is not high enough to comply with current grid requirements. Variable speed turbines all have power electronic components that convert the variable frequency current of the generator into the grid frequency. By means of this converter the output can satisfy all quality requirements with respect to frequency and voltage control, active and reactive power control. Grid requirements have had a major impact on the technology. All turbines in the multi-MW size have adopted the concept of variable speed operation. A second advantage of variable speed operation is that besides the electrical output, also the turbine loading can be controlled to a high extent. With the increasing size of the turbines, this becomes increasingly important in order to reduce turbine weight.

A second change in the concept is the safety system. In the past the design of the safety system was deterministic: an aerodynamic brake, by pitching the blade tip or the full span of the blade, and a full power mechanical brake at the low speed axis (between hub and gearbox) or fast speed axis (between gearbox and generator). The loads associated with mechanical breaking have become too large for modern large turbines, so the partial-load mechanical brake is now only used to stop an idling rotor. The full-load break is designed probabilistically. When the collective blade pitching should fail, at least two of the three blades will pitch to braking position. A detailed reliability analysis of common failure modes and built-in redundancy provides the level of safety prescribed by the safety standards such as IEC 61400-1 [8].

The most visible difference between wind turbine types is in the drive train. Most turbines use a three-stage gearbox to accelerate the low rotational speed of the rotor to the high speed required by the generator. The gearbox is one of the most critical components in a wind turbine, for several reasons. Unlike for gearboxes in other applications, the loading is highly variable, not at all deterministic and most of the time partial. This has led to the development of new tooth shapes, bearings, lubrication, dynamic analysis and dimensioning. A few manufacturers have removed the gearbox: the rotor directly drives a generator having an equally low rotational speed. These direct drive generators are specifically made for wind turbines. In order to produce sufficient power at a low rotational speed, the torque has to be very high. This is achieved in two ways: a large diameter and/or a high magnetic flux density. Figure 4.9 shows the nacelle for the largest direct drive generator at this time, developed by Enercon. Enercon is so far the only commercially successful manufacturer of this type of turbines. The generator power is 6 MW, the diameter of the aerodynamic rotor is 114 m. The rotational speed is extremely low: 8–13 rpm.

In general, elementary design rules dictate that the bigger the turbines are, the lower the rotational speed is. This makes the design of a direct drive generator for very large turbines increasingly difficult. For that reason, a hybrid concept has been developed by some manufacturers: a drive train with a single-stage gearbox, and therefore moderate transmission ratio, that drives a medium speed generator. Prototypes of a 5 MW version are now being tested. In general, there is not a clear winning drive train concept. The main advantage of a direct drive turbine, less drive train maintenance, has to show itself in the market where maintenance has to be extremely limited: offshore wind energy. Until now, all offshore turbines have gearboxes.

Fig. 4.9 The nacelle of the Enercon E112. The rotor of the generator and the aerodynamic rotor still have to be mounted. The front side of the nacelle shows the housing for the stator of the generator. The diameter is 10 m.

4.3.2 Offshore Applications

All countries with Baltic- or North-Sea shores are developing offshore wind energy. There are several reasons for this. Good sites with high wind speeds in the north-west of Europe are often located in densely populated areas, which prohibits a further increase of installations. Furthermore, both the Baltic and North Sea are shallow, with good harbour infrastructure nearby, and with high wind speeds. The cables to transport the electricity from the power station to the consumers do not need to be very long, since the densely populated areas are close to the shores. This makes both seas ideal breeding chambers for offshore wind power, the Baltic Sea in particular. This is sheltered water, where the North Sea is much more demanding. By the end of 2006 the total installed capacity was almost 900 MW in water depths ranging from 2 to 20 m. One of the most recent wind parks is the Dutch Offshore Wind Park Egmond aan Zee (OWEZ). The turbines have been installed in summer 2006 (see Figure 4.10). Furthermore, the Scottish Beatrice wind project was planned to go on line in summer 2007. Although limited in size with "only" two 5 MW tur-

Fig. 4.10 The installation of wind turbines at the Offshore Wind Park Egmond aan Zee (the Netherlands). The supporting monopiles have already been driven into the sea bed, with the transition piece above the water level. Tower and nacelle with two blades are mounted successively; the third blade is mounted separately.

bines, this is a very interesting project. It is the first one in deeper waters, 45 m, with a four-legged truss structure instead of a monopile as underwater support structure. Also with respect to installation a jump forward is realised. Unlike all offshore projects so far, the turbine is not installed in parts as shown in Figure 4.10, but as a whole in one single lift. This decreases the installation time, which is of paramount importance when calm water periods are scarce. Furthermore, the turbines are larger than ever installed offshore before. The larger the better is certainly true here, since many costs are size-independent as mentioned above.

The development of offshore wind power lags behind the onshore development, since it is younger and much more demanding. All wind turbines that have been installed until now are onshore turbines that have been adapted to offshore operation. A true merger of wind turbine technology and offshore technology has not yet been established, but will be required to increase the reliability of the turbines to a level where unscheduled maintenance does not occur, and scheduled maintenance is limited to once a year. Even the simplest repair can have a major influence on the performance of the turbine in case the machine is not accessible for months due to weather conditions. Improving the accessibility of the turbines in bad weather alleviates this problem. The turbines in the first Danish wind park in the North Sea, Horns Rev, have a platform at the nacelle upon which helicopters can lower maintenance crews. More conventional access systems have also been developed. An innovative one is the Dutch Ampelmann project where, like an inversed flight sim-

Fig. 4.11 Scale-model tests and animation of Ampelmann in action for access to offshore wind turbines.

ulator, a stable platform is created by hydraulic cylinders that compensate for the movement of the vessel (see Figure 4.11). In such a way the platform can be kept at a fixed position with respect to the turbine tower, allowing easy access under varying weather conditions. The design and feasibility study for the Ampelmann indicates that for sites near the Dutch shore the turbines can be accessed 90% of the year, where conventional vessels limit this to 75%. A first prototype is under construction and is planned to be put in service in 2007.

In several countries R&D programmes to support the development of offshore wind power are ongoing. As an example, the Dutch programme We@Sea is illustrative [9]. In this five-year programme more than 30 companies, research institutions, utilities, nature and environmental organisations, manufacturers and offshore contractors cooperate to enable an efficient realisation of the Dutch offshore ambition: 6000 MW in 2020.

4.3.3 Are Wind Turbines Ready?

It is often stated that wind turbines are "ready" and should no longer take a prominent place in the R&D agenda on renewable energy sources. A comparison with the development of aerospace technology shows that such a remark is out of place. When the pioneering phase of aircraft development was over and resulted in the first mass-produced, commercially successful airplane, the DC3 Dakota, no-one considered this to be the end of development. On the contrary, many similar milestones followed by breakthroughs in aerodynamics, materials, propulsion, control and logistics. The same holds for wind turbine technology. It is clear that (onshore) wind energy has left the phase of teething troubles, and has become mature. Nevertheless, new challenges drive further developments. The economics of wind power ask for turbines that are very large. The offshore market is almost endless when wind turbine technology follows offshore oil and gas technology to very deep waters, with floating constructions as the ultimate goal.

Within the wind energy community a long-standing debate on the optimal or maximum sizes of a wind turbine has come to a smooth end, with the conclusion that these do not exist. An optimal turbine size can be calculated for a given state of

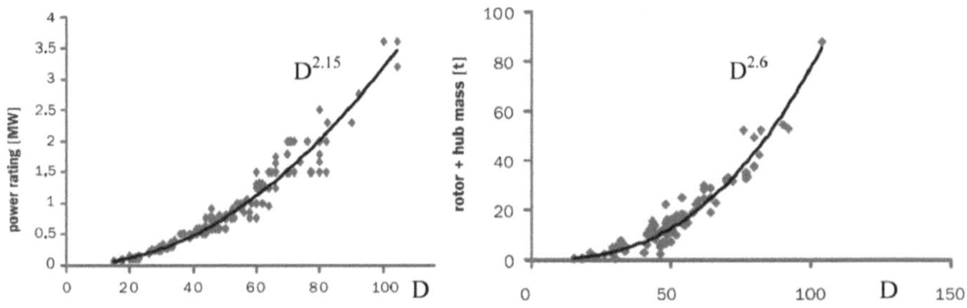

Fig. 4.12 Power rating and tower top mass as function of rotor diameter. Straightforward upscaling would give D^2 for the left- and D^3 for the right-hand graph [10].

technology and given infrastructure required for installation and grid connection. In Germany, the tower height was at some point limited by the absence of cranes that could reach the increasing tower top height. Since the wind energy market of Germany is the largest of all countries, new cranes were developed and a next step in tower height could be taken. For the turbine itself, technological developments continue to allow further upscaling by extending the design limits. Figure 4.12 shows the historical development of the installed power P and the tower top mass M as a function of the rotor diameter [10]. Scaling laws dictate that $P \sim D^2$ and $M \sim D^3$, so the power increases more, and the mass is lower than unchanged technology allows. The improvements in technology depend on two major aspects: (1) market size, which enables the manufacturers to learn from experience, re-engineering and operation & maintenance (O&M), and to invest in highly qualified engineering staff and test facilities, and (2) scientific progress that results in new design knowledge. This will be discussed in the section on R&D.

4.4 Grid Integration

Very often, discussions about the pros and cons of wind energy boil down to the question whether a reliable electricity supply is possible with a large amount of wind energy. A member of the Dutch parliament once stated that under those circumstances "we can only watch football on television when the wind blows". Reality is much more differentiating. Wind power has a number of impacts on power systems, which can be differentiated as local impacts and system-wide impacts. The local impacts have to do with grid requirements and turbine technology, whereas the system-wide impacts can only be analysed when the entire electricity supply system is taken into account. The problems and possibilities of large-scale integration of wind power in the electricity supply system are subject to many studies. A comprehensive description of all aspects is given in [11].

4.4.1 Local Impacts

Local impacts of wind power on the (distribution) system are mainly dependent on local grid conditions and the connected wind turbine type, and the effects become less noticeable when the (electrical) distance from the origin increases. The phenomena include changed branch flows, increased fault currents, the risk of electrical islanding, which all complicate system protection, and a number of power quality issues, such as voltage levels, harmonics and flicker. Modern wind turbines are equipped with versatile power electronics that, if controlled correctly, can mitigate most problems from the latter category.

Historically, wind turbines were considered as "insignificant" from a power system perspective: the penetration of wind power was small and the impacts on the system likewise. Therefore, technical requirements for wind power were nonexistent or the same as for distributed generation. Notably, in Denmark, distributed power sources used to be required to disconnect themselves in case of a voltage dip, and wind turbines were developed to do exactly this. As the penetration level of wind power in Denmark increased, it could happen that a significant amount of wind power disconnected itself as a result of a voltage problem, aggravating the voltage problem further. This has led to the development of grid codes specifically designed for the integration of wind turbines.

At the moment, many countries have adopted such grid connection rules for wind power, which list a number of technical requirements that any connected wind turbine or farm should comply with. Different codes are applicable in different countries, but most include requirements on reactive power capability, voltage dip ride-through capability, harmonics, fault current levels and voltage and frequency ranges in which the turbine should be able to operate. The development of grid codes has gone hand in hand with the technical developments of wind turbines themselves, and modern wind turbines include power electronics in order to enable the fulfillment of technical requirements. Interestingly, because of the variety of grid codes, wind turbines must often be designed to comply with all grid codes, which in practice means that wind turbines may be capable of more features than strictly required by the grid code in place.

4.4.2 System-Wide Impacts

With the growth of wind power in power systems, the system-wide impacts have grown as well. These impacts are a result of the variability and partial predictability of wind power and mainly depend on the wind power penetration level in the system, geographical dispersion of wind power and the size and characteristics of the power system. Most importantly, wind power has impacts on power system dynamics, secondary (load-frequency) control and tertiary reserves, the amount of power reserves and on the operation of other units in the system (unit commitment and dispatch).

Furthermore, with the partial predictability of wind power, load flows in systems with significant amounts of wind power may become more difficult to anticipate.

Wind power challenges the operation of power systems. Because electrical energy cannot be stored in substantial amounts, generation and load must be balanced at all times. This is the primary responsibility of the transmission system operator (TSO). Historically, it has always been the case that generation follows demand, by adjusting generation unit output. Thus, some excess reserves are always needed for power system balancing. With the liberalisation of energy markets in Europe, unit ownership became decoupled from system operation and market parties were allowed to freely make arrangements for trading electrical energy. The largest energy volumes are traded in the period before day-ahead: large blocks of electrical energy are traded for longer time periods to allow market parties to settle their physical positions (i.e., scheduling of generation unit maintenance, base-load power, etc.). One day ahead of operation, trade can be done on the spot market if needed, in order to allow a more accurate "tuning" of energy portfolios taking into account forecast demands. This tuning continues during intra-day trading, in which market participants may further optimise their unit operation taking into account updated information on units and load (i.e., unscheduled outages or wind power forecasts).

At gate closure, all trading for the physical delivery of electrical energy ceases: market participants submit their final energy schedules to the TSO. At this moment, the control over the power system is passed to a central imbalance settlement administrator, usually the TSO. Even though trading, generation and load are scheduled to match, this is not sufficient to guarantee a physical power balance in real-time. On top of the automatically summoned primary reserves, the TSO therefore continuously manages secondary (available within 15 minutes) and tertiary reserves (available after 15 minutes in order to free up secondary reserves) in order to maintain the balance in the system in real-time. These reserves are in general made available by market parties submitting bids for operating reserves to the TSO, usually up to the hour ahead. The balancing costs encountered by the TSO are then passed on to those market parties deviating from their submitted energy schedules.

Wind power complicates the process of power system balancing in the sense that it introduces additional power variations to the system (which can only be predicted to a certain extent) while at the same time it partially replaces the conventional units used for providing power reserves [13]. As an extreme example of wind power variability, consider Figure 4.13, showing the output of the Horns Rev A wind farm (160 MW installed capacity, off the Danish North Sea coast) on a summer's day. The variations may be as high as 80 MW within 30 minutes. Of course, wind power variations are not the only source of power imbalances in the system, as variations in load and generation must also be taken care of. Actually, the "fast" variations of this particular wind farm may be counteracted by sufficient geographical spread of wind power: a similar wind farm located 30 km ahead would balance out a significant part of the fluctuations. Figure 4.14 shows the effect of geographical aggregation of wind power in Germany. A 20% change in power in one hour occurs 100 times a year for a single, inland, wind farm, and less than once a year for 350 MW aggregated wind power in Germany.

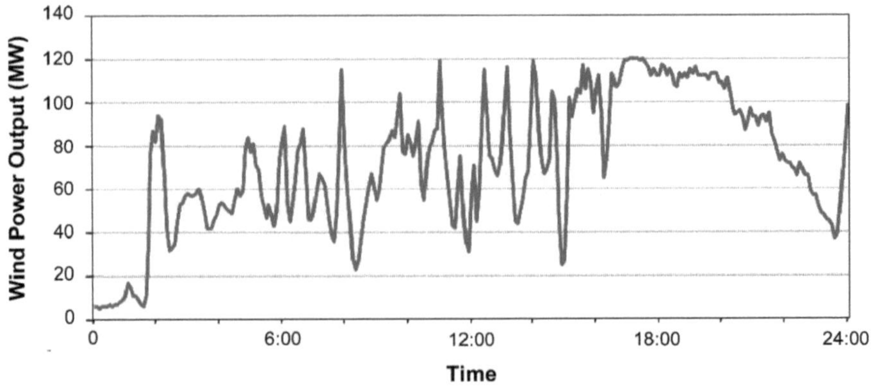

Fig. 4.13 Wind park output variations at Horns Rev A offshore wind farm, Denmark [12].

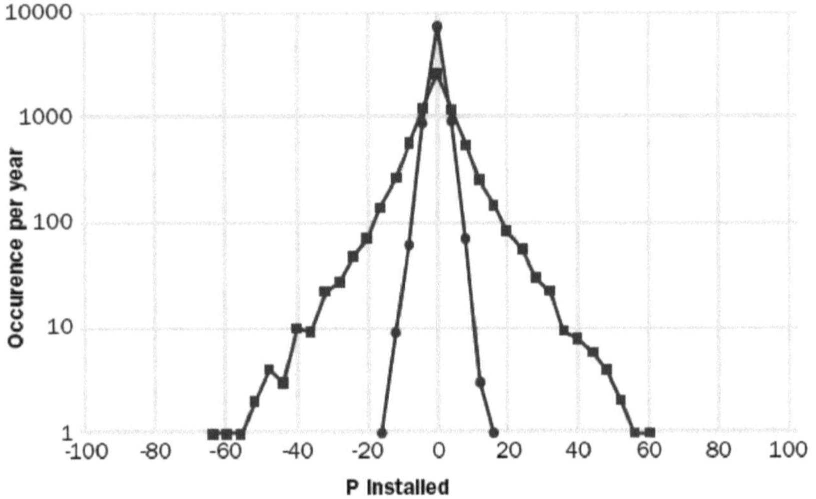

Fig. 4.14 Frequency of occurrence of hourly wind power variations (in %) and the effect of geographical aggregation. ● 350 MW aggregated wind power, ■ single inland windfarm, both in Germany [11].

As an example of the variations in load, wind power, international power exchanges and generation, consider Figure 4.15. From this figure, the load weekly pattern can be clearly differentiated. First of all, it can be concluded that for Western-Denmark, daily variations of load are in the same order of magnitude as wind power variations. Second, some variations of wind power and load counterbalance while others increase the overall power variation. The geographical spread of wind power in Denmark results in a relatively smooth overall wind power output. The need for power reserves at the system level depends on the combination of all power variations (load and wind power), to be balanced by power exchanges with neighboring

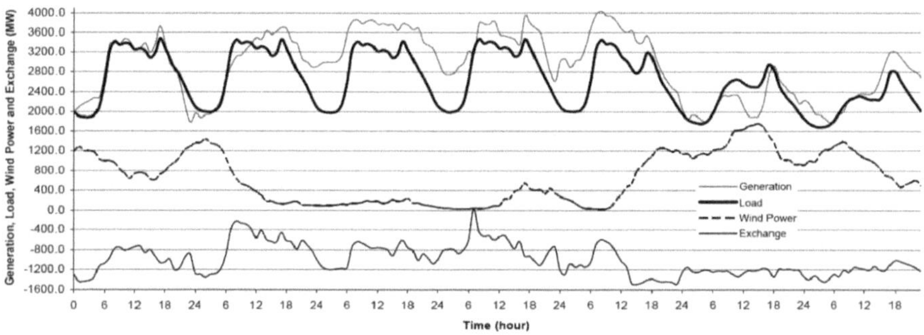

Fig. 4.15 Power system balancing with wind power in Western-Denmark for a typical week, based on data from Danish TSO Energinet.dk (previously Eltra).

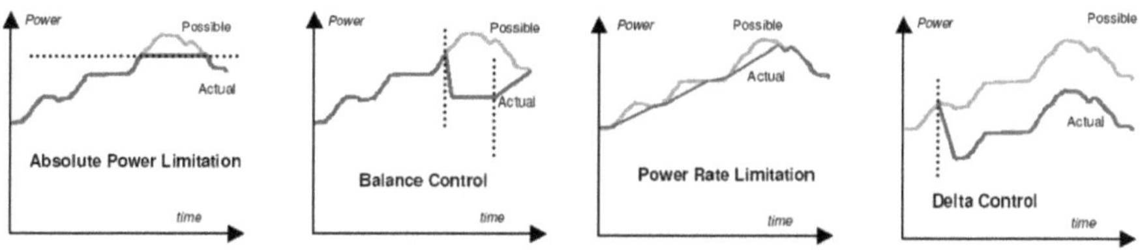

Fig. 4.16 Outline of the active power control functions as applied in the Horns Rev A wind farm, Denmark. The plot shows the possible power and the actual achieved power with the different control functions [14].

areas (trade) or by adjusting generation levels. For the latter, wind farms increasingly incorporate technical abilities to do so, as can be seen in Figure 4.16. Modern wind farms like Horns Rev A are very well controllable: the maximum power level, the actual power, the power change rate can be set by the operator, as well as the amount of reserve power by a Delta control.

Wind power is both variable and only partly predictable. In Figure 4.17, the standard deviation of the forecast error associated with the prediction of large-scale wind power in the Netherlands is approximated, based on wind speed forecasts and actual measured wind speeds for 15-minute time intervals up to one and a half day ahead. From the figure it follows that the forecast error exponentially decreases as the hour of operation draws nearer. However, a significant forecast error is present at the 12–36 hours ahead time-scale. This means that at the time of scheduling power trades (market closure 12–36 hours ahead of operation) the wind power forecast error is on average about 30% of the actual production. Of course, optimisation of the scheduling of generating capacity continues after market closure, and some generation units may be capable of fast start-up or shut down, down to some hours for gas plant and even minutes for hydro-power plants. Improved forecasts reduce the amount of additional reserves, however, wind power variability still requires power reserves for balancing.

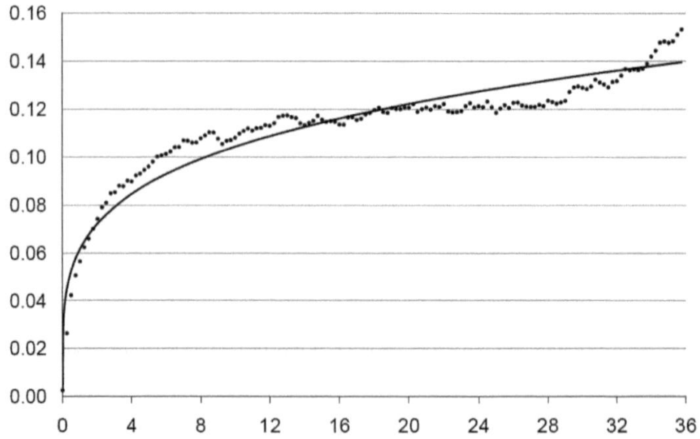

Fig. 4.17 Standard deviation (in %/100) of the forecast error of large-scale wind power in the Netherlands. The horizontal axis gives hours ahead of operation.

4.4.3 Balancing Costs

The cost of balancing wind power depends on a number of factors, including wind power penetration level, power system size, marginal cost of the units in operation, trading possibilities and market design [12] (who is responsible, the TSO or the wind plant owner?). It is therefore difficult to draw a general picture of the costs for integrating wind power into power system operation. In Figure 4.18 some estimates from different reports are given, most of these from the US. The IEA-NEA report [3] incorporates some European studies, and concludes that balancing costs depend highly on access to cheap or expensive regulating resources like hydro power or thermal power. The report indicates actual balancing costs in the range of 2–3 €/MWh for wind power. A Dutch study, based on a 15% electricity supply by wind, calculates ~1 €/MWh [15].

The balancing costs are not taken into account in Figures 4.3 and 4.4. The avoided costs to compensate for the environmental aspects of fossil fuel generation (30–40 €/MWh according to [6], see Section 4.2) are also not included. How to account for these cost aspects is subject to political decisions. In Denmark the balancing costs are considered as part of the costs for system services in general and therefore are socialised. In the Netherlands, these costs are charged to the wind power projects, i.e., added to the sales price per kWh. The environmental benefit for wind energy is only partially taken into account by charging fossil-generated electricity with CO_2 certificates. As a rule of thumb, it applies that the balancing costs for wind power are only 10% of the avoided external or environmental costs.

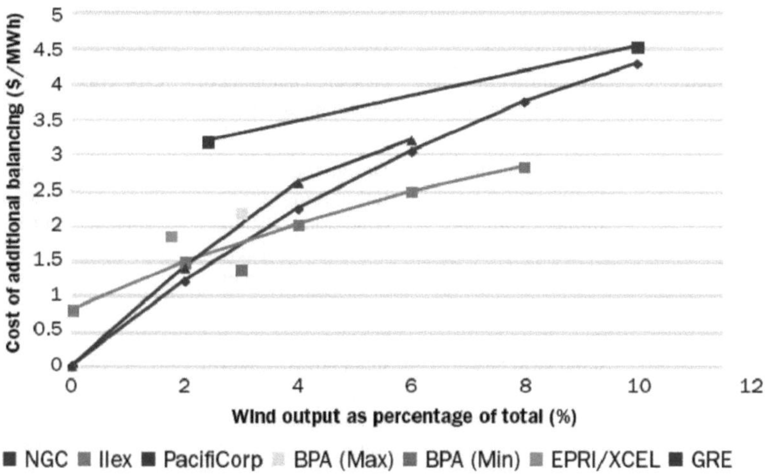

Fig. 4.18 Estimates of additional balancing costs from six studies [11].

4.5 The Research Agenda

Initiated by the European Wind Energy Association, the European wind energy sector has discussed the research agenda towards 2020, resulting in a well-supported document [16]. This agenda was launched because the wind energy sector was facing barriers in further up-scaling of the turbines [17]. Historically, the R&D efforts focused mainly on technology development and the scientific disciplines required for this. The present R&D agenda has a wider scope, incorporating also grid integration issues, public support and environmental issues. Priorities are addressed by distinguishing:

- "Show stoppers", which is to say that it is considered to be an issue (technological/societal) of such importance that failure to address it could halt progress altogether. Thus it needs special and urgent attention.
- "Barriers", defined as being a principal physical limitation in current technology, which may be overcome through the opening up of new horizons through generic/basic research over the medium to long term.
- "Bottlenecks". Not to be confused with the above, Bottlenecks are problems which can be relatively quickly overcome through additional short or medium term R&D, i.e., through the application of targeted funding and/or other resources.

Table 4.1 provides a brief summary of the subjects in these categories, setting clear priorities in the research agenda. Based on this agenda, several large European R&D programmes have started, supported by the European Commission in its 7th Framework Programme. Two programmes are mentioned here specifically: UpWind and EWIS.

Table 4.1 R&D priorities of the European wind energy sector [16].

Topic	Show stopper	Barrier	Bottleneck
Wind resource estimation and mapping	Availability of wind data in the public domain	Resource mapping of high wind sites that are yet unexplored	Cost-effective measurement units
Wind turbine technology	Robust, low maintenance offshore turbines with increased reliability	Design tools for turbines in extreme conditions – laboratories for accelerated testing of large components	Component level design tools and multi-parameter control strategies
Wind farms	Storage systems at wind farm level	Wind flow in/around wind farms – control of wind farm – risk assessment methodology	–
Grid integration	Initiatives for a Trans-European grid with connection points for wind power plants	Control strategies and requirements for grid compatibility, and to support and maintain a stable grid	Components and technologies for grid connection
Public support and the environment	Effects of large-scale wind power on ecology; communication to public and policy makers	Monitoring of effects near wind farms; Equipment to monitor, e.g., bird collisions and sea mammals behaviour	Exchange and communication of R&D results
Standards and certification	–	Developments of new standards on energy yield calculation, grid connection, risk assessment, design criteria, O&M standardisation	Accelerated finalisation of ongoing standards

4.5.1 UpWind

UpWind [18] is a 5-year project that started on the 1st of March 2006. The aim of the project is to develop and verify substantially improved models of the principal wind turbine components, which the industry needs for the design and manufacture of wind turbines for future very large-scale applications, e.g., offshore wind farms of several hundred MW. The wind turbines needed will be very large (>8–10 MW and rotor diameter >120 m). Present design methods and the available components and materials do not allow such up-scaling. In order to achieve the necessary up-scaling before 2020, full understanding of external design conditions, innovative materials with a sufficient strength-to-mass ratio, and advanced control and measuring systems are essential. In order to achieve this up-scaling in the most efficient way, several critical areas have been identified to be addressed in this Integrated Project. Aerodynamics, aero-elasticity, structural and material design of rotors, critical analysis of drive train components, and support structures (for offshore applications), are to be analysed, and new design approaches and concepts are to be developed, as well as supporting technology. As the characteristics of present monitoring and measuring techniques and control concepts are insufficient, the project will improve those techniques with the focus on large wind turbine structures. More than 30 companies, R&D institutes, utilities, and project developers cooperate in UpWind.

4.5.2 EWIS

EWIS, the European Wind Integration Study, started in 2006, and is undertaken by many European TSOs. The scope of work covers all the technical, operational and market/regulatory aspects related to the integration of wind power in Europe at a large scale. The project will gather both technical and market/legal aspects in the four main synchronous electricity systems in Europe. The overarching goal of the project is to address especially the network issues arising from large-scale wind power plants, which are particularly relevant to European TSOs, and to make proposals for a generic and harmonised European-wide approach towards wind energy issues. The final objective is to set up a model for the integration of the capacities of RES, and more specifically wind power within Europe, as forecasted in different scenarios to be covered by the study.

The UpWind and EWIS projects address most of the show stoppers mentioned in Table 4.1. As an example of the development of wind turbine technology, the possibilities for a next generation of wind turbine rotors are discussed here (this is only one of the topics in UpWind).

4.5.3 Developments in Wind Turbine Rotor Technology

The up-scaling of turbines has, until now, not led to significant changes in the blade structure: all blades are constructed as one single component, with the blade skin as a load carrying element. Most modern large wind turbines run at variable rotational speed, combined with the adjustment of the collective pitch angle of the blades to optimise energy yield. This blade pitch control has not only led to a perfect power regulation, but also to a significantly lighter blade construction due to the lower load spectrum and a lighter gear box due to shaved torque peaks. A next step in blade load control is almost ready for commercial application: pitch angle adjustment per blade instead of collective. This will further alleviate the rotor loads, especially the periodic loading due to yaw and wind shear. Not only will the blades benefit from this, but also the drive train and nacelle structure.

A possible further step, probably for the 2020 wind turbine generation with even larger rotor sizes, is a much more detailed and faster control of the loads. Control should be possible for each blade at any azimuthal position and any spanwise station, by aerodynamic control devices with embedded intelligence distributed along the span. The correspondence with the control devices at airplane wings (flaps at leading and trailing edges, ailerons) is apparent, but the requirements for blade control devices are probably much more severe. Modern blades are very reliable, and require only limited maintenance at the blade pitch bearing. Future blades with distributed control devices should be as reliable, without adding maintenance requirements.

The development of this kind of technology, often named in popular terms "smart structures" or "smart technology", is an interdisciplinary development par excellence. It requires a joint effort in many disciplines:

- Aerodynamics of aerofoils with control elements. Several options are available for the adjustment of lift and drag: flaps, microtabs, boundary layer suction or blowing or other means of influencing it, variable camber.
- Actuators. The activation of the aerodynamic devices has to be fast and reliable with as little as possible power use. Well known options are piezo-electric elements and shape-memory alloys.
- Control. Algorithms for this type of control are not yet available. Fast, real time load identification algorithms, allowing application of predictive control techniques is a challenging task. Algorithms like self-learning and adaptive algorithms will be used to design a fault-tolerant controller.
- Communication and power supply. The power supply and communication between the control devices should not increase sensitivity to lightning strikes.
- Blade material and construction. Preferably all devices should be embedded in the blade material, without creating slots in the blade surface to avoid contamination of the inner structure. The embedding can lead to new blade constructions, like the use of spars and ribs.
- Blade design tools. None of the available design tools include distributed control options, nor allow for totally different blade constructions.

It will take several years before this development is mature enough to be applied on commercial machines. After the present phase of analysis and laboratory testing, full-scale testing on an experimental turbine is the next step, followed by prototype testing.

4.6 Conclusions

History and perspectives of wind energy development have been described. Wind energy has left the phase of pioneering and teething troubles, and has the potential to become a mainstream electricity source. Several barriers need to be taken in order to achieve this. The generating cost approaches the cost of coal-fired generation. Continued cost reduction and dedicated measures to integrate wind energy in the European electricity supply system dictate the European wind energy research agenda. Several large European research projects address these topics, and will provide new knowledge to achieve the goal of a significant, sustainable, contribution to the electricity supply system.

References

1. World Market Update 2006, BTM Consult, March 2007.
2. NoFuel, Briefing of the European Wind Energy Association, February 2006.
3. Projected Costs of Generating Electricity, 2005 Update, IEA-NEA Report, 2005, Paris.
4. Wind Power Economics, Brochure of the European Wind Energy Association, available at www.ewea.org (accessed June 2007).
5. van Kuik, G.A.M., Where is wind?, *Journal for Public Administration*, June/July 2006, 18–22 [in Dutch].
6. European Commission DG Research, External Costs, research results on socio-economic damages due to electricity and transport EUR 20198, 2003.
7. Junginger, M., Faaij; A. and Turkenburg, W.C., Cost reduction prospects for offshore wind farms, *Wind Engineering* **1**, 2004, 97–118.
8. IEC 61400-1 – Edn. 3.0, *Wind Turbines, Part 1: Design Requirements.*
9. Large-scale wind power generation offshore, towards an innovative and sustainable business, R&D programme of the consortium We@Sea, available at www.we-at-sea.org.
10. Wind Energy: The Facts; An analysis of wind energy in the EU-25, European Wind Energy Association, 2004, available at www.ewea.org (accessed June 2007).
11. Large scale integration of wind energy in the European power supply, European Wind Energy Association, 2005, available at www.ewea.org (accessed June 2007).
12. Ummels, B.C., Gibescu, M., Kling, W.L. and Paap, G.C., Integration of wind power in the liberalized Dutch electricity market, *Wind Energy* **9**(6), November/December 2006, 579–590.
13. Ummels, B.C., Gibescu, M., Kling, W.L. and. Pelgrum, E., System integration of large-scale wind power in the Netherlands, in *Proceedings of the IEEE PES General Meeting*, Montreal, June 18–22, 2006, 8 pp.
14. Kristoffersen, J.R., The Horns Rev wind farm and the operational experience with the wind farm main controller, in *Proceedings of Copenhagen Offshore Wind Conference*, Copenhagen, October 25–28, 2005.
15. de Vries, H.J. and van Sambeek, E.J.W., Schatting van de kostenontwikkeling van offshore windenergie in Nederland en de benodigde Rijksbijdrage voor het behalen van 6.000 MW in 2020, ECN-C–04-045 [in Dutch].
16. Prioritising Wind Energy Research; Strategic Research Agenda of the Wind Energy Sector, European Wind Energy Association, 2005, available at www.ewea.org (accessed June 2007).
17. van Kuik, G.A.M., Are wind turbines growing too fast?, in *Proceedings EWEC2001 Wind Energy Conference*, July 2001, Copenhagen.
18. UpWind website: www.upwind.eu.

Chapter 5
Photovoltaic Cells for Sustainable Energy

Augustin J. McEvoy and Michael Grätzel

Abstract. Incident solar radiation is the most fundamental of the sustainable energy sources, being at the origin of all others – wind and wave, biomass and even fossil fuels which represent stored solar energy over a geological time scale. Only nuclear and geothermal sources are non-solar. The direct conversion of incident solar radiation into the most convenient energy vector, electricity, motivates the development of lower cost, higher efficiency and larger area photovoltaic devices and systems. Their history, technology and applications are here reviewed.

Key words: Incident radiation, solar spectrum, photovoltaics, semiconductor devices, photo-electrochemistry, sensitisation.

5.1 The Solar Energy Resource

Incident solar radiation is the most fundamental of the sustainable energy sources, the others being its consequences – wind, wave, biomass and even the fossil fuels which represent solar energy stored over a geological time scale. Only nuclear, geothermal and tidal sources are essentially non-solar. In consequence an informed awareness of the physical realities of solar energy input is essential to anyone interested in sustainable energy. On the basis of such knowledge, an appreciation of the utility and effectiveness of a solar energy exploitation technology can arise.

The "solar constant" is the energy flux per unit area impinging on a surface normal to the sun's radiation at a position on the earth's orbit, and has a value of 1.367 kW/m^2 outside the atmosphere, an incidence which is conventionally referred to as AM0 (for "air mass zero"). On the earth's surface however, this figure is significantly reduced by absorption and scattering, as well as diurnal and seasonal variations. Therefore while the whole-earth solar energy input is approximately 1.2×10^{17} W, or 1.56×10^{18} kW·h/year and only 0.13% of earth's cross-section could supply our total energy requirements through solar panels of 10% efficiency,

K. Hanjalić et al., Sustainable Energy Technologies: Options and Prospects, 99–119.

only a fraction can be usefully harvested by fixed installations on land. The solar energy flux on a plane surface at the equator under clear sky conditions is just over 1 kW/m^2, or AM1. Typical midday conditions in western Europe approximate to AM1.5. Practical experience in Germany – which has the highest installed solar photovoltaic park in Europe, in great part due to government incentives – demonstrates an annual electrical energy yield of approximately 850 kW·h per rated kW of photovoltaic generating capacity. The insolation (i.e., the incident solar energy, usually expressed in kW·h·m^{-2}day^{-1}) is therefore a rather dilute energy source, so the essential problem is to develop and fabricate a significant area of photovoltaic surface with an acceptable conversion efficiency of incident solar light to electricity at a competitive cost. A related issue is the storage of energy on daily and seasonal scales. A further limitation is shown in a comparison of the global distribution of population and therefore localisation of energy demand, and the areas of intense insolation in Figure 5.1: there is a significant mismatch.

The most significant factor for the operation of any photo-conversion system is the spectral distribution characteristic of solar light. The solar emission spectrum approximates that of a black body with a temperature of 5800 K, and is shown in Figure 5.2. The figure also presents the direct solar flux, utilisable in focussing concentrator systems, and also the total when scattered light is included, under AM1.5 conditions. It is of interest in passing to note the infra-red absorption bands due to water vapour and carbon dioxide which are responsible for the greenhouse effect by intercepting thermal re-emission in those spectral ranges from the earth's surface. The irradiance is the integral with respect to wavelength of each curve, giving a result in W/m^2. Evidently most of the incident light is in the visible and near infra-red region of the spectrum, wavelengths 400–1500 nm, corresponding to photon energies in the range 3–0.8 electron volts (eV). The two AM1.5 spectra define a standard total (global, hemispherical, within a 2π steradian field of view of the tilted plane) spectral irradiance and a standard direct normal spectral irradiance. The direct normal spectrum is the direct component contributing to the total global (hemispherical) insolation. AM1.5 irradiance is 970 W/m^2.

5.2 Principles of Photovoltaic Solar Energy Conversion

The basic requirement of any solar-to-electric conversion system is obviously the absorption of a significant part of the incident solar radiation with a consequent utilisation of the energy of the photons in an electrical process. The first reported observation of such an effect was by Edmond Becquerel in 1839, reported in his *Mémoire sur les effets électriques produits sous l'influence des rayons solaires* in which he used a contact to what we now recognise as a semiconductor, a silver halide. The sensitivity to light of these salts was simultaneously attracting attention in the work which gave rise to photography. The first solid-state photocell followed in 1877 using selenium [2]. When the modern era of photovoltaics began, essentially with Rappaport and others at the RCA Labs in Princeton, NJ [3] silicon

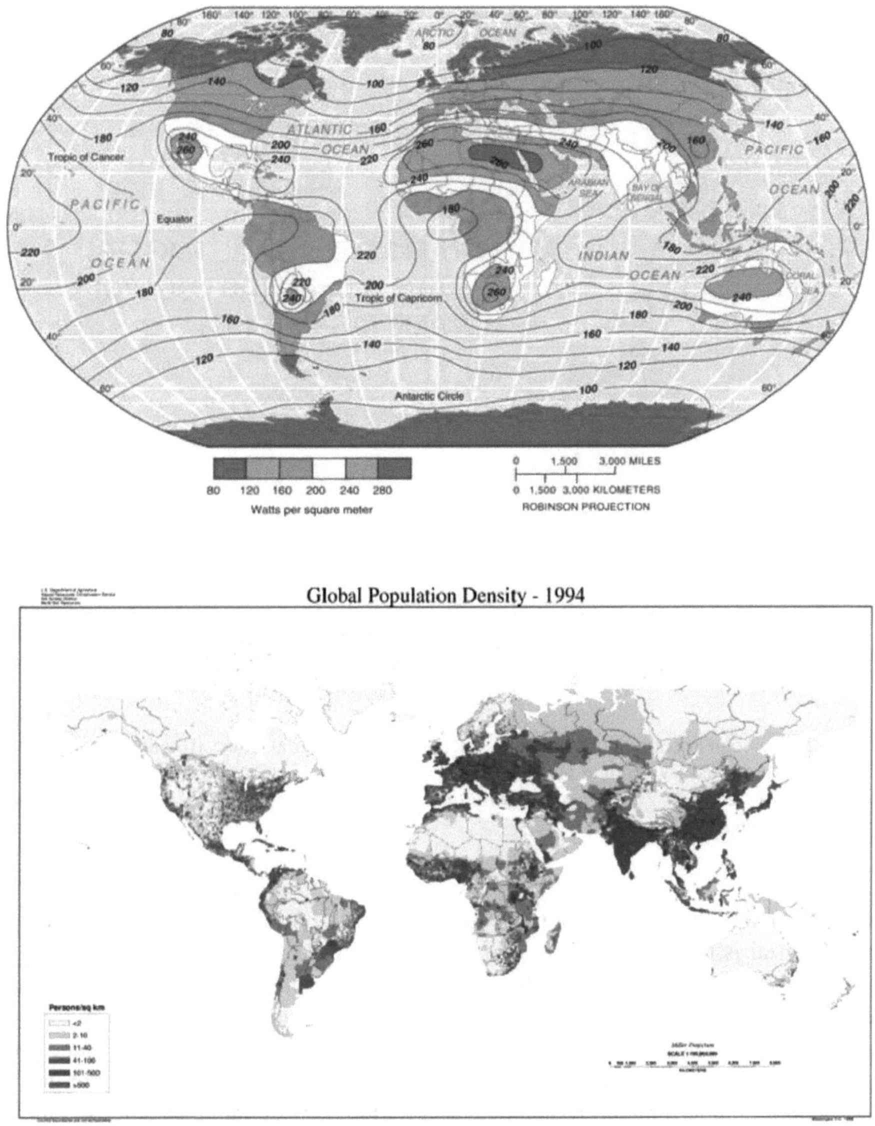

Fig. 5.1 (a) The solar resource – insolation chart; (b) population distribution. (Technical Report TR-95-6. National Center for Geographic Information analysis. University of Santa Barbara, USA).

was the preferred semiconductor. The essential characteristic of a semiconductor is that at normal temperature it has only a very few mobile charges able to conduct electricity, but on absorption of a photon of appropriate energy an electron can be promoted from immobility in the valence band to mobility in the conduction band, which leaves a corresponding hole, equivalent to a positive mobile charge, in the

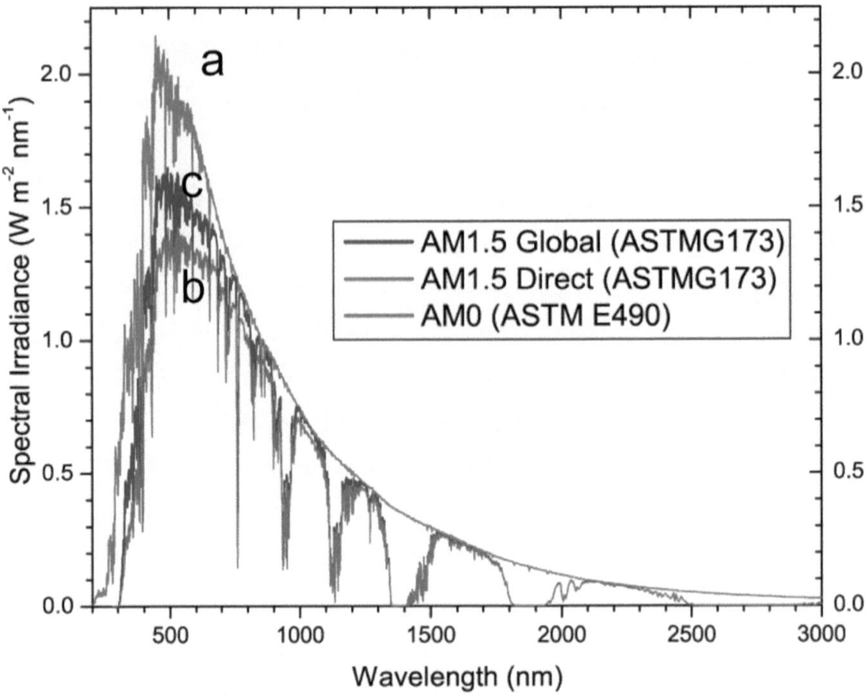

Fig. 5.2 The solar spectrum, AM0 (a) in space and the terrestrial direct (b) and global normal insolation on a tilted PV module plane (c) incident radiation under AM1.5 conditions, according to the ASTM standards. *Direct = small solid angle about the sun, global = direct + diffuse light from the remainder of the sky.*

valence band. Since this excitation requires a certain minimum amount of energy, the material is transparent to the longer wavelength lower energy photons but fits the requirement stated above for higher energy photons in that, once absorbed, their energy provides mobile charges in the semiconductor. The minimum energy for optical absorption by silicon, for example, is 1.1 eV, corresponding to the band gap between valence and conduction states. As well as silicon, a periodic table group IV element, modern materials science has provided a wide range of other semiconductors, so that electronic device designers have available for selection a range of band gaps, charge carrier mobilities and other optoelectronic properties for specific advantages in photovoltaics, and equally and more recently in light-emitting diodes which use the converse of the photovoltaic process. One such family of materials is the III–V semiconductors in which adjacent sites in a silicon-like cubic lattice are occupied alternately by elements from groups III and V of the periodic table. The available binary III–V semiconductors are presented, with their band gaps and structural lattice dimensions, in Figure 5.3. Along the lines joining the binary compounds it is possible to form solid solutions, in which different group III or group V elements may be substituted for the majority element of the same group. As a

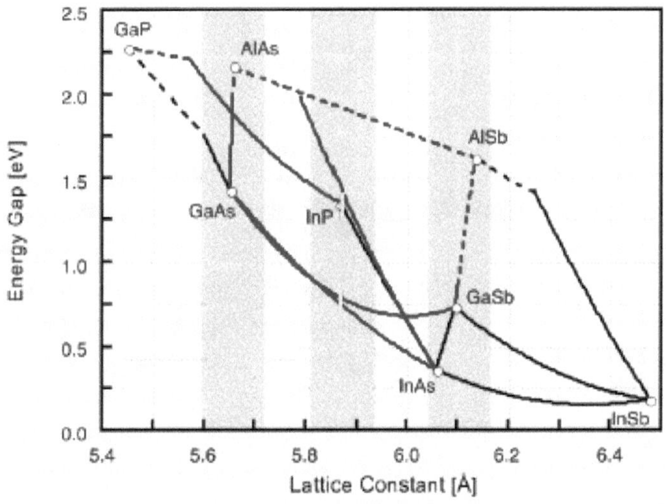

Fig. 5.3 The III–V family of compound semiconductors. The binary compounds AB are positioned according to their band gaps and lattice dimensions. Solid solutions with intermediate properties can be obtained as shown by suitable substitution on the A or B sites in the lattice. The dashed and solid lines represent indirect and direct semiconductors, respectively, with direct semiconductors showing higher absorption coefficients.

consequence, any band gap from 0.3 to 2.8 eV, and any lattice dimension from 5.4 to 6.2 Å can be obtained. The II–VI series also provide binary semiconductors with significant applications in optoelectronics.

In a similar process a further, ternary, semiconductor family can be generated, the I–III–VI_2 compounds (Figure 5.4). All these are under investigation and development, the III–V's in particular for radiation resistance in aerospace applications and for multijunction devices as will be described later, and the I–III–VI_2 series for thin-film solar cells because of their particularly high optical absorption coefficient. In a similar notation, Becquerel's silver halide is a I–VII compound.

The electrical energy represented by the photo-generated carriers must be made available to an external electrical circuit. This charge separation action is achieved by contacting the semiconductor with another material of different conduction mechanism – a metal to form a Schottky junction, an electrolyte in a photo-electrochemical cell, a chemically different semiconductor to form a heterojunction, or the same material but with a majority of carriers of the opposite polarity for a homojunction. Since it is the dominant commercial photovoltaic device, the discussion of device operation will focus on silicon p-n homojunction solar cells. The physics and operating principles of a photo-sensitive semiconductor device will now be presented.

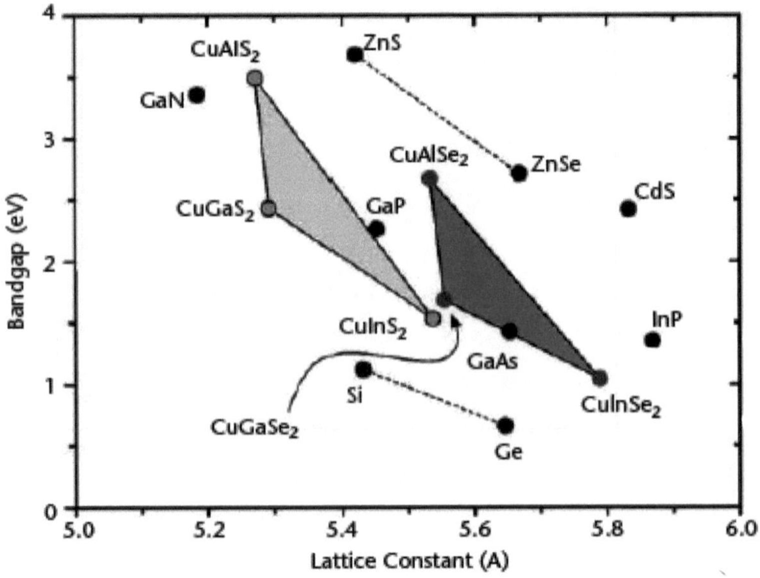

Fig. 5.4 Band gap/lattice constant relation for some II–VI and I–III–VI_2 semiconductors, compared with elemental (group IV, Si and Ge), and III–V materials.

5.3 Fundamentals of Solar Cell Operation

In a crystal of pure silicon, the 4-coordinated atoms are bonded to each nearest neighbour covalently, so that no free mobile electrons are available. To control the electrical properties of the material, a small proportion of lattice sites are substituted, or "doped", with elements of a different valence. If a pentavalent dopant is employed, only four electrons for each atom inserted on a silicon lattice site are incorporated in covalent bonding, and the excess electrons (one per dopant atom) provide mobile electrons as charge carriers; this type of donor dopant therefore makes the semiconductor n-type. Where positive charge carriers are required, an acceptor dopant such as boron may be employed, with only three valence electrons when fourfold covalency is required in the silicon crystal; the electron deficiency or hole then effectively provides a mobile positive charge carrier. One consequence of the excess or deficiency induced by doping is that the Fermi level, the energy of the highest occupied electron state, in the silicon is determined by the small proportion of dopant inserted, rather than by the bulk properties of the host material itself. With excess electrons in an n-type material, the Fermi level is close to the conduction band edge, and conversely for a p-type material the Fermi level descends towards the valence band edge. On contact between two such materials, p- and n-, the Fermi levels equilibrate, so the electronic band structure is displaced, positively in the n-type material and negatively in the p-type, as shown in Figure 5.5. This then provides an electric field capable of separating oppositely-charged carriers, as

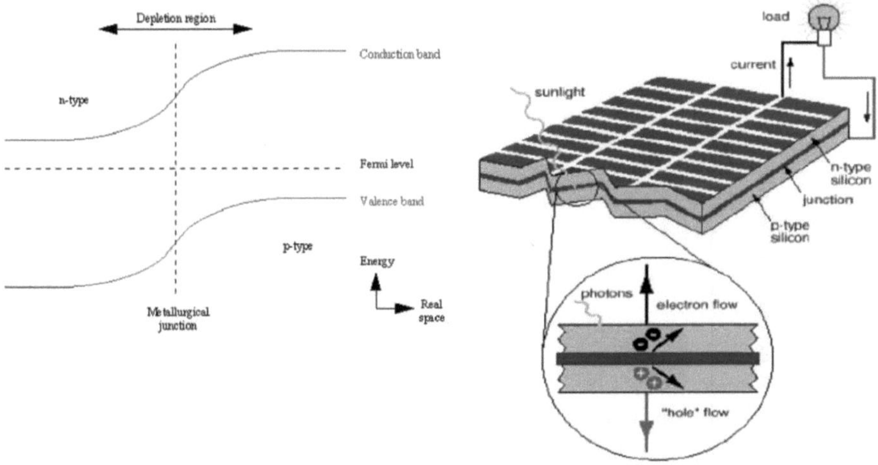

Fig. 5.5 Electron band structure at a p-n junction in darkness, showing equilibration of the Fermi level and consequent "band-bending" representing an internal electric field. An electron with higher energy is at a more negative potential. (right) schematic of a p-n junction photovoltaic cell.

required for the exploitation of the converted solar energy represented by the photo-generated carrier pair. Electrons can descend to the n-type side of the junction where they accumulate, while holes rise to and accumulate on the p-type side; the electrons are therefore available to transit an external circuit, returning to the p-type zone to combine with and neutralise holes. The maximum potential difference generated under illumination, that is under open-circuit conditions, is therefore somewhat less than the band gap of the semiconductor material. In the silicon case, for example, V_{OC} is approximately 600 mV, for a band gap of 1.1 eV. Figure 5.6 shows the electrical characteristics of a silicon solar cell, which shows only slight dependence of V_{OC} on the intensity of illumination. As for the maximum available current under short-circuit conditions, it corresponds to harvesting of the carrier pairs photo-generated by absorption of sufficiently energetic, or "actinic" photons, and therefore is directly proportional to the incident light intensity.

5.4 Materials and Performance of Solar Cells

Given the range of potential photovoltaic semiconductor materials available, the criteria for choice can now be addressed. Clearly, materials with narrower band gaps will be sensitive to a wider range of the spectrum, as the longer wavelength photons of lower energy are still actinic for them. The photo-current density is therefore increased. However, a contrary trend applies to the photo-induced voltage, which must be reduced with a narrower band gap. In consequence, for a single-junction

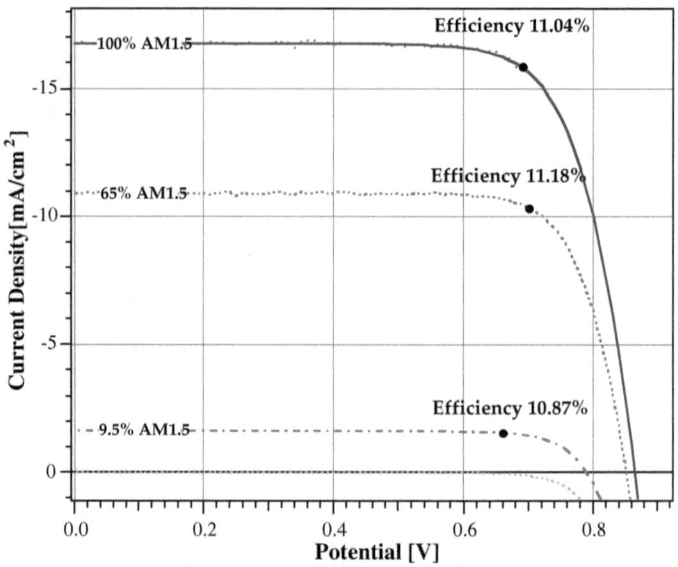

Fig. 5.6 Current-voltage characteristics of an actual solar cell, under different levels of illumination, and indicating the maximum power point and efficiency in each case.

cell, there is an optimal band gap, as shown in Figure 5.7. This in practice indicates a band gap of 1.4 eV as optimal, corresponding for example to the III–V compound GaAs. However, the loss of efficiency in using silicon is obviously relatively minor, and the questions of cost, engineering experience and other materials parameters then dominate the choice.

On the basis of this information, it is now possible to estimate the efficiency with which a single-junction solar cell can be expected to convert incident light. When the cell operates without delivery of current to an external circuit, a steady state situation develops in which the photo-generated carriers are lost at the same rate by recombination, in which electrons occupy available holes and dissipate their energy as heat. When current is drawn, some recombination still occurs, representing an efficiency loss, and the output voltage correspondingly decreases (near the open-circuit voltage, generated carriers are shared between current in the external circuit and the remainder which recombine). When most generated carriers are accepted by the external circuit, a point is reached when in turn the voltage delivered drops rapidly. Along the performance curve there is evidently a point at which the current/voltage product is optimised, the maximum power point. These features are identifiable in the performance curves of Figure 5.6, obtained on an actual solar cell. It should also be noted that where an incident photon has an energy significantly above the minimum necessary for carrier pair generation, the electron drops to the lowest level of the conduction band and the excess energy is again lost as heat. Finally the internal resistance of the cell provides a further loss factor. In consequence the idealised efficiency of about 25% indicated for silicon in Figure 5.7 is

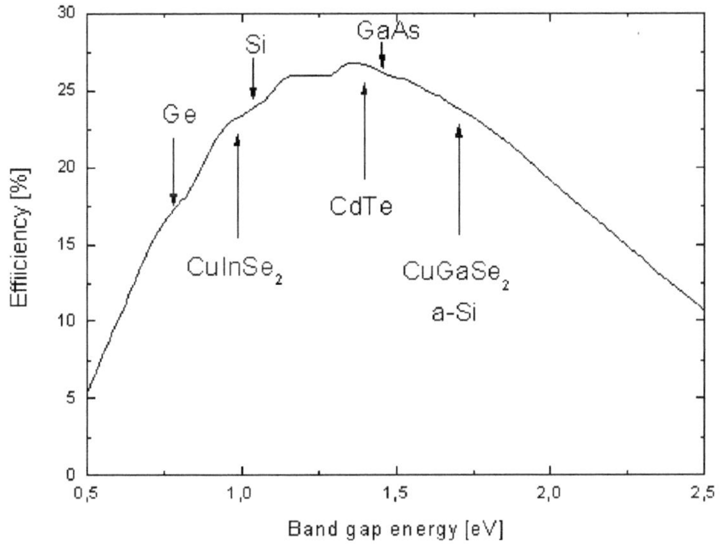

Fig. 5.7 Dependence of efficiency on the band gap of the selected semiconductor, and consequent indication of optimal band gap, 1.4 eV, for a single junction PV cell.

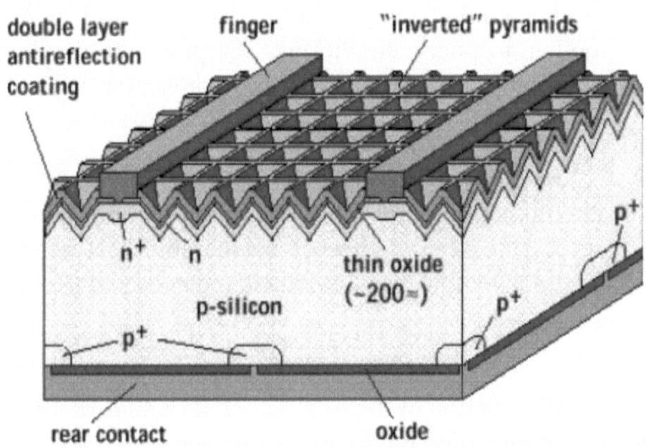

Fig. 5.8 Silicon solar cell, indicating electrical and optical features necessary for highest efficiency. These include graded doping (n^+ and p^+ regions) to accelerate charge carrier separation, metallic current collecting fingers to minimise series resistance, a thin 200 μm oxide passivation layer to reduce carrier loss through surface recombination, as well as surface texturing and antireflection coating to ensure absorption of incident light.

not realisable in practice; the very best silicon cell, as in Figure 5.8, fabricated as a technical exercise rather than as a production device, may closely approach that figure [4]. Certain production cells using high-quality single-crystal silicon reach 20%, however, while in a trade-off of efficiency against unit cost, polycrystalline silicon cells are widely available with an efficiency of 15%. On the basis of previous data,

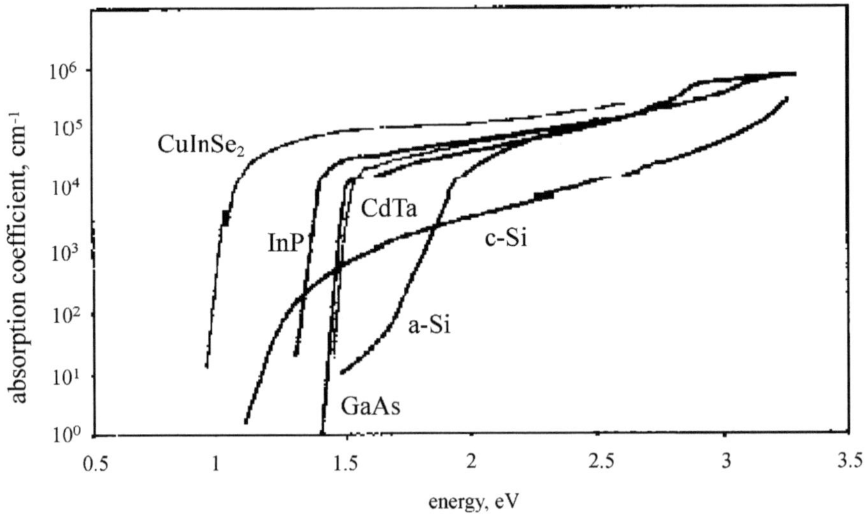

Fig. 5.9 Optical absorption coefficients of some practical semiconductors as a function of incident photon energy; c-Si = monocrystalline silicon; a-Si = amorphous silicon.

therefore, a 1 m^2 module could be expected to provide some 150 W under midday insolation under good atmospheric conditions. The power output per unit area of a solar cell under given lighting conditions is usually presented as:

$$P = V_{oc} \cdot J_{sc} \cdot \text{ff}.$$

The factor "ff" is the fill factor, representing the fraction of the area of the rectangle delimited by V_{oc} and J_{sc}, the open-circuit voltage and the short-circuit current density, formed by that defined by the current and voltage axes and the maximum power point.

$$\text{ff} = V_{mpp} \cdot J_{mpp}/V_{oc} \cdot J_{sc}.$$

The cell efficiency is the fraction of the total incident radiant energy represented by the electrical power output at the maximum power point.

The highest energy conversion efficiency achieved with silicon, as already mentioned, uses high-purity single crystal material. The minimisation of recombination losses is required, and this occurs principally at dislocations and defects in the semiconductor. Each cell has a thickness of some 100 microns, not only for the mechanical strength required when the bulk silicon ingot is sawn into the individual slices, but also because of an intrinsic optoelectronic property of silicon, as shown in Figure 5.9. Due to the indirect character of its band gap, the optical absorption coefficient of crystalline silicon is an order of magnitude lower than that of other, "direct", semiconductors. A much thicker absorbent structure is therefore required, as is a longer diffusion of charge carriers within the material, placing a further demand on the purity and crystallinity of the silicon. Given that the prepar-

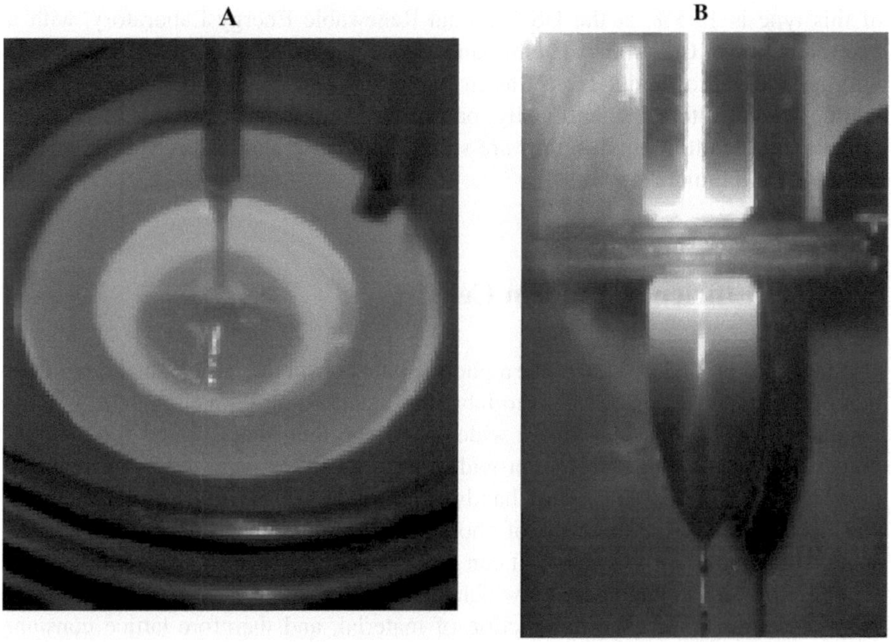

Fig. 5.10 (a) Drawing a single-crystal silicon ingot from the melt ("Czochralski growth"); b) zone refining. These illustrate the high temperature energy-intensive processing of single crystal silicon.

ation of such a material is a high-technology, cost- and energy-intensive process involving the drawing of a single crystal from a high-temperature melt and zone-refining (Figure 5.10) before sawing it into wafers and diffusion of the donor and acceptor dopants, again at high temperature, it is not surprising that in most silicon PV applications the energy recovery time is optimistically estimated as 2 years [5]. Cost and energy considerations therefore encourage research on alternatives.

The earliest credible candidate for a thin-film cell, requiring a significantly smaller amount of semiconductor material and with an effective direct band gap and therefore higher optical absorption, was paradoxically a structural modification of silicon [6]. This amorphous silicon, in which the long-range order of the crystalline form is not developed, is produced by plasma cracking of a gaseous silicon hydride, silane, SiH_4, which deposits the silicon on a transparent conducting substrate. As with crystalline silicon, it can incorporate donor or acceptor dopants, and can therefore form photovoltaically active p-n junctions. The expected recombination losses are minimised through passivation of defects by residual hydrogen incorporated in the semiconductor structure. Unfortunately the practical application of amorphous silicon devices has been limited by the fact that their efficiency degrades over the first few months of service life through loss of a part of the incorporated hydrogen.

There is at present considerable research and industrial interest in cells using the I–III–VI_2 semiconductors, particularly compounds or solid solutions of copper, gallium and indium and sulfur or selenium. The highest efficiency reported for cells of this type is 19.5%, at the US National Renewable Energy Laboratory, with a heterojunction of CdS, a II–VI compound, contacting a $Cu(In_{0.3}Ga_{0.7})Se_2$ thin film, with a band gap of about 1.2 eV as in Figure 5.4 [7]. As for organic solar cells, while of great interest scientifically, particularly due to their affinity to OLEDs, organic light emitting diodes, they are still very much in the research stage, of low efficiency and limited stability.

5.5 High Efficiency Tandem Cells

The favoured procedure to achieve a photovoltaic conversion efficiency higher than that of a single junction device is to fabricate a stacked multijunction cell. A semiconductor can be selected with a wider band gap than that optimal for a single junction device. It will therefore provide a higher voltage, but a lower current due to its restricted optical absorption bandwidth. It is, however, transparent to light of wavelength longer, and therefore of photon energy lower, than its band gap and associated optical absorption edge. It can therefore be followed in optical series by a second cell of lower band gap, for which the longer wavelength transmitted photons are still actinic. With proper selection of material, and therefore lattice constant and band gap, the current generated in each cell can be equalised so that the cells may also be in electrical series, and grown monolithically for the high performance and low recombination loss of a single crystal device. Appropriate materials in the III–V series are to be found in Figure 5.3, with band gaps of 1.75 and 1.1 eV respectively. The concept can be extended to triple-junction devices with an expected efficiency of 39% [8]. Evidently, however, such multiple thin layer devices are extremely costly, and can be considered only for space applications, as well as terrestrial applications in concentrator systems. Fewer cells are required if they are of higher efficiency, and so costly aerospace payload weight is reduced. Concentrator systems, focusing more light on a given solar cell, could limit system cost by increasing the total power production of a single cell but on the other hand require mechanical tracking systems to follow the sun and are effective only for direct sunlight in clear sky conditions, being ineffective for diffuse light.

5.6 Photo-Electrochemical Solar Cells

The nonlinear junction in a photovoltaic device is not necessarily formed by a semiconductor or even by a metal as in the Schottky barrier diode; it should not be forgotten that in the original Becquerel device of 1839 the phase contacting the photoactive semiconductor was a liquid electrolyte. The first photo-cell therefore was a photo-electrochemical device. If the more recent developments and practical applications generally use solid state devices, it is not least because an electrolyte, particularly under illumination, is a very hostile environment for a semiconductor.

When the charged carrier pair is photo-generated in an n-type semiconductor, the electron is available for an external current. However a valence band hole is in fact a powerful oxidising agent, so at the semiconductor surface its reaction with the electrolyte finds itself in competition with a damaging side reaction – the electrochemical photo-corrosion of the semiconductor itself. If we take the example of a common II–VI semiconductor, CdS, the following reaction occurs:

$$\mathrm{CdS} + 2\mathrm{h}^+ \Rightarrow \mathrm{Cd}^{2+} + \mathrm{S} \qquad (\mathrm{h}^+ = \text{hole, positively charged})$$

with cadmium ions dissolving in the electrolyte and elemental sulfur coating the electrode surface. As a result only wide band gap fully oxidised materials, typically titanium dioxide, TiO_2, are acceptable for stable photo-electrodes. Unfortunately only photons in the ultraviolet spectral region are actinic for such a wide band gap material. In consequence and despite long and intensive research, stability is recognised as incompatible with photo-conversion efficiency in a simple semiconductor – aqueous electrolyte system.

The resolution of this dilemma lies in the separation of light absorption and charge separation functions, by sub-band gap sensitisation of the semiconductor with an electroactive dye. This has been a long-term research and development effort in our Laboratory. The wide band gap, intrinsically stable semiconductor, such as titanium dioxide with its band gap of 3.1 eV, can then photo-respond to visible light of wavelength 400–750 nm, or 1.6–3.0 eV photons. The sensitisation principle has been known in photo-electrochemistry for over a century [9]. and has been applied in photography to secure wide-spectrum response, at first in panchromatic black-and-white emulsions, and more recently with spectrally selective dyes in colour photography [10]. The work of Gerischer and Tributsch [11, 12] on ZnO definitively established the mechanism of dye sensitisation and indicated its significance for photo-electrochemistry by finally rendering compatible effective wideband visible spectral absorption with the stability of a semiconductor substrate which would otherwise respond only to ultraviolet light. It is now evident that the process involves the excitation of the dye from its charge-neutral ground state to an excited state by the absorption of the energy of a photon, followed by relaxation through electron loss to the semiconductor substrate. The dye is left as a positively-charged surface-adsorbed cation. This process is presented schematically in Figure 5.11. In a redox electrolyte two dissolved ion species are interconvertible by exchange of electrons. Here the dye cation recovers charge neutrality by such a redox reaction with an anion in solution. The redox system in turn is restored by an electron from a counterelectrode. A closed regenerative cycle for the conversion of incident light into an electric current is therefore established. The standard redox system is the iodide/tri-iodide couple, I^-/I_3^-:

$$I_3^- + 2e^- \leftrightarrow 3I^-.$$

This regenerative photo-electrochemical device is evidently functionally equivalent to a conventional solid-state photovoltaic cell. However, as only a monomolecular adsorbed film of the sensitising dye can transfer charge to the substrate, the original

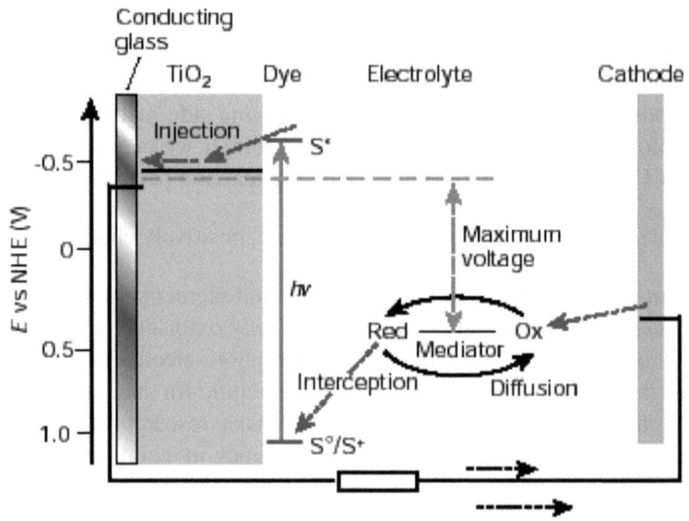

Fig. 5.11 Schematic of operation of the dye-sensitised electrochemical photovoltaic cell.

sensitised devices had low optical absorption, and therefore low photovoltaic efficiency. To provide a device capable of challenging the efficiency of its solid-state counterpart, the dye was adsorbed on a nano-structured semiconductor, so that the active interface area for light absorption and charge transfer greatly exceeded the projected geometrical area of the surface and gave the required opacity and light absorption. Since the first presentation of this device [13], further development work has enabled these dye-sensitised cells to reach a white-light conversion efficiency exceeding 11%, as shown by the cell characteristics in Figure 5.6.

That such a performance is attainable with a nano-structured semiconductor is due to the ultra-fast kinetics of electron injection from the excited dye into the solid. That separation of charge carriers, associated with the majority carrier nature of the device where the electrons enter an n-type material, strongly inhibits charge carrier recombination losses. Electrons can be lost only through recapture by the dye cation or by the redox electrolyte after crossing a phase boundary – an inherently slow process.

It is evident that to achieve this level of photovoltaic performance and maintain it over an adequate in-service life requires optimisation of each component in the system and attention to their synergies. Of particular importance is the choice of a suitable electroactive dye; in this aspect the development was inspired by the natural prototypes, the porphyrins and the chlorophyll molecule with its magnesium ion within a nitrogen "cage" in particular [14]. As the basis for photo-synthesis in plants, chlorophyll requires a photo-excited state attainable by the absorption of photons widely across the visible spectrum, exactly as does a photo-sensitising dye. The energy level of the excited state, or LUMO (lowest unoccupied molecular orbital) must lie higher than the conduction band edge of the semiconductor, so that

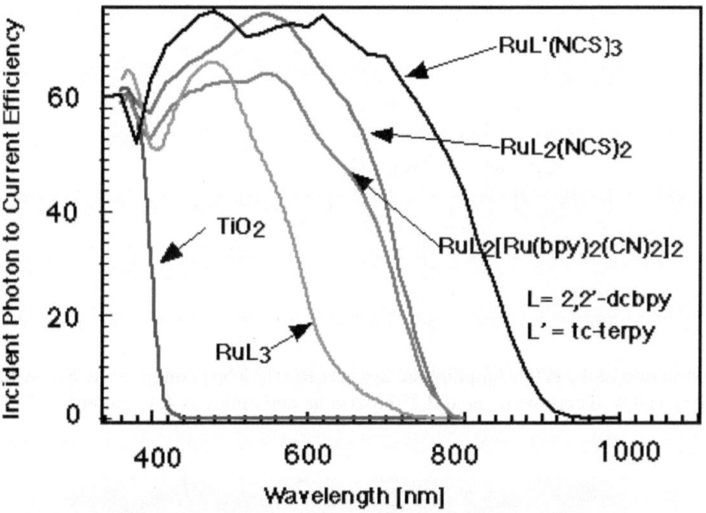

Fig. 5.12 the photo-current action spectra of cells containing various sensitisers, where the incident photon to current conversion efficiency is plotted as a function of wavelength.

an electron can be injected during the relaxation process. For optimum absorption of white light, therefore, the HOMO (highest occupied molecular level) should be about 1.5 eV lower. In terms of photo-excitation, the molecular HOMO and LUMO levels are analogous to the valence and conduction features in the band structure of a semiconductor. An appropriate molecular structure for the sensitisation of a titanium dioxide semiconductor had been identified already [15] by Clark and Suttin, using a ruthenium tris-bipyridyl complex. By substitution of carboxylic groups on the pyridyl rings, the chemical attachment of the molecule to a titanium dioxide surface can be enhanced, producing a stable photoanode. However with this sensitiser, the HOMO–LUMO gap is about 2.0 eV, and as a result the photovoltaic response is limited to wavelengths below 600 nm, as shown in Figure 5.12 (curve RuL_3; L = dicarboxylbipyridyl, or dcbpy). With detailed molecular engineering, the electronic structure of the molecule was modified to raise the HOMO level and thereby extend the response towards the red end of the spectrum. Figure 5.12 presents this progress, first using cyanide (CN) and later thiocyanide (SCN) substituents. Ultimately, a "black dye", absorbing across the whole visible spectrum and providing a panchromatic response, was evolved by associating the ruthenium "core" with a terpyridyl complex (L'), and three spectrum-modifying thiocyanide groups. As an example of dye molecular engineering, the molecule (laboratory code Z-907) shown in Figure 5.13 has of course the two bipyridyl ligands bonded to ruthenium to establish photo-sensitivity; two thiocyanide ligands to modify the spectral response; two carboxyl groups to chemically adsorb to the semiconductor oxide surface; and finally two hydrocarbon chains to present a hydrophobic surface to the electrolyte, thereby rejecting contaminant water. The consequence of this structure is the self-

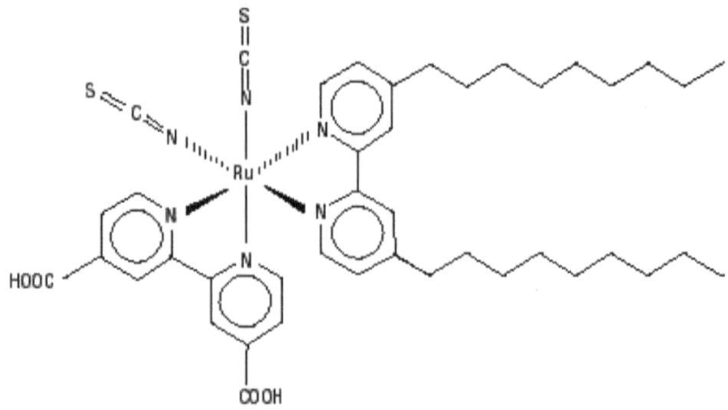

Fig. 5.13 structure of the Z-907 amphiphilic dye (cis-Ru(H2dcbpy)(dnbpy) (NCS)2, where the ligand H2dcbpy is 4,4'-dicarboxylic acid-2,2'-bipyridine and dnbpy is 4,4'-dinonyl-2,2'-bipyridine) [16].

Fig. 5.14 Scanning electron micrograph of the surface of a mesoporous anatase film prepared from a hydrothermally processed TiO_2 colloid.

assembly of a monomolecular dye layer whose specific orientation provides the required physical properties, uniformly chemisorbed on the nano-structured semiconductor surface.

A similar development process and attention to detail was required for the semiconductor substrate. On the transparent support – glass or polymer – a layer of transparent conducting oxide (TCO) such as tin-doped indium oxide (ITO) provides the electrical contact to the active semiconductor. In the earliest prototypes of the cell this was a fractured and fractal layer deposited by a sol-gel process. In current

Fig. 5.15 (a) Large-area roof installation, CSIRO, Australia. (b) flexible dye-sensitised cell on polymer substrate (Konarka Inc.).

practice a hydrothermal process can be applied [17] for grain ripening of nano-crystalline particles in the anatase structure, as shown in the micrograph in Figure 5.14. A colloidal suspension of the nano-particles is then deposited by spin coating or tape casting on the conducting substrate and fired at a relatively low temperature (450°C in the example shown, on a glass substrate) to establish electrical continuity. The process clearly avoids the repeated high temperature processing steps required in silicon PV cell manufacturing, a significant energy saving. It is very satisfactory to note that the dye-sensitised device is now emerging from the research laboratory and that pilot production and field evaluation can be reported. Figure 5.15a shows an installation on the roof of a CSIRO (Commonwealth Scientific and Industrial Research Organisation) building in Australia (courtesy of Sustainable Technologies International Ltd.). At the same time, cells on flexible substrates appropriate for application in consumer goods and under low-level illumination are emerging from industrial development laboratories to realise the "artificial leaf" in practice, Figure 5.15b (Konarka Inc., USA).

5.7 Balance of Plant

Operating as it does on the basis of the absorption of photons of actinic light with conversion of their energy into a separated charge-carrier pair in a semiconductor, the solar cell is necessarily a low-voltage DC device, and therefore not directly compatible with commercial electricity generation and distribution practice, which is a high-voltage low-current AC system. Individual solar cells are therefore employed in series-connected modules, which in turn form interconnected arrays. For optimum performance all the cells in a module should have characteristics as closely matched as possible and operate under uniform illumination. If one cell in the array is defective or shaded, for example, the series current may even impose an inverse voltage, leading to power dissipation, a "hot spot" and even module failure. Module

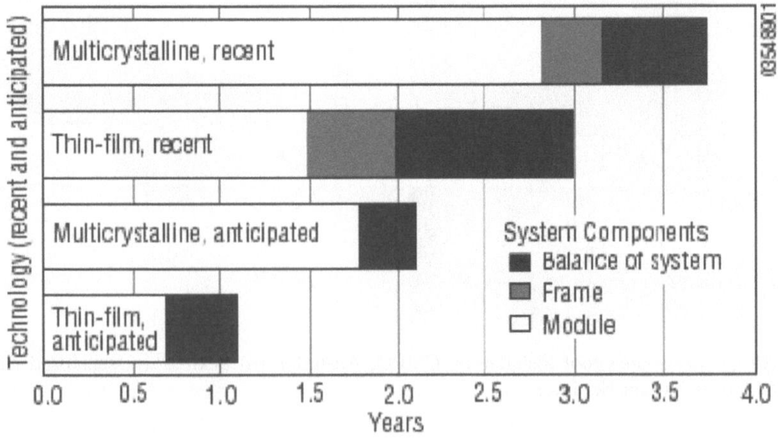

Fig. 5.16 Energy payback time for silicon PV devices, according to US Dept. of Energy.

engineering requires safety systems such as shunt diodes to avoid such events. It should also be remembered that a module or array, with a large number of cells in series, can present a safety hazard to personnel; as long as the sun shines, a solar PV array cannot be "switched off" (this section is generic to PV devices, many of which are stable under illuminated open-circuit conditions. The dye cell is of course less stable in these conditions). System efficiency also requires operation as close as possible to the maximum power point, so an electronic power tracking control unit is required. When integrated into an AC distribution system an inverter must be employed, as well as an appropriate metering system for accounting purposes. Given the diurnal, seasonal and climatic variations, mains-connected PV must be associated with reserve generation capacity, either conventional or renewable. When used in an isolated system where mains backup is not available, electrical storage by a battery bank is required. Obviously therefore, photovoltaic system engineering is multidisciplinary, with efficiency and economy dependent not only on choice of appropriate cells and modules and correct installation, but also on the associated electronics and power management equipment. As a result the photovoltaic devices themselves often represent only 30% of the overall system cost. The key issue of energy payback time already mentioned and future prospects for its reduction on the cell and system levels are indicated in Figure 5.16.

5.8 Conclusions

This presentation aims at an objective information to energy specialists in other disciplines on photovoltaic technologies and their place in a renewable energy scenario. It is intended that the sustainable energy option represented by photovoltaic

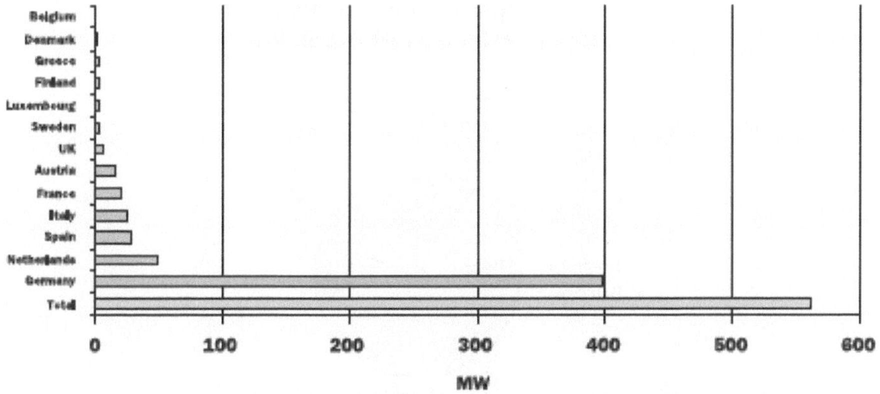

Fig. 5.17 Total PV installed power in selected European countries in December 2003.

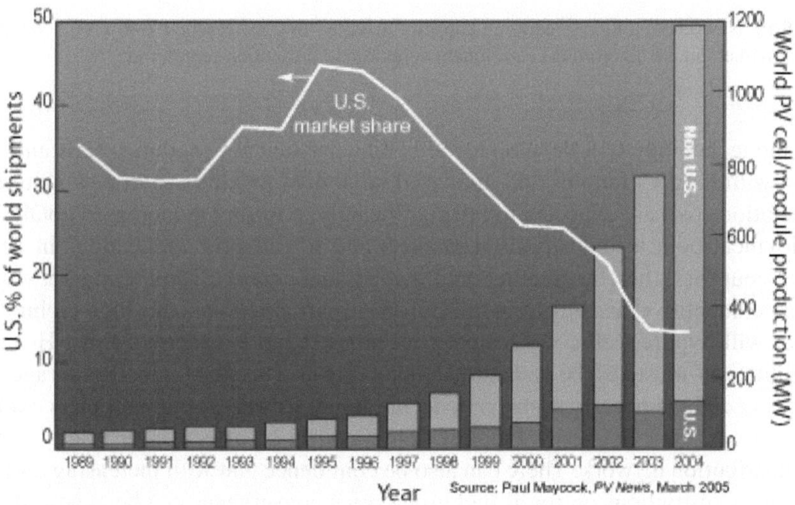

Fig. 5.18 Development of the world market for photovoltaic devices over the past 10 years. It is expected that the trend can be maintained.

technology should be appreciated at its real worth. By 2004 for example an annual production of photovoltaic devices with a rated power under AM1.5 illumination of 1.2 GW has been achieved [20]. Presently, Germany leads Europe in installation of solar PV systems, due to a vigorous programme of government incentives, and has some 12% of world PV generating capacity; installation in other European countries is presented in Figure 5.17. A European perspective is to be found in the report of the European Union PV Technology Advisory Council (PV-TRAC) [19]. The Council asserts that PV potentially can deliver electricity on a large scale at competitive cost. In 2030 PV could generate 4% of electricity worldwide. It is envisaged

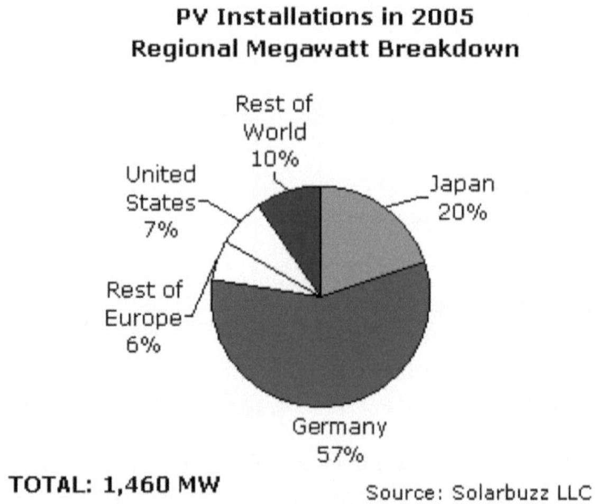

Fig. 5.19 New photovoltaic generating capacity installed in 2005 reached 1.46 GW, 57% of the world market, due to a vigorous Government support and promotion programme.

that the technology will develop towards higher efficiency modules, cells and systems, with longer lifetimes and improved reliability, making use of new materials. Generation costs are expected to fall significantly, resulting in increased uptake and deployment both in industrialised markets and for off-grid applications in developing countries, thereby creating employment and exports. The PV market will be highly competitive, and ensuring full European participation in this high-technology sector will require well-coordinated, concentrated and long-term efforts. However its limitations must also be understood, involving grid backup or energy storage, and requiring complementary sustainable technologies as well as more efficient exploitation of consumable energy supplies, integrated in an effective electricity supply and distribution network. There can also be confidence that with increasing cost and regulatory restrictions on fossil fuel use, a price convergence can be expected. The trend in market size over the past few years is provided in Figure 5.18. As for prices, there has been a slight rise in the past twelve months, as the increased demand for solar-quality silicon has depleted available stocks surplus to the electronics industry requirements. Currently (October 2006) module prices in Europe are on average over €5.00/Wp [21] and are unlikely to decrease significantly until new silicon production facilities are in operation. As a guide to longer-term trends, the industry expects that module prices can be reduced to about €1.50 over the next ten years, a target which must be attained if solar photovoltaic systems are to significantly penetrate the grid-connected market without subsidy and thereby maintain the expected growth rate. The impact of an intensive government promotion programme can be judged from Figure 5.19, showing that over half the PV generating capacity installed in 2005 was placed in Germany. It is estimated that worldwide the total installed photovoltaic peak power is some 5 GW.

Acknowledgements

Thanks are due to our funding agencies, the Swiss National Fund for Scientific Research, the European Union and others which have supported this work; the industrial organisations which have licensed the corresponding patents, and of course the research team members, past and present, for their commitment.

References

1. Becquerel, A.E., Mémoire sur les effets électriques produits sous l'influence des rayons solaires, *C.R. Acad. Sci., Paris* **9**, 1839, 561.
2. Adams, W.G. and Day, R.E., The action of light on selenium, *Proc. Roy. Soc.* **A25**, 1877, 113.
3. Rappaport, P., The photovoltaic effect and its utilization, *RCA Review* **20**, 1959, 373.
4. Zhao, J-H., Wang, A-H. and Green, M.A., High efficiency PERT cells on a variety of single crystalline silicon substractes, in *16th European PV Solar Energy Conference*, Glasgow, Scotland, 2000, p. 1100.
5. Kato, K., Murata, A. and Sakuta, K., Energy Payback Time and Life-Cycle CO_2 Emission of Residential PV Power System with Silicon PV Modules, Appendix B, *Environmental Aspects of PV Power Systems*, Report No. 97072, Utrecht University, the Netherlands, 1997.
6. LeComber, P.G. Madan, A. and Spear, W.E., Electronic transport and state distribution in amorphous Si films, *J. Non Crystalline Solids* **11**, 1972, 219.
7. Ramanathan, K., Teeter, G., Keane, J.C. and Noufi, R., Properties of high-efficiency $CuInGaSe_2$ thin film solar cells, *Thin Solid Films* **480/481**, 2005, 499.
8. http://www.nrel.gov/ncpv/higheff.html
9. Moser, J., Notiz über Verstärkung photoelektrischer Ströme durch optische Sensibilisierung, *Monatsh. Chem. (Vienna)* **8**, 1887, 373.
10. West, W., First hundred years of spectral sensitization, Vogel Centenary Symposium, *Photogr. Sci. Eng.* **18**, 1974, 35.
11. Gerischer, H. and Tributsch, H., Eletrochemische Untersuchungen zur spectrale Sensibilisierung von ZnO-Einkristallen, *Ber. Bunsenges. Phys. Chem.* **72**, 1968, 437.
12. Tributsch, H., Eine elektrochemische Methode zum Studium der spektralen Sensibilisierung und heterogener photochemischer Reaktionen an ZnO-Elektroden, Ph.D. Thesis, Technische Hochschule München, Germany, 1968.
13. O'Regan, B. and Grätzel, M., A low-cost, high-efficiency solar cell based on dye-sensitized colloidal TiO_2 films, *Nature* **353**, 1991, 737.
14. Grätzel, M., The artificial leaf, bio-mimetic photocatalysis, *Cattech* **3**, 1999, 3.
15. Clark, W.D.K. and Suttin, N., Spectral sensitisation of n-type titanium dioxide by polypyridine-ruthenium complexes, *J. Amer. Chem. Soc.* **99**, 1977, 4676.
16. Wang, P., Zakeeruddin, S.M., Moser, J.E., Nazeeruddin, M.K., Sekiguchi T. and Grätzel, M., A stable quasi-solid-state dye-sensitized solar cell with an amphiphilic ruthenium sensitizer and polymer gel electrolyte, *Nature Mater.* **2**, 2003, 402.
17. Brooks, K.G., Burnside, S.D., Shklover, V., Comte, P., Arendse, F., McEvoy, A.J. and Grïzel, M., Mesoporous nanocrystalline thin films for electrochemical applications, in *Proc. Am. Ceram. Soc.*, Indianapolis, USA, 1999/*Electrochem. Mater. Divices* (MRS), 2000, 115.
18. Global Investor Focus, Credit Suisse, November 2005.
19. http://europa.eu.int/comm/research/energy/index_en.htm
20. Solar Annual 2006, Photon Consulting Inc.
21. www.solarbuzz.com

Chapter 6
Photo-Electrochemical Production of Hydrogen

Roel van de Krol and Joop Schoonman

Abstract. The direct conversion of solar energy into hydrogen represents an attractive but challenging alternative for photovoltaic solar cells. Several metal oxide semiconductors are able to split water into hydrogen and oxygen upon illumination, but the efficiencies are still quite low. The operating principles of the photoelectrochemical device, the materials requirements, main bottlenecks, and the various device concepts will be discussed, and some promising directions for future research will be indicated.

Key words: Solar hydrogen production, semiconductors, photocatalysis, photoelectrochemistry, metal oxides, devices.

6.1 Introduction

In 2001, the global energy consumption rate was 13.5 TW, of which more than 85% originated from fossil fuels. Growing environmental concerns and a projected doubling of the energy demand in 2050 drive the search for new, sustainable sources of energy [1]. While many different scenarios and opinions exist on how to meet future energy demands, there is general agreement that the challenges involved are indeed formidable. Clean fossil fuels will undoubtedly play an important role in the foreseeable future, but the energy penalty for carbon capture and storage is high ($\sim$20%) and the technology is still in an early stage of development. Large-scale utilisation of nuclear fission energy would require the construction of a 1 GW nuclear power plant every 1.6 days for the next 45 years. These need to be highly efficient breeder reactors to avoid fast depletion of the low-cost terrestrial uranium supplies. Renewable energy sources, such as hydroelectric power, tidal power, geothermal energy, and wind energy are attractive options, but the total amount of energy that can be exploited from these sources is less than 10 TW. The same is true for biomass: to generate 10 TW of energy, all remaining cultivable land on earth would have to

K. Hanjalić et al., Sustainable Energy Technologies: Options and Prospects, 121–142.

be used for growing high-energy crops [2]. These examples suggest that there is no single solution that can solve the entire energy problem, and that all available options need to be explored.

One particularly attractive option that has not been mentioned yet is solar energy. With 120,000 TW of solar energy striking the surface of the earth at any given moment, it is by far the most abundant energy source available. Most of the energy sources mentioned above (fossil fuels, wind, tide, hydroelectric, biomass) originate from solar energy. However, direct use of solar energy by e.g., solar thermal power or photovoltaic solar cells is still limited. At this time, only ∼0.04% of all energy is generated by photovoltaics.

By covering 0.16% of the surface of the earth with 10% efficient solar cells, 20 TW_{peak} of power could be generated [2]. While 0.16% seems only a small number, it should be realised that this corresponds to the surface areas of France and Germany combined. Covering such a large area with solar cells presents a daunting task, even when this is undertaken on a (de-centralised) global scale. Interestingly, the solar cell market currently grows by 35–40% per year, and is one of today's fastest growing markets. With a total produced capacity of 1200 MW_{peak} in 2004 [2], a sustained 35% annual growth rate would lead to a total installed capacity of 20 TW_{peak} in 2036.

One of the challenges with solar cells is that they only generate electricity during daytime. Hence, large-scale use of solar energy requires an efficient energy storage solution. So far, the only practical way to store such large amounts of energy is in the form of a chemical energy carrier, i.e., a fuel. Hydrogen is one of the prime candidates as a future energy carrier. It can be made by electrolysis of water, and upon combustion in, e.g., a fuel cell it returns to its original form (water) without generating any harmful by-products. The future prospect of a so-called Hydrogen Economy has attracted much interest, and many research efforts are currently underway to develop new technologies for the production, storage, utilisation, and transport of hydrogen. This chapter addresses the sustainable production of hydrogen from water by using sunlight.

The most straightforward method to produce hydrogen from water and sunlight is by coupling an electrolyser to a solar cell array. Current commercially available electrolysers have current-to-hydrogen conversion efficiencies of up to 85%, which seems to make this an attractive route. However, to achieve the ∼1 A/cm^2 current density normally used in these electrolysers, an electrolyser cell voltage of 1.9 V is required. Since the thermodynamic potential required for splitting water into hydrogen and oxygen is 1.23 eV, the overall energy conversion efficiency is limited to 65% (1.23/1.9). When combining the electrolyser with a 12% solar cell, the overall conversion efficiency is limited to ∼8% [3].

A potentially more attractive route is the use of photo-electrochemical cells. Here, the photo-active semiconductor, usually a transition metal oxide, is immersed in water and the photo-generated electrons and electron-holes are directly used to reduce and oxidise water, respectively. Since the photo-active area is the same as the electrochemically active area, the current densities are much lower (10–30 mA/cm^2). This results in lower overpotentials and higher overall energy con-

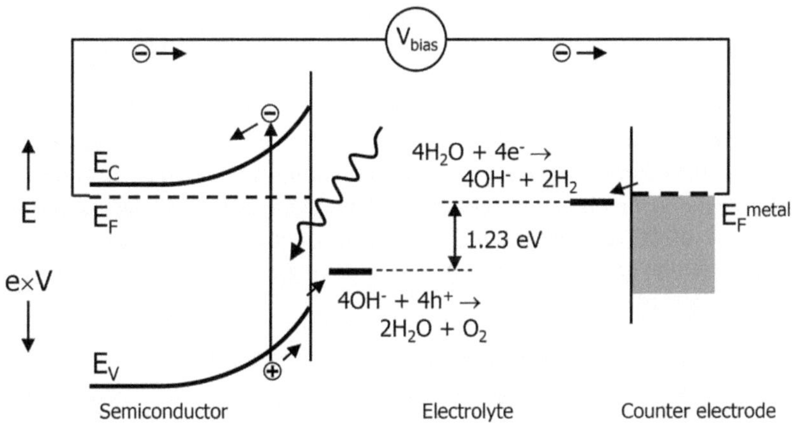

Fig. 6.1 Energy diagram for a photo-electrochemical cell (PEC) for water splitting using an n-type semiconductor as a photo-anode. The chemical reactions are written for an alkaline solution. A bias potential can be applied to aid the reduction reaction if E_C is positioned below the reduction potential of water.

version efficiencies [3]. Moreover, the functions of solar cell and electrolyser are now combined in a single device, which is expected to lead to much lower costs.

In this chapter, the operating principles of photo-electrochemical devices for water splitting will be explained. The requirements on the materials properties will be outlined, and several approaches and device concepts to overcome some of the limitations imposed by the currently available materials will be discussed. Finally, some recent work performed at Delft University of Technology on new materials for photo-electrochemical hydrogen production will be presented.

6.2 Principle of Operation

The principle of operation for a photo-electrochemical cell (PEC) for water splitting is shown in the diagram of Figure 6.1. In this case, an n-type semiconductor is chosen as the photo-active material (photo-anode). When the semiconductor is illuminated with photons that have an energy that is equal to or larger than the band gap of the material, electrons are excited from the valence band into the conduction band. The electrons travel to the back contact and are transported to the counter electrode where they reduce water to form hydrogen gas. The electron-holes that remain in the valence band migrate to the surface, where they oxidise water to form oxygen gas.

The driving force for the transport of photo-generated electrons and holes towards the bulk and the surface, respectively, is the presence of an electric field at the surface of the semiconductor. This electric field causes a bending of the energy bands, as shown in Figure 6.1, and is formed when free electrons from the bulk of the semiconductor are trapped into surface states. The resulting negative surface charge is compensated by the remaining positive charges in the bulk, which are ionised shallow donors (e.g., impurities or oxygen vacancies in the case of metal oxides). The concentration of electrons in the positive space charge is negligible, and this region is therefore often called the *depletion layer*. By applying an external potential to the semiconductor, the width of the depletion layer can be varied by extraction or injection of additional electrons from or into the material. The width of the depletion layer, W, depends on the donor concentration, N_D, and the applied potential, V, according to

$$W = \left(\frac{2\varepsilon_0\varepsilon_r}{eN_D}\right)^{1/2}\left(V - V_{\mathrm{fb}} - \frac{kT}{e}\right)^{1/2}. \tag{6.1}$$

Here, ε_0 is the vacuum permittivity, ε_r is the static dielectric constant of the semiconductor, V_{fb} is the flatband potential (see below), e is the elementary charge (1.6022×10^{-19} C), k is Boltzmann's constant (1.38×10^{-23} J/K), and T is the temperature in Kelvin.

The width of the depletion layer generally varies between 10 nm and ~500 nm. The flatband potential is the applied potential at which there is no electric field present in the material and W is negligible. Its value depends on the relative amounts of surface-adsorbed OH^- and H^+ species from the electrolyte since this affects the charge distribution at the interface [4, 5]. The ratio of surface-adsorbed OH^- and H^+ depends on the pH of the solution, and the relation between V_{fb} and the pH of the solution is given by [4]:

$$V_{\mathrm{fb}} = V_{\mathrm{fb}}^{\mathrm{PZC}} + \frac{2.3RT}{F}(\mathrm{PZC} - \mathrm{pH}). \tag{6.2}$$

Here, F is Faraday's constant (96485 C/mole), and $V_{\mathrm{fb}}^{\mathrm{PZC}}$ is the potential at the *point of zero charge* (PZC), i.e., the pH at which the net amount of adsorbed surface charges is zero. At room temperature V_{fb} changes with -59 mV per unit of pH.

At distances larger than W from the surface, no electric field is present to separate the photo-generated electrons and holes. As a result, the minority carriers (holes in the case of an n-type semiconductor) move in random directions until they are trapped at some impurity or native defect where they recombine with majority carriers (electrons). The average distance that the holes travel before trapping and recombination occurs is referred to as the minority carrier diffusion length, L_p. Compared to conventional electronics-grade semiconductors such as silicon, metal oxides usually have much larger impurity concentrations. As a consequence, their minority carrier diffusion lengths are very short, in the order of 1–20 nm. This implies that only light absorbed within a distance of $W + L_p \approx W$ contributes to the photo-response.

Ideally, all light should be absorbed within the space charge region to minimise electron transport losses in the bulk. This can be achieved by matching the space charge width with the absorption coefficient α, so that $W \approx \alpha^{-1}$. The absorption coefficient, α, depends on the photon energy according to [6]

$$\alpha = \frac{A(h\nu - E_g)^m}{h\nu}, \tag{6.3}$$

where A is a constant, $h\nu$ is the photon energy ($= hc/\lambda$), E_g is the band gap, and m is a constant that depends on the nature of the optical transition; $m = 2$ for an indirect semiconductor and $m = 1/2$ for a direct semiconductor. Compared to direct semiconductors, indirect semiconductors have much smaller absorption coefficients because they require a change in the crystal momentum upon absorption of a photon with $h\nu > E_g$ [7]. Unfortunately, most metal oxide semiconductors have an indirect band gap and their reciprocal absorption coefficients are usually too small for optimal matching of W and α^{-1} ($\alpha^{-1} \gg W$).

The overall solar-to-hydrogen conversion efficiency of the device can be determined from the electrical characteristics of the device, according to

$$\eta_{\mathrm{STH}} = \frac{j(V_{\mathrm{redox}} - V_B)}{P_{\mathrm{light}}}. \tag{6.4}$$

Here, j is the photocurrent density (A/m^2), P_{light} is the incident light intensity in W/m^2, V_{redox} is the potential required for water splitting, and V_B is the bias voltage that can be added in series with the two electrodes in order to assist the water splitting reaction (Figure 6.1). Equation (6.4) assumes that all the current is used for the water splitting reaction, i.e., no side reactions occur. V_{redox} is usually taken to be 1.23 eV (at room temperature) based on a Gibbs free energy change for water splitting of 237 kJ/mole. Alternatively, the enthalpy change (286 kJ/mole) is sometimes used, which corresponds to a redox potential of 1.48 eV. There is some debate on which of these values is the more appropriate one for photo-electrolysis efficiency calculations. Usually, the Gibbs free energy value is used which results in more conservative efficiency numbers. In the literature, the Gibbs free energy and the enthalpy values are sometimes referred to as the lower and high heating values (LHV and HHV) of hydrogen, respectively. These terms originate from the mechanical engineering of steam engines, and their use in (photo-)electrochemistry often leads to confusion and should be discouraged.

Suitable photo-electrode materials for efficient solar hydrogen generation have to fulfil the following requirements:

1. Sufficiently high (visible) light absorption.
2. High chemical stability in the dark and under illumination (no photo-corrosion).
3. Suitable band edge positions to enable reduction/oxidation of water by the photogenerated electrons/holes.
4. Efficient charge transport in the semiconductor.
5. Low overpotentials for the reduction/oxidation reaction (high catalytic activity).

The spectral region in which the semiconductor absorbs light is determined by the band gap of the material. The minimum band gap is determined by the energy required to split water (1.23 eV) plus the thermodynamic losses ($\sim$0.4 eV) [8] and the

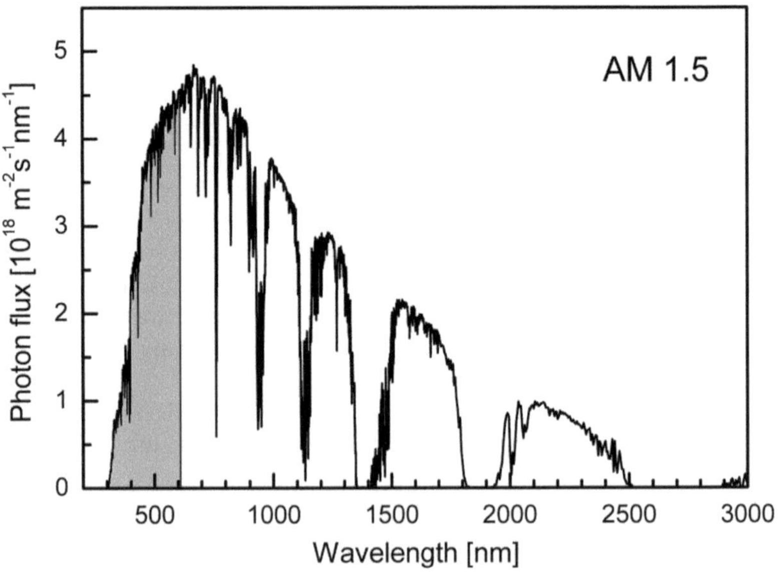

Fig. 6.2 Intensity of sunlight *versus* wavelength for AM1.5 conditions. The grey area respresents the part of the spectrum that can be absorbed by a hypothetically ideal band gap of 2.03 eV [9].

overpotentials that are required at various points in the system to ensure sufficiently fast reaction kinetics (~0.3–0.4 eV) [9, 10]. As a result, the band gap should be at least ~1.9 eV, which corresponds to a wavelength of ~650 nm. The maximum value of the band gap is determined by the solar spectrum shown in Figure 6.2. Below 400 nm the intensity of sunlight drops rapidly, imposing an upper limit of ~3.1 eV on the band gap. Hence, the optimum value of the band gap should be somewhere between 1.9 and 3.1 eV, which is within the visible range of the solar spectrum.

Several authors have made predictions for the maximum attainable efficiency based on the band gap of the material, the solar spectrum, and estimated values for the thermodynamic losses, overpotentials, optical reflection losses, etc. [9–11]. Most recently, Murphy et al. predicted a maximum efficiency of 16.8% for a hypothetically ideal material with a band gap of 2.03 eV [9]. Higher efficiencies can be obtained by using multiple layers, each with its own band gap and corresponding optical absorption range [10].

The second requirement, the stability against (photo-)corrosion, is a severe one that limits the usefulness of many photo-active materials. Most non-oxide semiconductors, such as Si, GaAs, GaP, CdS, etc., either dissolve or form a thin oxide film which prevents electron transfer across the semiconductor/electrolyte interface. Almost all metal oxide photo-anodes are thermodynamically unstable, which means that the photo-generated holes are able to oxidise the semiconductor [5]. However, if the kinetics of charge transfer across the interface (oxidation of water) are faster than the anodic decomposition reaction, photo-corrosion is inhibited, although it

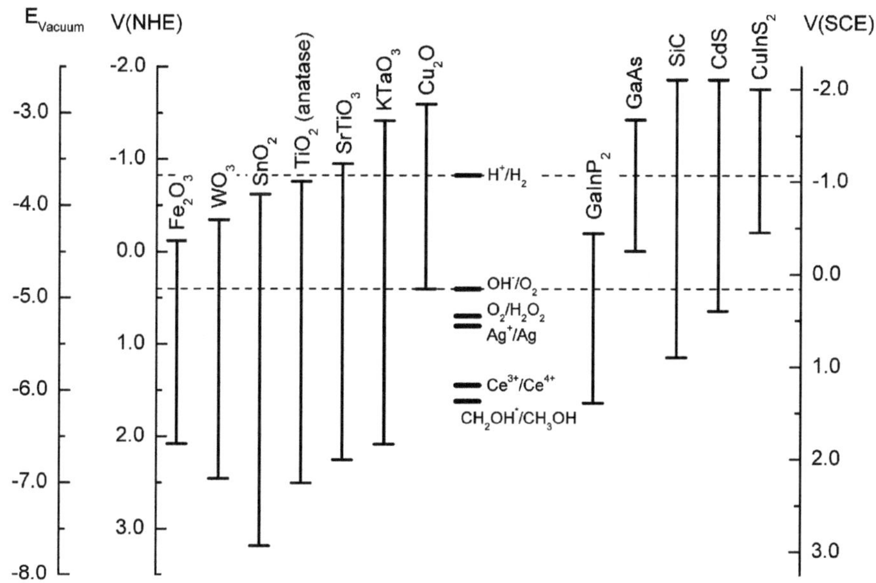

Fig. 6.3 Energy band positions for various semiconductors at pH 14 [5, 38, 66].

can remain as a side-reaction. For example, TiO_2 and SnO_2 show excellent stability over a wide range of pH values and applied potentials, while ZnO always decomposes in aqueous environments upon illumination. Fe_2O_3 represents an intermediate case, for which the stability appears to depend on the presence of dopants, pH, and oxygen stoichiometry [12,13]. The general trend is that wide-band gap metal oxide semiconductors are stable against photo-corrosion, while small-band gap semiconductors are not. While this is in obvious conflict with the requirement of visible light absorption, it does not represent a fundamental limitation.

The third requirement implies that the conduction and valence band edges should "straddle" the reduction and oxidation potentials of water. Specifically, E_C should be above $E_{\mathrm{red}}(H_2/H^+)$ and E_V should be below $E_{\mathrm{ox}}(OH^-/O_2)$. Figure 6.3 shows the band edge positions of various semiconductors, along with the reduction and oxidation potentials of water. Most non-oxide semiconductors are able to reduce, but not oxidise water. Conversely, most oxide semiconductors are able to oxidise, but not reduce water. Since the stability criterion favors oxide semiconductors, the reduction of water appears to be challenging. In most cases, an externally applied voltage or a separate electrode compartment with a different pH [14] is necessary to assist the reduction reaction. This is, however, equivalent to a galvanic cell in series and can therefore not be sustained. Indeed, only a few metal oxides have been demonstrated to be able to evolve both oxygen and hydrogen, and only $SrTiO_3$ has been shown to photo-cleave water in a two-electrode system without any assistance [15]. The efficiency was, however, less than 1% due to the large band gap of the material (3.3 eV).

An alternative and exciting possibility to overcome the limitations imposed by the band energies is by exploiting quantum-size effects. When reducing the size of a semiconducting particle to values that approach the characteristic wavelength of the electrons and holes, the band gap of the material increases. Hence, E_C and E_V increase and decrease, respectively, the amount of which is determined by the effective masses of the electrons and holes [16]. For example, by decreasing the particle size of Fe_2O_3 one might be able to raise the conduction band to a level just above the reduction potential of water [17]. Since metal oxide semiconductors usually have comparatively large effective electron and hole masses, the particle size needs to be reduced to less than $\sim$10 nm before quantum-size effects become noticeable. The consequence of using quantum-confined structures is the blue-shift of the absorption spectrum due to the increased band gap value. This should be taken into account when aiming for an optimal match with the solar spectrum. While the diagram in Figure 6.3 is drawn for a pH value of 14, it remains valid for other pH values so long as the variation of the flatband potential of the semiconductor (see below) corresponds to the change in the reduction and oxidation potentials of water, i.e., -59 mV per pH unit. While this is indeed observed for most, even non-oxide, semiconductors, it should be borne in mind that there are exceptions (e.g., WSe_2) [18]. Moreover, the positions of the energy bands with respect to the water redox potentials may shift during illumination (band edge "unpinning") [5].

The fifth requirement implies that hole transfer across the n-type semiconductor/electrolyte interface should be sufficiently fast in order to compete with the anodic oxidative photocorrosion. Furthermore, it should also be sufficiently fast in order to avoid the accumulation of holes at the surface, as this would lead to a decrease of the electric field and a concomitant increase in electron-hole recombination. This is especially important for very small (e.g., quantum-confined) structures, since they cannot sustain an electric field that is large enough to effectively separate the photo-generated electrons and holes. For such systems, the interfacial charge transfer kinetics need to be sufficiently fast in order to avoid recombination. To improve the kinetics of hole transfer, catalytically active surface species can be added. Perhaps the best known example of an effective oxygen evolution catalyst is RuO_2, whereas Pt is usually employed as a catalyst for hydrogen evolution [5].

6.3 Device Concepts

To overcome some of the limitations outlined above, several different device concepts have been reported in the literature. The following overview will be limited to some of the most promising and/or interesting ones. It aims to give the reader a brief and representative (but necessarily incomplete) overview of current developments in the field.

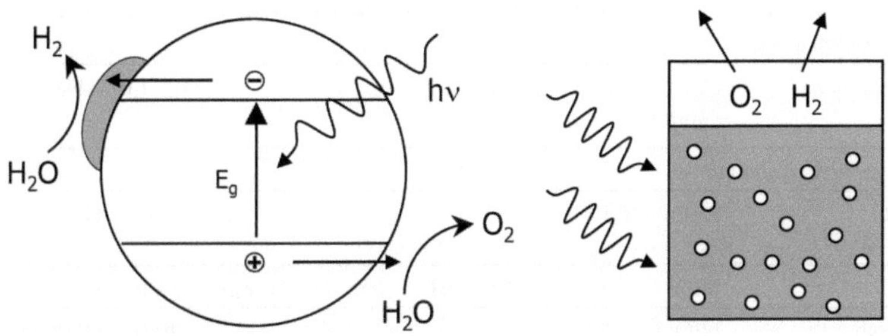

Fig. 6.4 Left: Energy diagram for a particle-based photo-catalyst for water splitting. For very small particles (<100 nm), the electric field inside the particles is negligible. To avoid electron-hole recombination, co-catalysts (here shown only for the reduction reaction) are added to enhance the surface reaction rates. Right: One-pot system containing a suspension of photo-catalyst particles in water. Illumination occurs either from the side or from the top of the reaction vessel. The hydrogen and oxygen need to be separated to avoid explosive mixtures.

6.3.1 Photo-Catalyst Particles in Suspension

The most straightforward approach towards solar water splitting is by dispersing a photo-catalyst powder in water. With this approach, the reduction and oxidation reactions both have to take place at the surface of the same particle, as illustrated in Figure 6.4. To enhance the kinetics and to avoid chemical recombination of H_2 and O_2, small amounts of co-catalysts, such as Pt, NiO_x or RuO_2 are deposited onto the photocatalyst's surface to ensure that H_2 and O_2 evolve at spatially separated sites. Typical loadings are 1–3 g of catalyst powder per 1000 ml H_2O, and co-catalyst loadings of 0.5–1% *versus* the amount of photo-catalyst. The aqueous slurry is contained in a closed pyrex glass cell which is illuminated using fluorescent tubes ("black-light"), or high-power xenon or high-pressure mercury lamps. By connecting the cell to a closed gas circulation system, the amounts of evolved hydrogen and oxygen can be determined on-line with gas chromatography.

In many cases, the photo-catalyst is only able to evolve either hydrogen or oxygen gas, not both at the same time. If only hydrogen (oxygen) evolves, a sacrificial electron donor (acceptor) must be present to ensure the stoichiometric consumption of electrons and holes. While sacrificial systems are of considerable technical interest as they permit the study of electrode reactions, they cannot be used as a commercial hydrogen source due to the consumption of high-value (in terms of energy and cost) compounds, such as methanol, EDTA, formic acid, etc. Examples of oxygen evolution photo-catalysts are $BiVO_4$ [19, 20] and WO_3 [21]. $AgNO_3$ or Ce^{4+}(aq) are often used as electron acceptors (or *electron scavengers*) in these systems. Oxygen gas is sometimes used as an electron scavenger in other photo-oxidation reactions. Hydrogen evolution photo-catalysts appear to be more common, examples include $InVO_4$ [22], Bi_2RNbO_7 (R = Y, rare earth) [23], and $BaCr_2O_4$ [24]. Methanol is almost always used as a sacrificial electron donor (hole scavenger), but other scavengers such as EDTA and formic acid may also work. The poor hydrogen evolution

Table 6.1 Literature overview of particle-based photo-catalyst systems that are able to split water in stoichiometric ratios.

Photo-catalyst	Co-catalyst	Lamp, electrolyte	η_{STH}	Stability	Absorption	Ref.
Single particle systems:						
$M_2Sb_2O_7$ M=Ca, Sr	RuO_2	L1	–	>4 h	UV	[55]
$NaTaO_3$	NiO	L2, 10^{-3} M NaOH	20% @ 270 nm	>14 h	UV	[56]
$BaIn_{0.5}Nb_{0.5}O_3$	NiO_x	L3		>40 h	UV/VIS	[57]
$In_{0.9}Ni_{0.1}TaO_4$	NiO_x	L3	0.66% >420 nm	>120 h	UV/VIS	[58]
$(Ga_{1-x}Zn_x)(N_{1-x}O_x)$	Cr-Rh oxide	L2/L3, pH 4.5 (H_2SO_4)	2.5% @ 420 nm	>35 h	UV/VIS	[59]
Layered materials:						
$Ba{:}La_2TiO_7$	NiO	L2, >12 mM NaOH	50% in UV	–	UV	[60]
$Sr_2Ta_2O_7$	NiO	L3	12% @ 270 nm	>20 h	UV	[61]
Composite particles:						
$Zn{:}Lu_2O_3/Ga_2O_3$	NiO	L2	–	>220 h	UV	[62]
$Cr{:}Ba_2In_2O_5{:}In_2O_3$	NiO_x	L2	4.2% @ 320 nm	>400 h	UV	[63]
Particle mixture, Z-scheme:						
WO_3/(Ca, Ta):$SrTiO_3$	Pt/Pt	L3, 100 mM NaI	0.1% @ 421 nm	>250 h	UV/VIS	[64]
an-TiO_2/ru-TiO_2	Pt/Pt	L2, 0.1 M NaI, pH 11	–	>100 h	UV	[65]

L1 = 200 W Hg-Xe; L2 = 400–500 W high pressure Hg; L3 = 300 W Xe

capability of $BiVO_4$ has been investigated with computational methods, and was attributed to the large distance between the protons from the water molecules and the vanadium reduction sites [25]. Bi_2RNbO_7 shows a somewhat unusual behavior in that it appears to evolve hydrogen without a hole scavenger being present. Apparently oxygen is formed at this photo-catalyst which is immediately photo-adsorbed at the particle's surface [23].

Table 6.1 presents an overview of recently reported suspension-based photo-catalyst systems that are able to split water in a $H_2{:}O_2$ ratio of 2:1 without the need for any sacrificial agents. Co-catalysts are required in all cases and the reported efficiencies are low, especially in the visible-light region. In addition to the straightforward single component systems, several other systems have been studied. In all cases, the main goals are to (i) spatially separate hydrogen from oxygen to prevent chemical recombination and (ii) provide different surfaces with different catalytic activities for hydrogen and oxygen evolution.

In the layered perovskite compounds (Table 6.1), water molecules are able to diffuse spontaneously into the crystal structure of the material where photo-oxidation takes place. The oxygen diffuses back to the surface through the interlayers, while photo-reduction takes place at the surface of the co-catalyst [26]. In the composite particles, an n-n junction forms between two different particles. Due to the different energetic positions of the conduction and valence band edges, the electrons are transported to one particle, while the holes move to the other particle. In the mixed-particle system, this concept is taken a step further by performing the oxidation and reduction of water on two different particles that are not in direct contact with each other. To ensure a complete redox reaction at each individual particle, an I^-/IO_3^- redox mediator is used to transport electrons from the oxygen evolving particles to their hydrogen evolving counterparts.

The use of powder-based photo-catalysts (Figure 6.4) has several advantages over the photo-electrochemical approach (Figure 6.1). Conventional high-temperature solid-state synthesis routes can be used, allowing complex, single-phase compounds to be made with relative ease. Co-catalysts can be easily added in the desired quantities by simple mixing and firing. Upscaling is readily accomplished and allows accurate chemical analysis of the evolved gasses, even at very low efficiencies. However, there are also some disadvantages. When reactivation of the catalyst is necessary, it must be separated from the water by filtration, which is not very convenient. Furthermore, hydrogen and oxygen are produced in the same reactor volume, forming an explosive mixture, and need to be separated immediately afterwards. This not only represents a safety issue, the separation also costs energy and lowers the overall efficiency of the process. Finally, it is often difficult to determine the actual number of photons that is absorbed by the suspension, which prohibits accurate efficiency measurements.

6.3.2 Tandem Cells

While the particle-based approach may offer a convenient and valuable screening method for selecting suitable materials, PEC systems are generally considered to be more promising for solar hydrogen generation. This is primarily due to the rather stringent requirement that hydrogen and oxygen should be produced in separate volumes. As mentioned above, $SrTiO_3$ has been the only material so far for which unassisted photo-cleavage of water has been observed in a PEC configuration. Other materials require the application of a bias voltage (Figure 6.1) to assist with the reduction of water. Since water splitting requires at least 1.9 eV of energy, photons with less energy are not absorbed by the photo-anode and may be used for generating the bias voltage in a tandem cell configuration.

Augustynski and Grätzel used a dye-sensitised solar cell (DSSC) to provide a bias voltage to a WO_3 or Fe_2O_3 photo-anode [27–29]. An efficiency of 4.5% has been published for the WO_3-based photo-anodes [27], close to the theoretical efficiency that has been reported for this material [9]. Recent work on doped α-Fe_2O_3 (hematite) photo-anodes indicates that overall solar-to-hydrogen efficiencies of 2.1% can now be reached (based on the enthalpy change for the water splitting reaction) [30]. The estimated theoretical efficiency for Fe_2O_3 is 12.9% [9], which shows that there is still plenty of room for improvement. In fact, Fe_2O_3 is currently being regarded as one of the most promising candidates for efficient photo-cleavage of water using a photo-electrochemical cell.

Although biasing with a separate solar cell is attractive from an efficiency point of view, the fact that it involves two separate devices, each with its own electrodes and electrolyte (in the case of a DSSC), complicates the system and increases the cost. Several efforts have been made to design and fabricate a monolithic tandem cell, where multiple functional layers are combined into a single plate. When this plate is immersed in a solution and illuminated, hydrogen should evolve from one side while oxygen is formed on the other side. Turner et al. have successfully fabricated a monolithic tandem cell based on a p-type $GaInP_2$ photo-cathode biased by a GaAs p-n junction solar cell [31]. The solar-to-chemical conversion efficiency is 12.4% (based on the Gibbs free energy change of water splitting) under an illumination intensity of $\sim$11 suns. As pointed out by the authors, the use of concentrated sunlight is necessary due to the high cost of the device. The main problem, however, is that the lifetime of the tandem device is measured in hours instead of years due to severe photo-corrosion.

Licht et al. addressed the photo-corrosion problem by using efficient catalysts for H_2 and O_2 evolution (Pt and RuO_2, respectively) [32, 33]. They used a complex multiple band gap system based on $Al_xGa_{1-x}As$ and Si p-n junctions to achieve a record 18.3% efficiency for water splitting. However, it should be noted that this system is comprised of a small area tandem cell coupled to a large-area electrode for electrolysis. Moreover, with more than 10 different high-purity semiconducting layers, the cost of such a device will be much too high for large-area terrestrial applications [27]. It is not a priori clear whether such a system will be cost effective in a solar concentrator system. Practical design issues and light scattering by evolving gas bubbles limit the illumination intensity to an estimated 10–20 suns [31]. For comparison, photovoltaic concentrator systems based on these materials usually operate at light intensities exceeding 100 suns [34].

Miller et al. used a triple-junction amorphous silicon photovoltaic cell in combination with efficient hydrogen and oxygen evolution catalysts [35]. Amorphous silicon (a-Si) is much cheaper than GaAs-based materials and requires less energy to process. Sputter-deposited NiMo and Fe:NiO_x were used as hydrogen and oxygen evolution catalysts, respectively. These materials are pure electrocatalysts, i.e., they are not photo-active. A solar-to-hydrogen conversion efficiency of 7.8% was achieved for a small-area prototype device (<1 cm^2). Both catalysts showed excellent resistance against (photo-)corrosion. It should be noted that the photo-active area is not the same as the electrochemically active area, since the catalysts are not transparent for visible light. To address this, one of the a-Si junctions is replaced by Fe_2O_3 or WO_3 [36]. So far, the a-Si/WO_3 hybrid photo-electrodes show solar-to-hydrogen conversion efficiencies of 1% in outdoor tests [37]. A particular challenge for these systems is to deposit the metal oxides at sufficiently low temperatures (<300°C) in order to avoid degradation of the underlying a-Si junctions. Aerosol spray pyrolysis requires a temperature of $\sim$400°C and cannot be used. With reactive sputtering the substrate temperature could be kept below 200°C at the cost of being a more expensive technique.

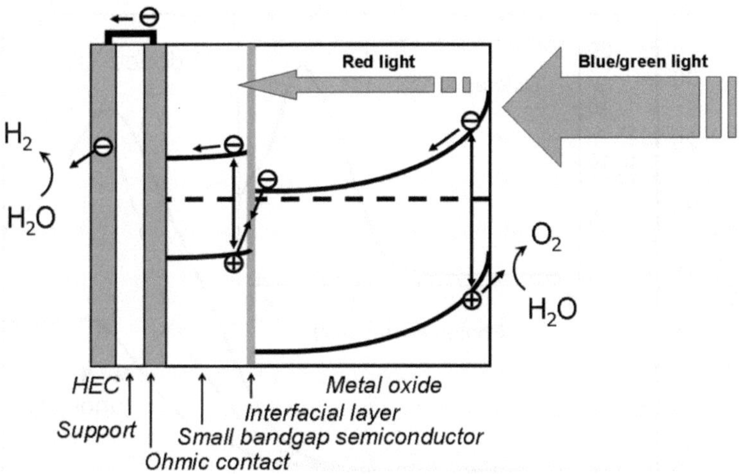

Fig. 6.5 Energy diagram for the proposed monolithic PEC device for water splitting using an n-type metal oxide semiconductor as a photo-anode, and a buried small-band gap semiconductor to boost the energy of the electrons. Oxygen evolves at the metal oxide photo-anode, while hydrogen is formed at the hydrogen evolution catalyst (HEC).

From the above it appears that (i) a viable large-area photo-electrochemical device for water splitting should be based on the monolithic tandem cell concept and (ii) the key ingredient for such a device is a photo-active metal oxide that is stable against photo-corrosion and that can be deposited using low-cost techniques. Most visible-light absorbing metal oxides have a conduction band minimum that is below the hydrogen evolution potential, cf. Figure 6.3. Therefore, we propose a PEC structure that uses an underlying small-band gap semiconductor to boost the energy of the electrons by absorbing the remaining low-energy photons. The proposed PEC device is shown in Figure 6.5 and is based on an n-n semiconductor junction. The metal oxide can be Fe_2O_3, WO_3, or another oxide with a band gap between ~2 and ~2.8 eV. The smaller band gap semiconductor should have a high enough conduction band edge for hydrogen evolution. Since it is protected by the metal oxide, it does not need to be photo-chemically stable in water. However, it should be near-intrinsic to ensure that the energetic distance between the Fermi level and the conduction band minimum is high enough. The critical component is the interfacial layer, which should be compatible with both semiconductors. The device is based on a reverse-operated version of the solid state TiO_2/$CuInS_2$ solar cell that was recently developed in our department by Nanu et al. [38, 39].

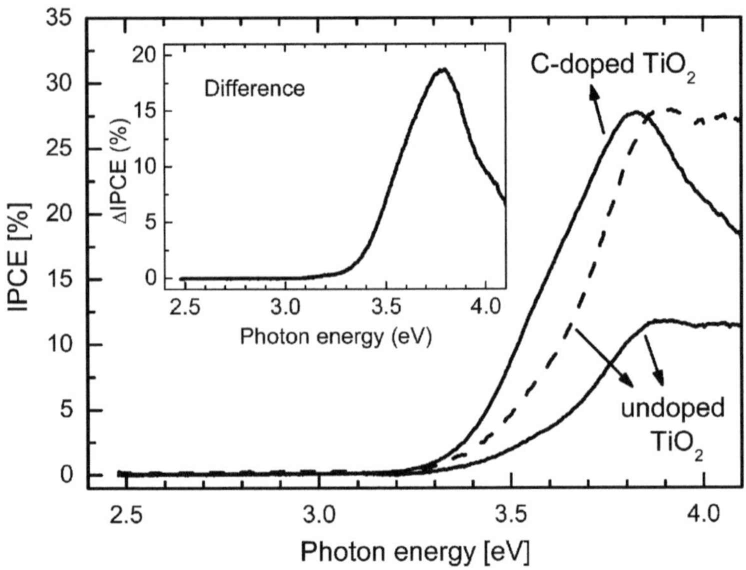

Fig. 6.6 Incident photon-to-current conversion efficiency (IPCE) for undoped and carbon-doped TiO_2 made by spray pyrolysis. The carbon was introduced by spraying under a CO_2 atmosphere. The spectra are recorded at a potential of 0 V *versus* SCE in a 0.1 M KOH aqueous electrolyte [46].

6.4 Materials for Photo-Anodes

After Fujishima and Honda's paper on water splitting with TiO_2 under UV illumination [14], many efforts were undertaken to sensitise TiO_2 to visible light by doping [40, 41]. While certain dopants, such as Cr and Fe, were indeed able to shift the absorption to the visible part of the solar spectrum, it did not result in significantly higher water splitting efficiencies. This is because the metal dopants act as efficient recombination centers for the photo-generated electron-hole pairs. In 2001, research interest in the sensitisation of wide-band gap metal oxides by doping returned after a paper by Asahi and co-workers on anion-doped TiO_2 [42]. Quantum-chemical calculations indicated that the wavefunctions of anion dopants, such as nitrogen and carbon, show significant overlap with the valence band wavefunctions. This implies that the sub-band gap defect levels are less localised than for cation dopants, and recombination should be less of a problem. Nitrogen- and carbon-doping of TiO_2 particles have indeed been shown to result in higher photo-catalytic activities [43–45].

To investigate the effect of anion dopants on TiO_2 photo-anodes, we have incorporated carbon into thin film anatase TiO_2 by spray pyrolysis under a CO_2 atmosphere [46]. Figure 6.6 shows the incident photon-to-current conversion efficiency (IPCE) as a function of photon energy. While the UV-response of the material increased, the amount of incorporated carbon was too small to cause any significant

Fig. 6.7 Scanning electron micrograph of fractal-shaped TiO_2 deposited by chemical vapor deposition.

shift towards the visible part of the spectrum. A high-temperature treatment of a porous nano-particulate film of anatase TiO_2 in a hexane-containing environment also did not result in a significant shift towards the visible part of the solar spectrum [47]. To increase the concentration of carbon in TiO_2, the oxidative annealing of TiC films may be a more promising route [48]. Sputter-deposition in a CO_2 containing atmosphere may also lead to enhanced carbon concentrations. Lindgren et al. published some interesting work on nitrogen-doped TiO_2 films deposited by reactive DC magnetron sputtering [49, 50]. They showed that, while an optimal concentration of nitrogen enhances the white-light response with a factor of 200 compared to undoped TiO_2, the presence of nitrogen also causes significant recombination.

To tackle the adverse effects of recombination, earlier work in our laboratory has focused on developing highly structured TiO_2 photo-electrodes [51]. Figure 6.7 shows an example of such a morphology made by Chemical Vapor Deposition (CVD). The high aspect ratio of these fractal-like structures ensures that the photo-generated holes only have to travel short distances before reaching the surface. This limits the chance of being trapped at a dopant site, which would ultimately result in recombination. These fractal-shaped TiO_2 electrodes did not generate any photo-current under visible light, although sensitisation with a Ru- based dye resulted in a solar cell with an efficiency of 2.7% [51].

In close collaboration with Everest Coatings, a spin-off company originating from our laboratory, we showed that fractal-shaped TiO_2 can be sensitised to visible light by co-doping with Fe and Cr [52]. Figure 6.8 shows the IPCE spectra for these films compared to those for thin, undoped films of TiO_2. It should be noted that this response could only be obtained for the fractal-shaped samples; flat films of (Cr,Fe)-doped TiO_2 made by spray pyrolysis exhibited a much smaller visible-light

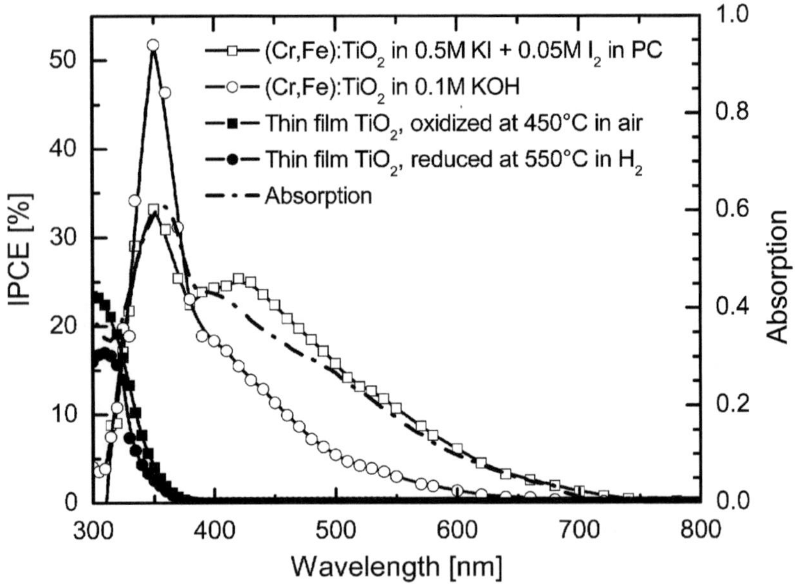

Fig. 6.8 IPCE spectrum of fractal-shaped TiO_2, co-doped with Fe and Cr and deposited with CVD.

response. Presumably, the nano-structured morphology of the leaf-shaped material stabilises the defects that are responsible for the visible-light photo-activity. The IPCE tracks quite well with the optical absorption, indicating that charge transport through these films is not a limiting factor.

Although these films absorb far more visible light than undoped TiO_2, the absorption in the UV part of the spectrum (<350 nm) is in fact much smaller. Photocurrent measurements under AM1.5 simulated sunlight with and without a UV filter show that visible light ($\lambda > 400$ nm) contributes more than 75% to the total AM1.5 photocurrent response. At the same time, the overall solar cell efficiency is $\sim$0.3%, which is similar to that of undoped TiO_2. This implies that the activity enhancement in the visible part of the solar spectrum is cancelled by the activity decrease in the UV part of the spectrum. Moreover, simultaneous exposure to light and water (or oxygen) causes these films to lose their visible light photo-activity. The color changes from cinnamon-brown to white and the remaining optical absorption corresponds to that of undoped TiO_2. Clearly, these films are of limited use for photo-electrochemical water splitting. More generally, it seems that all reported efforts to enhance the visible-light photo-activity of wide-band gap metal oxides for water splitting are unsuccessful. Taking into account the recent developments on, e.g., Fe_2O_3 and WO_3 [27, 30, 53] it appears that smaller-band gap metal oxides offer a more promising route towards solar water splitting.

As a final example, we will discuss some recent work on InVO4 that was recently performed in our group [54]. Motivated by the high visible-light photo-activity of $InVO_4$ powders reported by Ye and co-workers [22], we have prepared and charac-

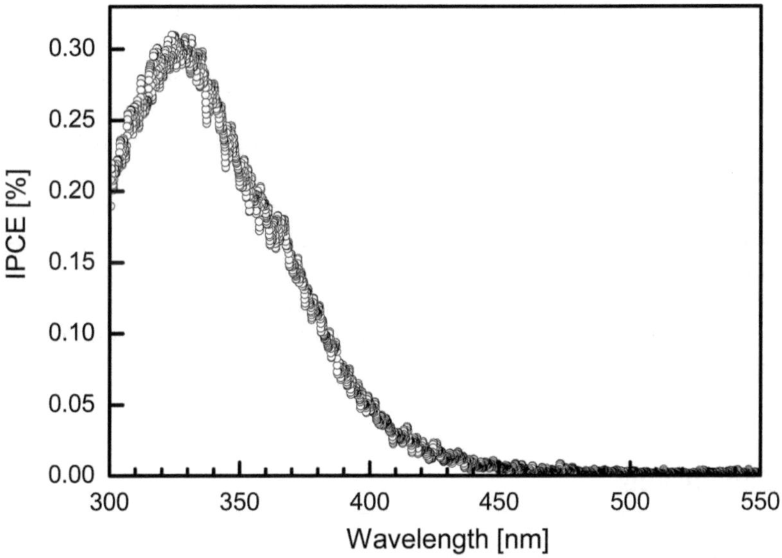

Fig. 6.9 IPCE of a ~70 nm spray-deposited orthorhombic $InVO_4$ film, recorded in an aqueous 0.1 M KOH electrolyte at a potential of 0.8 V *versus* Ag/AgCl. A lock-in amplifier was used to measure the photo-current and a calibrated photodiode was used to determine the light intensity.

terised thin film $InVO_4$ photo-electrodes. The films were deposited by spray pyrolysis from an ethanolic solution of $In(NO_3)_3$ and VCl_3. As-deposited films exhibit the monoclinic structure, which could be converted to the desired orthorhombic structure by a post-deposition anneal in air at 550°C. The IPCE of the resulting films is shown in Figure 6.9. The photocurrent starts at ~470 nm (~2.6 eV). This agrees well with the optical absorption spectra, from which a band gap of 3.2 eV and a sub-band gap absorption onset of 2.5 eV were derived (not shown). Interestingly, these results are rather different from the absorption characteristics of $InVO_4$ powders that were made by high-temperature solid state synthesis [22]. Considering the different synthesis methods, this is attributed to differences in the nature and concentration of bulk and/or surface defects. It should be noted that the IPCE in Figure 6.9 is very low, less than 0.3%. Analysis of the electrochemical impedance spectra for these films reveals that the low photo-response is caused by a high concentration of deep donors. According to Equation (6.1), a high donor density results in a small space charge width in which the photo-generated electrons and holes are efficiently separated by the electric field. Moreover, the fact that these donor states are located fairly deep within the band gap implies that they are efficient traps for charge carriers, which further increases the amount of recombination.

The nature of the defects that cause the high donor density is not yet clear. With $InVO_4$ being a ternary oxide, its defect chemistry is much more difficult to control than that of simple binary oxides. Specifically, small deviations from the desired 1:1 ratio of In:V may lead to relatively large concentrations of cation vacancies. The oxygen stoichiometry can be controlled by high-temperature thermal treatments in, e.g., air. However, the temperature range is limited to at most 550°C due to the conducting glass substrate. This is not the case for the synthesis of photo-catalyst powders, where temperatures up to 1200°C or more can be used. The limited control over the relative cation concentrations and the low upper limit of the processing temperature are the main reasons why the exciting new ternary and even more complex metal oxide photo-catalysts are difficult to prepare in the form of a photo-electrode for a PEC device. This calls for the development of new low-cost and low-temperature techniques for the deposition of thin film complex metal oxides that allow accurate control over the defect chemistry of these materials. Progress in this area is one of the key challenges for future research efforts on the photo-electrochemical production of hydrogen.

6.5 Conclusions

A brief overview of the main materials requirements and device concepts for solar hydrogen production from water using semiconducting photo-electrodes has been presented. Compared to the suspension-based photo-catalytic route for water splitting, the use of thin film photo-electrodes offers several crucial advantages: (i) the ability to produce hydrogen and oxygen in separate compartments, (ii) the higher solar-to-hydrogen conversion efficiencies, and (iii) the ease of removing the photo-active material from water. Due to the stringent and often conflicting demands on the photo-electrode material, it appears that a tandem configuration that absorbs high- and low-energy photons in two separate layers represents the most promising concept. The key component in this configuration is the metal oxide semiconductor that is in contact with water. Numerous attempts have been made over the past three decades to shift the photo-response of wide-band gap semiconductors, such as TiO_2, to the visible part of the solar spectrum by doping with anions or cations. While a significant shift of the optical absorption towards the visible part of the spectrum has often been observed, it does not lead to an enhanced conversion efficiency due to a concomitant increase in recombination. Recent developments indicate that smaller-band gap metal oxide semiconductors, such as Si-doped α-Fe_2O_3 and WO_3, offer a more promising direction of research. In addition, work on powder-based photo-catalysts suggests that small-band gap ternary and even more complex metal oxides may be suitable photo-electrode materials. The key challenges for all these materials are to achieve low concentrations of recombination centers and high charge carrier mobilities. This requires accurate control over the composition and the defect chemistry, which is especially important in the case of ternary oxides. Nano-structured materials generally have different defect distributions than bulk materials and may offer new possibilities to engineer photo-active metal oxides for water splitting.

Acknowledgements

The authors gratefully acknowledge the contributions of the following postdocs, students, and other co-workers to this work: Dr. J.C. van 't Spijker, E.L. Maloney (Everest Coatings), Dr. Y. Liang, C.S. Enache, D.A. Lloyd, and M.R. Damen. The authors are grateful to Dr. A.J. McEvoy (EPFL, Switzerland) for helpful comments. Financial support for this work was provided by the Delft Institute for Sustainable Energy and by SenterNovem.

RvdK gratefully acknowledges the Netherlands Organization for Scientific Research (NWO) for a VENI grant which made part of these investigations possible.

References

1. Lewis, N.S. and Nocera, D.G., Powering the planet: Chemical challenges in solar energy utilization, *Proc. Nat. Acad. Sci. USA* **103**, 2006, 15729).
2. Basic Research Needs for Solar Energy Utilization, Report of the US Department of Energy, Office of Basic Energy Sciences, 2005.
3. Khaselev, O., Bansal, A. and Turner, J.A., High-efficiency integrated multi-junction photovoltaic/electrolysis systems for hydrogen production, *Int. J. Hydrogen Energy* **26**, 127 (2001).
4. Morrison, S.R., *Electrochemistry of Semiconductor and Oxidized Metal Electrodes*, Plenum Press, New York, 1980.
5. Memming, R., *Semiconductor Electrochemistry*, Wiley, 2000.
6. Basu, P.K., *Theory of Optical Processes in Semiconductors*, Oxford University Press, 1997.
7. Kittel, C., *Introduction to Solid State Physics*, Wiley, New York, 1986.
8. Weber, M.F. and Dignam, M.J., Efficiency of splitting water with semiconducting photoelectrodes, *J. Electrochem. Soc.* **131**, 1984, 1258.
9. Murphy, A.B., Barnes, P.R.F., Randeniya, L.K., Plumb, I.C., Grey, I.E., Horne, M.D. and Glasscock, J.A., Efficiency of solar water splitting using semiconductor electrodes, *Int. J. Hydrogen Energy* **31**, 2006, 1999.
10. Bolton, J.R., Strickler, S.J. and Connolly, J.S., Limiting and realizable efficiencies of solar photolysis of water, *Nature* **316**, 1985, 495.
11. Weber, M.F. and Dignam, M.J., Splitting water with semiconducting photoelectrodes – Efficiency considerations, *Int. J. Hydrogen Energy* **11**, 1986, 225.
12. Shinar, R. and Kennedy, J.H., Competition reactions at alpha-Fe_2O_3 photo-anodes, *J. Electrochem. Soc.* **130**, 1983, 860.
13. Shinar, R. and Kennedy, J.H., Photoactivity of doped alpha-Fe_2O_3 electrodes, *Solar Energy Mater.* **6**, 1982, 323.
14. Fujishima, A. and Honda, K., Electrochemical photolysis of water at a semiconductor electrode, *Nature* **238**, 1972, 37.
15. Mavroides, J.G., Kafalas, J.A. and Kolesar, D.F., Photoelectrolysis of water in cells with $SrTiO_3$ anodes, *Appl. Phys. Lett.* **28**, 1976, 241.
16. Brus, L., Electronic wave-functions in semiconductor clusters – Experiment and theory, *J. Phys. Chem.* **90**, 1986, 2555.
17. Vayssieres, L., Sathe, C., Butorin, S.M., Shuh, D.K., Nordgren, J. and Guo, J.H., One-dimensional quantum-confinement effect in alpha-Fe_2O_3 ultrafine nanorod arrays, *Adv. Mater.* **17**, 2005, 2320.
18. McEvoy, A.J., Etman, M. and Memming, R., Interface charging and intercalation effects on d-band transition-metal diselenide photoelectrodes, *J. Electroanal. Chem.* **190**, 1985, 225.

19. Kudo, A., Omori, K. and Kato, H., A novel aqueous process for preparation of crystal form-controlled and highly crystalline $BiVO_4$ powder from layered vanadates at room temperature and its photocatalytic and photophysical properties, *J. Am. Chem. Soc.* **121**, 1999, 11459.
20. Sayama, K., Nomura, A., Arai, T., Sugita, T., Abe, R., Yanagida, M., Oi, T., Iwasaki, Y., Abe, Y. and Sugihara, H., Photoelectrochemical decomposition of water into H_2 and O_2 on porous $BiVO_4$ thin-film electrodes under visible light and significant effect of Ag ion treatment, *J. Phys. Chem. B* **110**, 2006, 11352.
21. Bamwenda, G.R. and Arakawa, H., The photoinduced evolution of O_2 and H_2 from a WO_3 aqueous suspension in the presence of Ce^{4+}/Ce^{3+}, *Solar Energy Mater. Solar Cells* **70**, 2001, 1.
22. Ye, J.H., Zou, Z.G., Oshikiri, M., Matsushita, A., Shimoda, M., Imai, M. and Shishido, T., A novel hydrogen-evolving photocatalyst $InVO_4$ active under visible light irradiation, *Chem. Phys. Lett.* **356**, 2002, 221.
23. Zou, Z.G., Ye, J.H. and Arakawa, H., Photocatalytic properties and electronic structure of a novel series of solid photocatalysts, Bi_2RNbO_7 (R = Y, rare earth), *Topics Catal.* **22**, 2003, 107.
24. Wang, D., Zhou, Z.G. and Ye, J.H., A new spinel-type photocatalyst $BaCr_2O_4$ for H_2 evolution under UV and visible light irradiation, *Chem. Phys. Lett.* **373**, 2003, 191.
25. Oshikiri, M. and Boero, M., Water molecule adsorption properties on the $BiVO_4$ (100) surface, *J. Phys. Chem. B* **110**, 2006, 9188.
26. Takata, T., Furumi, Y., Shinohara, K., Tanaka, A., Hara, M., Kondo, J.N. and Domen, K., Photocatalytic decomposition of water on spontaneously hydrated layered perovskites, *Chem. Mater.* **9**, 1997, 1063.
27. Grätzel, M., Photoelectrochemical cells, *Nature* **414**, 2001, 338.
28. Santato, C., Ulmann, M. and Augustynski, J., Photoelectrochemical properties of nanostructured tungsten trioxide films, *J. Phys. Chem. B* **105**, 2001, 936.
29. Khan, S.U.M. and Akikusa, J., Photoelectrochemical splitting of water at nanocrystalline n-Fe_2O_3 thin-film electrodes, *J. Phys. Chem. B* **103**, 1999, 7184.
30. Cesar, I., Kay, A., Martinez, J.A.G. and Grätzel, M., Translucent thin film Fe_2O_3 photoanodes for efficient water splitting by sunlight: Nanostructure-directing effect of Si-doping, *J. Am. Chem. Soc.* **128**, 2006, 4582.
31. Khaselev, O. and Turner, J.A., A monolithic photovoltaic-photoelectrochemical device for hydrogen production via water splitting, *Science* **280**, 1998, 425.
32. Licht, S., Multiple band gap semiconductor/electrolyte solar energy conversion, *J. Phys. Chem. B* **105**, 2001, 6281.
33. Licht, S., Wang, B., Mukerji, S., Soga, T., Umeno, M. and Tributsch, H., Efficient solar water splitting, exemplified by RuO_2-catalyzed AlGaAs/Si photoelectrolysis, *J. Phys. Chem. B* **104**, 2000, 8920.
34. Algora, C., Ortiz, E., Rey-Stolle, I., Diaz, V., Pena, R., Andreev, V.M., Khvostikov, V.P. and Rumyantsev, V.D., A GaAs solar cell with an efficiency of 26.2% at 1000 suns and 25.0% at 2000 suns, *IEEE Trans. Electron Devices* **48**, 2001, 840.
35. Rocheleau, R.E., Miller, E.L. and Misra, A., High-efficiency photoelectrochemical hydrogen production using multijunction amorphous silicon photoelectrodes, *Energy & Fuels* **12**, 1998, 3.
36. Miller, E.L., Rocheleau, R.E. and Deng, X.M., Design considerations for a hybrid amorphous silicon/photoelectrochemical multijunction cell for hydrogen production, *Int. J. Hydrogen Energy* **28**, 2003, 615.
37. Miller, E.L., Paluselli, D., Marsen, B. and Rocheleau, R.E., Development of reactively sputtered metal oxide films for hydrogen-producing hybrid multijunction photoelectrodes, *Solar Energy Mater. Solar Cells* **88**, 2005, 131.
38. Nanu, M., Schoonman, J. and Goossens, A., Inorganic nanocomposites of n- and p-type semiconductors: A new type of three-dimensional solar cell, *Adv. Mater.* **16**, 2004, 453.
39. Nanu, M., Schoonman, J. and Goossens, A., Solar-energy conversion in $TiO_2/CuInS_2$ nanocomposites, *Adv. Funct. Mater.* **15**, 2005, 95.

40. Maruska, H.P. and Ghosh, A.K., Transition-metal dopants for extending the response of titanate photoelectrolysis anodes, *Solar Energy Mater.* **1**, 1979, 237.
41. Borgarello, E., Kiwi, J., Grätzel, M., Pelizzetti, E. and Visca, M., Visible-light induced water cleavage in colloidal solutions of chromium-doped titanium-dioxide particles, *J. Am. Chem. Soc.* **104**, 1982, 2996.
42. Asahi, R., Morikawa, T., Ohwaki, T., Aoki, K. and Taga, Y., Visible-light photocatalysis in nitrogen-doped titanium oxides, *Science* **293**, 2001, 269.
43. Sakthivel, S. and Kisch, H., Daylight photocatalysis by carbon-modified titanium dioxide, *Angew. Chem. Int. Ed.* **42**, 2003, 4908.
44. Sakthivel, S. and Kisch, H., Photocatalytic and photoelectrochemical properties of nitrogen-doped titanium dioxide, *Chem. Phys. Chem.* **4**, 2003, 487.
45. Chen, X.B., Lou, Y.B., Samia, A.C.S., Burda, C. and Gole, J.L., Formation of oxynitride as the photocatalytic enhancing site in nitrogen-doped titania nanocatalysts: Comparison to a commercial nanopowder, *Adv. Funct. Mater.* **15**, 2005, 41.
46. Enache, C.S., Schoonman, J. and van de Krol, R., The photoresponse of iron- and carbon-doped TiO_2 (anatase) photoelectrodes, *J. Electroceram.* **13**, 2004, 177.
47. Enache, C.S., Schoonman, J. and van de Krol, R., Addition of carbon to anatase TiO_2 by n-hexane treatment – Surface or bulk doping?, *Appl. Surf. Sci.* **252**, 2006, 6342.
48. Irie, H., Watanabe, Y. and Hashimoto, K., Carbon-doped anatase TiO_2 powders as a visible-light sensitive photocatalyst, *Chem. Lett.* **32**, 2003, 772.
49. Lindgren, T., Mwabora, J.M., Avendano, E., Jonsson, J., Hoel, A., Granqvist, C.G. and Lindquist, S.E., Photoelectrochemical and optical properties of nitrogen doped titanium dioxide films prepared by reactive DC magnetron sputtering, *J. Phys. Chem. B* **107**, 2003, 5709.
50. Lindgren, T., Lu, J., Hoel, A., Granqvist, C.G., Torres, G.R. and Lindquist, S.E., Photo-electrochemical study of sputtered nitrogen-doped titanium dioxide thin films in aqueous electrolyte, *Solar Energy Mater. Solar Cells* **84**, 2004, 145.
51. Goossens, A., Maloney, E.L. and Schoonman, J., Gas-phase synthesis of nanostructured anatase TiO_2, *Chem. Vapor Depos.* **4**, 1998, 109.
52. van 't Spijker, J.C., Maloney, E.L., Schoonman, J. and van de Krol, R., Unusually high photocurrents in anatase TiO_2 thin films with fractal morphologies, Presented at the *16th International Conference on Photochemical Conversion and Storage of Solar Energy (IPS-16)*, Uppsala, Sweden, 2006.
53. Santato, C., Ulmann, M. and Augustynski, J., Enhanced visible light conversion efficiency using nanocrystalline WO_3 films, *Adv. Mater.* **13**, 2001, 511.
54. Enache, C.S., Schoonman, J. and Van de Krol, R., Manuscript in preparation.
55. Sato, J., Saito, N., Nishiyama, H. and Inoue, Y., Photocatalytic water decomposition by RuO_2-loaded antimonates, $M_2Sb_2O_7$ (M = Ca, Sr), $CaSb_2O_6$ and $NaSbO_3$, with d(10) configuration, *J. Photochem. Photobiol. A* **148**, 2002, 85.
56. Kato, H. and Kudo, A., Water splitting into H_2 and O_2 on alkali tantalate photocatalysts $ATaO_3$ (A = Li, Na, and K), *J. Phys. Chem. B* **105**, 2001, 4285.
57. Yin, J., Zou, Z.G. and Ye, J.H., Photophysical and photocatalytic properties of $MIn_{0.5}Nb_{0.5}O_3$ (M = Ca, Sr, and Ba), *J. Phys. Chem. B* **107**, 2003, 61.
58. Zou, Z.G., Ye, J.H., Sayama, K. and Arakawa, H., Direct splitting of water under visible light irradiation with an oxide semiconductor photocatalyst, *Nature* **414**, 2001, 625.
59. Maeda, K., Teramura, K., Lu, D.L., Takata, T., Saito, N., Inoue, Y. and Domen, K., Photocatalyst releasing hydrogen from water – Enhancing catalytic performance holds promise for hydrogen production by water splitting in sunlight, *Nature* **440**, 2006, 295.
60. Kim, J., Hwang, D.W., Kim, H.G., Bae, S.W., Lee, J.S., Li, W. and Oh, S.H., Highly efficient overall water splitting through optimization of preparation and operation conditions of layered perovskite photocatalysts, *Topics Catal.* **35**, 2005, 295.
61. Kudo, A., Kato, H. and Nakagawa, S., Water splitting into H_2 and O_2 on new $Sr_2M_2O_7$ (M = Nb and Ta) photocatalysts with layered perovskite structures: Factors affecting the photocatalytic activity, *J. Phys. Chem. B* **104**, 2000, 571.
62. Wang, D.F., Zou, Z.G. and Ye, J.H., A noval Zn-doped Lu_2O_3/Ga_2O_3 composite photocatalyst for stoichiometric water splitting under UV light irradiation, *Chem. Phys. Lett.* **384**, 2004, 139.

63. Wang, D.F., Zou, Z.G. and Ye, J.H., Photocatalytic water splitting with the Cr-doped $Ba_2In_2O_5/In_2O_3$ composite oxide semiconductors, *Chem. Mater.* **17**, 2005, 3255.
64. Sayama, K., Mukasa, K., Abe, R., Abe, Y. and Arakawa, H., Stoichiometric water splitting into H_2 and O_2 using a mixture of two different photocatalysts and an IO_3^-/I^- shuttle redox mediator under visible light irradiation, *Chem. Commun.* **23**, 2001, 2416.
65. Abe, R., Sayama, K., Domen, K. and Arakawa, H., A new type of water splitting system composed of two different TiO_2 photocatalysts (anatase, rutile) and a IO_3^-/I^- shuttle redox mediator, *Chem. Phys. Lett.* **344**, 2001, 339.
66. Harris, L.A. and Wilson, R.H., Semiconductors for photoelectrolysis, *Ann. Rev. Mater. Sci.* **8**, 1978, 99.

Chapter 7
Distributed Energy Generation, The Fuel Cell and Its Hybrid Systems

Kenjiro Suzuki†, Hideo Yoshida and Hiroshi Iwai

Abstract. In anticipation that massive production of hydrogen from renewable energy sources will not be achieved soon, we consider a distributed energy generation system using the Solid Oxide Fuel Cell (SOFC), and its hybrid system, which are characterised by fuel flexibility and high electricity generation efficiency. It is demonstrated that this system competes well in efficiency with the most advanced oil-fired large-scale combined-cycle electricity generation plants. The SOFC is very effective in reducing the emission of carbon dioxide and thus it is seen as a potential energy generation method that does not contribute to global warming. A short overview is given of the state-of-the-art of commercial type home cogeneration systems in Japan, using an SOFC or another type of fuel cell, Polymer Electrolyte Fuel Cells (PEFC). We also discuss numerical simulations of an SOFC single unit and of an SOFC stack module. These can be used as a tool for optimum design and thermal management either of an SOFC single unit or of an SOFC stacked module aimed at improving their durability and at lowering their cost.

Key words: Distributed energy generation, fuel cells, hydrogen economy, SOFC, hybrid system, simulator, optimum design, thermal management.

7.1 Introduction

Weather statistics show that the globe became warmer by 0.7°C during the last 130 years, or by 0.5°C during the last 50 years. Since the yearly rate of warming is less than 0.01°C, we cannot sense this warming and this is the reason why most people do not take the issue seriously. However, the prediction by the Intergovernmental Panel on Climate Change (IPCC) shows that according to a pessimistic but probable

† Professor Kenjiro Suzuki passed away shortly after submitting this manuscript.

K. Hanjalić et al., Sustainable Energy Technologies: Options and Prospects, 143–158.

scenario[5] the global temperature may further increase by 2 to 4°C by the end of this century. Such large global temperature increases would seriously affect the climate, ecosystem and agriculture on our planet. It would also cause further serious health problems in non-tropical countries by, e.g., expanding the habitation area of malaria mosquitoes. It would also lead to large-scale melting of the polar icecaps which, in turn, would cause rising of the sea level, endangering seriously the existence of the coastal areas of many sea-level and island countries. Suppression of global warming has, therefore, become a very important issue for all the people living on Earth.

Global warming is the consequence of an imbalance between the solar incidence onto the globe and the thermal radiation from the Earth into the outer space. The increased concentration of some of the chemical species in the atmosphere intensifies the opacity of the Earth's atmosphere in the far infra-red range of electromagnetic waves, leading to a decrease of thermal radiation from the Earth's surface into space, resulting in a surplus of solar energy incidence to the globe. Carbon dioxide and methane are the two major global warming chemical species. Major emission sources of carbon dioxide are coal and oil fired energy conversion systems, as well as automobiles and aircraft. Therefore, a reduction of the emission rate of energy-conversion originated carbon dioxide is urgently required. In order to meet this goal, we need to develop new energy conversion systems of higher efficiency and to use renewable energy sources much more than hitherto. The ideal solution would be the replacement of coal and oil with hydrogen, but at present the desired scale of realisation of a real hydrogen economy is still very distant and virtually impossible to achieve in a short period of time. A real hydrogen economy means that enough hydrogen is produced from renewable energy sources to fill a significant part of the world's needs, or at least to fill this need at an acceptable low emission rate of carbon dioxide. We still have a long way to go before we can establish such an economy. In the meantime, there are many tasks to do as the responsibility of the ones living in this transient society.

Since there is no single path to complete suppression of global warming, we have to combine multiple options to reduce the emission rate of global warming chemical species. We consider here some options based on fuel cells.

7.2 Distributed Energy Generation

In populous countries like Brazil, Russia, India and China (referred to often as BRICs) that have rapidly developing economies, the demand for electricity supply is growing very fast. Similar trends exist in several larger Asian countries such as Indonesia, Vietnam and Malaysia, as well as in Europe, North America and Japan. However, large-scale centralised power stations cannot be built in short periods of time and their construction cost is very high. Therefore, the shortage of electricity will inevitably be a great problem in those countries. Furthermore, extremely

[5] See also http://www.gcrio.org/OnLnDoc/pdf/wg2spm.pdf

high demands for electricity are often localised to areas with high concentrations of population, industry and business. In such regions, the energy supply has become a primary issue and a topic of intense discussion, such as for instance in the Shanghai area in China. In most critical regions the introduction of distributed energy generation systems is regarded as the most viable interim solution. Such distributed energy generation systems have other advantages. Large-scale blackouts occurred in the northeastern part of the United States and in Canada on August 14, 2003 and shortly later in large areas of Switzerland, France and Italy on September 28, 2003. Since in most of these urban areas electricity was supplied only from the grids of large electric companies, it took a long time to recover from the blackouts. Therefore, introduction of distributed energy generation systems is effective in preventing large-scale blackouts and at least mitigating the effects of a large-scale blackout if it occurs.

Some Asian countries including Japan recently suffered serious disasters caused by strong earthquakes. Recovery from such disasters is slow and people suffer from collapse of the water, gas and electricity supply network. Natural disasters due to strong winds such as Typhoons and Hurricanes or heavy rains can occur in other parts of the world causing also serious disruption of life lines. Distributed energy generation systems are usually less prone to serious damages and are more flexible in mitigating such disasters.

Distributed energy generation is characterised by much smaller unit capacity as compared to the most advanced centralised power station. A wind power system is a typical example of a distributed energy generation system; a biomass-fired gas turbine system is another. Fuel cells and fuel cell – gas turbine hybrid systems are also considered to be suitable for distributed energy generation. Among others, SOFC (Solid Oxide Fuel Cells) and SOFC-MGT (Micro Gas Turbine) hybrid systems are promising for this purpose because of high electricity generation efficiency and fuel flexibility as discussed below.

7.3 Hydrogen Production

Hydrogen is not as primary an energy source as coal, petroleum, solar and biomass and wind power, because it needs to be produced by electrolysis of water or by reforming of hydrocarbons. Such production processes need additional energy and produce carbon dioxide. In this sense, it is not completely clean unless the process is energy effective and the produced carbon dioxide is separated in the process and finally sequestered in rock or in the depths of the sea. Needless to say, for massive use of hydrogen as fuel, production processes must be cost effective too. Otherwise no one would be interested in using hydrogen as fuel.

The present production rate of hydrogen in Japan is 27 billion Nm^3 per year. At present, all of it is used for industrial purposes and almost none as fuel for electricity production. Hydrogen is now produced mainly by reforming of naphtha and plant off-gas, and by decomposition of ethylene. Purity of such hydrogen is normally

60 to 70% and at best 98% [1]. However, in order to use hydrogen as fuel for a PEFC, as discussed in detail below, it is necessary to raise its purity up to a value higher than 99.99%. For both of the primary processes and purification of hydrogen, additional energy input is necessary. Electrolysis of water produces hydrogen of high purity but consumes electricity. Electricity to be used in this process should be produced by making use of renewable energy. Otherwise, no improvement is made in the emitting rate of carbon dioxide as compared with the conventional fossil-fuel electricity generation. The use of nuclear energy for this purpose is another option since additional carbon dioxide is not emitted in running a nuclear power plant.

The predictions issued by the New Energy and Industrial Technology Development Organization (NEDO)[6] and World Energy Network (WE-NET)[7] show that 38 billion Nm^3 of hydrogen per year will be used in Japan as fuel in 2030, but this amount is equivalent to only 7% of the forecasted total energy demand in that year.[8] This means that non-hydrogen fuel will still be widely used in energy generation 25 years from now. Taking this into account, another path to reduce the emission of carbon dioxide has to be developed in parallel with the effort to establish a hydrogen economy.

In relation to the above discussed points, we consider first the fuel cells including SOFC and then the distributed energy generation systems using SOFC and its hybrid. Finally, we present briefly a numerical simulator to be used as a tool for optimum design and thermal management of an SOFC single unit and its stacked electricity generation module.

7.4 Fuel Cells and Superior Features of SOFC

Fuel cells are devices for direct conversion of the chemical energy of fuel into electricity. They are classified into several categories and, among others, three types of fuel cells are tabulated in Table 7.1. A PEFC is a low-temperature proton-type fuel cell using a polymer membrane as an electrolyte. It can be started in a reasonably short period of time. Therefore, it is expected to be used for an FCV, Fuel Cell Vehicle. There are still several problems to be solved but basic technologies for FCVs have been developed. A number of FCVs are currently on field test around the world. Cost control and durability of products are among the most important R&D tasks, and success in achieving acceptable targets is a prerequisite for mass introduction of those products. For stationary use, several companies have been making serious efforts in Japan to develop PEFC co-generation systems for home use and two utility companies started to install a 1kW PEFC home-cogeneration system to limited numbers of contracted homes in the Tokyo area on a half-commercial and half-test base as outlined in Table 7.2. One is fuelled by town gas and another by

---

[6] See also http://www.nedo.go.jp/english/index.html

[7] See also http://www.enaa.or.jp/WE-NET/

[8] See also http://www.env.go.jp/earth/gijyutsu_k/03/ref_04.pdf

Table 7.1 Three typical types of fuel cells.

	Polymer Electrolyte Fuel Cell (PEFC)	Molten Carbonate Fuel Cell (MCFC)	Solid Oxide Fuel Cell (SOFC)
Temperature °C	60–80	600–700	600–1000
Fuel	H_2, CH_3OH	H_2, CO	H_2, CO, C_xH_y
Electrolyte	Polymer Membrane	Li_2CO_3+Na_2CO_3	YSZ, SSZ, Ceria
Conducting ion	H^+	CO_3^{2-}	O^{2-}
Oxidiser	Air	Air with CO_2	Air
Catalyst	Pt	Pt, Pt-Ru	–

Table 7.2 1 kW PEFC home cogeneration system.

	Tokyo Gas	ENEOS
Fuel	Town gas	Liquid propane, kerosene
Electricity generation efficiency (LHV)	37%	35%
Efficiency of rejected heat recovery (LHV)	50%	46%
Hot water storage capacity	200 l	200 l
Size	$1260D \times 2650H \times 3400W$	$640D \times 1990H \times 740W$

liquid propane. They have external fuel reformers upstream of the fuel cells. Its electricity generation efficiency is publicly claimed to be higher than 35% and efficiency of the recovery of the rejected thermal energy (hot water storage) is higher than 45% (both LHV base). Hot water storage capacity is 200l. This installation has initiated a new era of home co-generation systems based on fuel cells.

The Molten Carbonate Fuel Cell (MCFC) and the SOFC are high temperature fuel cells. In an MCFC, carbon dioxide mixed with air is changed into its ion CO_3^{2-} at an air electrode and the produced ion is conducted through an electrolyte, the layer of molten carbonate. The MCFC has been developed for stationary power stations with capacity from a few hundred kilowatts to a few megawatts and its commercial-based installation has started a few years ago in the US, Japan, Korea and Europe. It can use carbon monoxide as well as hydrogen as fuel, and the installed MCFC plants use town gas, natural gas, methanol, coal synthesis gas and sewage, sludge and other anaerobic digester gases. Its electricity generation efficiency is expected to be as high as 45%. Unfortunately, the capacity of MCFC plants can not be made so small as to be used for a home co-generation system.

An SOFC is operated at high temperature in the range from 700 to 1000°C. Its quick start is difficult but this is not a serious impediment in case of stationary electricity generation systems. In fact, it is superior to other types of fuel cells in several features: high electricity generation efficiency, fuel flexibility, no necessity to enhance the electrochemical reaction with a catalyst, possibility of internal fuel reforming, viability of a small capacity electricity generation system and various possibilities to recover the high temperature effluent thermal energy. Due to these superior characteristics, the SOFC is regarded to be prospective as a unit for dis-

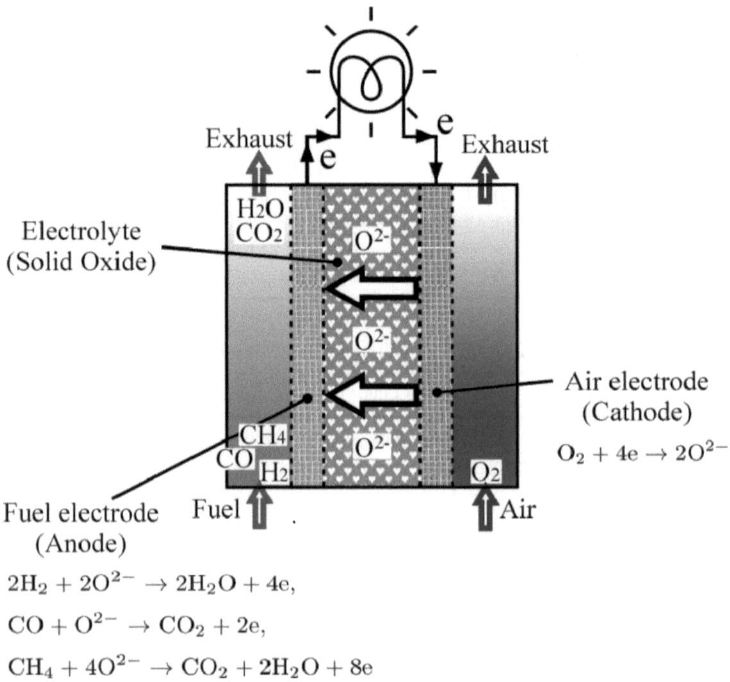

Fig. 7.1 Structure of SOFC cell and mechanism of electricity generation.

tributed energy generation in the near future. In the last two to three years, several companies announced in their press releases or issued technical reports that electricity generation efficiency reached 50% or higher with a 1 or 5 kW module developed for a home cogeneration system (see Table 7.3).[9]

In fact, a 1 kW town gas fuelled SOFC module installed in the Energy Flow Research Center, Shibaura Institute of Technology, in 2005 proved that 50% DC electricity generation efficiency (LHV) can easily be attained.

An SOFC uses solid oxides like YSZ (Yttria-stabilised Zirconium), SSZ (Scandia-stabilised Zirconium), Ceria-based oxides and Sr-doped $LaGaO_3$ as the material for an electrolyte [2]. These are oxygen ion conducting materials. As shown schematically in Figure 7.1, the electrolyte is sandwiched by two electrodes, a cathode or air electrode and an anode or fuel electrode. The sandwich separates the air and fuel flow. In the case of an SOFC, oxygen reduction occurs at the cathode and the produced oxygen ion is conducted through the electrolyte. At the anode, oxygen ion reacts with a fuel releasing electron. A released electron at the anode returns to the cathode through the external circuit, consuming electricity at the electric devices

[9] See also the following websites: http://www.kyocera.com.vn/news/2003/1205.html; http://www.performance-materials.net/htm/f20040201.817666.htm; http://www.fuelcelltoday.com/FuelCellToday/IndustryInformation/IndustryInformationExternal/NewsDisplayArticle/0,1602,5445,00.html

Table 7.3 Recent achievement of Japanese industries for home co-generation SOFC stack module.

Companies	Cell geometry	Capacity	Fuel	Efficiency (LHV)
Kyocera[a]	Flat-tubular	1 kW	Town gas	54% DC
Kyocera/Osaka Gas[b]	Flat-tubular	1.5 kW	Town gas	44% AC
Toto[c]	Tubular	1 kW	Town gas	55% DC
KEPCO/MMC[d]	Disk	1 kW	Town gas	58% DC

[a]http://www.kyocera.co.jp/news/2003/1206.html
[b]Press release, November 24, 2004
[c]SOFC IX, Quebec, May 2005
[d]SOFC IX, Quebec, May 2005: KEPCO (The Kansai Electric Power Co., Inc.), MMC (Mitsubishi Material Corporation)

connected in the circuit. Produced chemical products are discharged together with remaining fuel from the outlet of the fuel passage.

In principle, oxygen can react at the anode with hydrogen, carbon monoxide or a variety of hydrocarbons. This means that an SOFC can operate with a variety of fuels, such as town gas, natural gas, biomass or coal synthesis gas. In fact, fuel flexibility is one of the important characteristics of an SOFC. However, hydrogen is less sluggish in electrochemical reaction compared to carbon monoxide, and this leads to higher efficiency and more stable performance. Hydrogen is certainly preferable to hydrocarbons as fuel since it does not produce carbon deposition at all. Carbon deposition blocks the micro-scale porous passages of the anode leading to deterioration of the fuel cell performance. Therefore, even with SOFC, fuel reforming is still popularly used and reformed hydrogen-rich fuel is introduced into the cell stack of the SOFC module. Because of high operation temperature, internal fuel reforming is possible. Heat transfer from the cell stack to the reformer, closely allocated to the cell stack, can provide heat to support strongly endothermic fuel reforming reaction.

At the operation temperature of an SOFC, oxygen reduction reaction and electrochemical reaction of hydrogen proceed at a reasonable rate without the help of a catalyst. The electrochemical catalyst is a platinum based material which is likely to be contaminated by the presence of a small portion of carbon monoxide likely to be left in the reformed fuel even after purification. This means that an SOFC does not need an expensive catalyst and is immune from poisoning by carbon monoxide.

Although fuel cells convert directly the fuel chemical energy into electricity, it does not mean that every bit of the fuel chemical energy can be utilised for electricity. The inefficient part of energy conversion is the generation of heat. Therefore, heat removal is needed to keep the fuel cell at an appropriate temperature, thus thermal management is important in any fuel cell. Cell temperature should be high enough on one hand to keep the ion conductivity of electricity at a reasonably level, since ion conductivity of the electrolyte is strongly dependent on temperature. On the other hand, cell temperature must be limited, since an excessively high temperature will cause thermal damage to the cell. In the case of SOFC the operation temperature is high, so elongation and strain due to thermal expansion occurring in

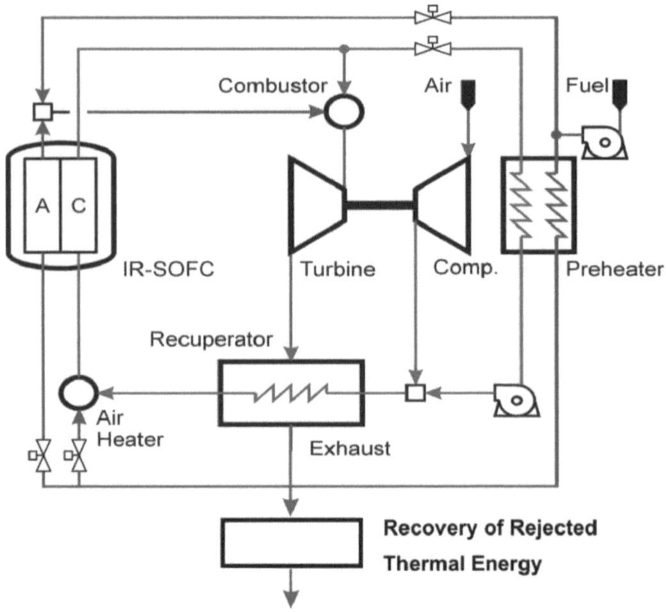

Fig. 7.2 Flow chart of SOFC-MGT hybrid system.

the cell can lead to cell thermal damage. For the same reason, the temperature gradient must also be kept as low as possible and the temperature distribution must be kept as uniform as possible. Thermal management is also important for the SOFC module. Heat must be transferred at an appropriate rate from the cell stack to the internal fuel reformer to support the strong endothermic fuel reforming reaction. These issues are discussed in more detail below in the section dealing with the numerical simulators of a single cell and of the SOFC stack module, which can be used as tools for thermal management and for optimum design of the cell and of the module.

7.5 SOFC-MGT Hybrid System

One possibility to further raise the electricity generation efficiency is a hybrid system fusing an SOFC with a gas turbine. A 220 kW SOFC-MGT hybrid system tested at NFCRC attached to UC Irvine demonstrates that this system is promising. The present authors carried out studies on the same hybrid system of a smaller electricity generation capacity 30 kW [26] under a CREST project supported by Japan Science and Technology Corporation[10] [3–6]. Figure 7.2 shows the flow chart of such a hybrid system.

[10] See also http://www.jst.go.jp/kisoken/en/index.html

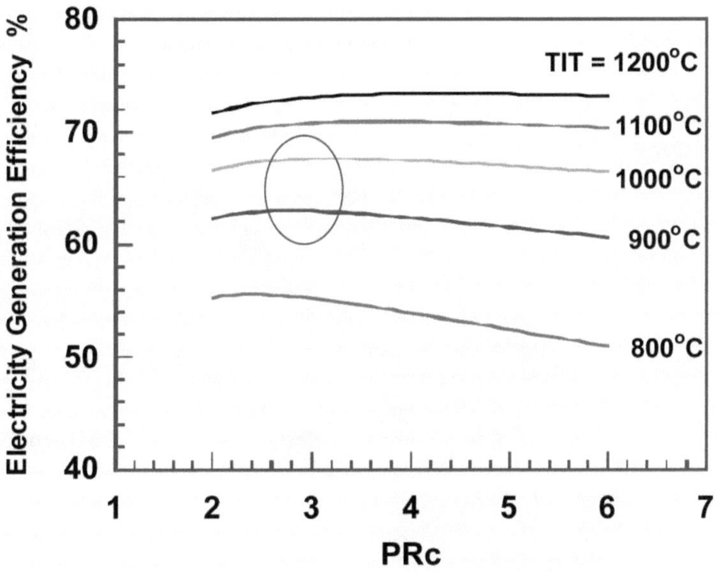

Fig. 7.3 Efficiency of SOFC-MGT hybrid system.

An SOFC is operated at an elevated pressure which is in the range from 1.5 to 3 bars and is optimised depending on the gas turbine inlet temperature TIT. The gas turbine is too small to introduce a blade cooling system so that a regenerative turbine should be used. For conventional metal turbines, the turbine inlet temperature (TIT) should be lower than 900°C. A possibility to utilise higher TIT is to use a ceramics turbine. Figure 7.3 shows the LHV-based electricity generation efficiency of an SOFC-MGT hybrid system fueled by methane and the thickness of the YSZ electrolyte is assumed to be 50 μm [7–11]. It is seen that the SOFC-MGT hybrid system with TIT equal to 900°C can compete in efficiency with the large-scale most advanced combined-cycle power station. This level of efficiency can be attained even at TIT equal to 800°C if the electrolyte is thinned down to 10 μm or if other materials like Cerium series materials or SSZ are used for the electrolyte [9]. The development of a 5 kW micro gas turbine for a 30 kW hybrid system is not easy but larger capacities are within the reach of the present technology. At present, cost reduction and durability are the two major issues in SOFC technology, and thus of the SOFC-MGT hybrid system.

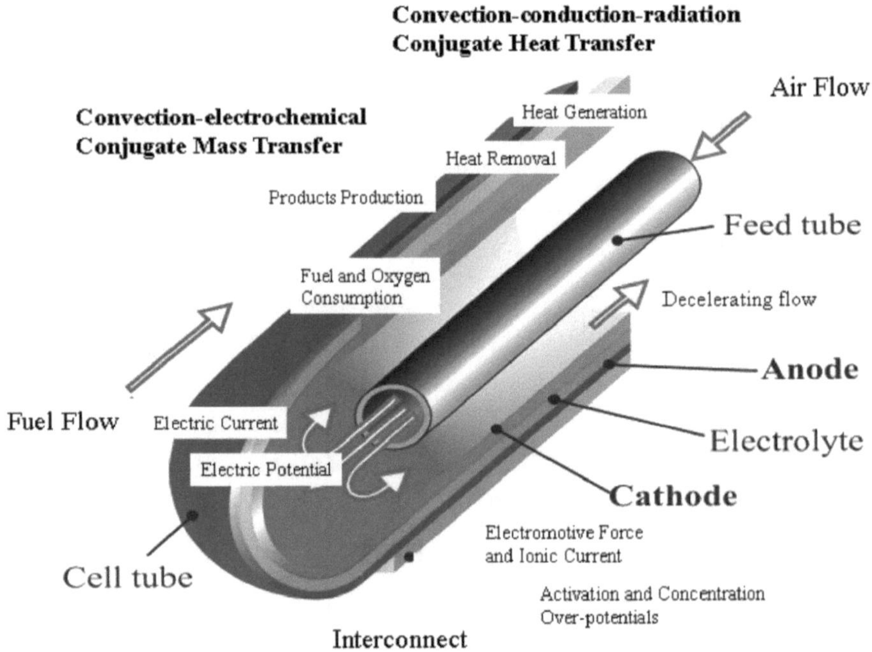

Fig. 7.4 Summary of the phenomena occurring in a tubular cell.

7.6 Numerical Simulators of SOFC Single Unit and SOFC Module

Durability of an SOFC is closely related to its local thermal condition in practical operation. However, the measurement of the local temperatures during the operation is very difficult, almost impossible. All that can be done is to provide a good guess by a numerical simulator.

Figure 7.4 illustrates the major phenomena occurring in a single cell. Numerical simulation should take into account all of the physical and chemical processes: air and fuel flows, consumption and production of chemical species participating in the electrochemical and reforming reactions, electric potential field in the cell, ionic and electric current, heat generation and heat removal. Figure 7.5 shows an outline of the iterative calculation procedure which accounts for all of these phenomena [12–14]. A full three-dimensional (3D) computation is too demanding on the CPU time but a reasonable approximation can be achieved with a quasi-2D approach for a tubular SOFC single unit. In this 2D approach, fluid flow, temperature and mass transfer fields are treated as axi-symmetric and integrated in a consistent manner with the treatment of the electric potential field allowing the electric current along the peripheral direction. Details of the computation scheme are given in [14, 15].

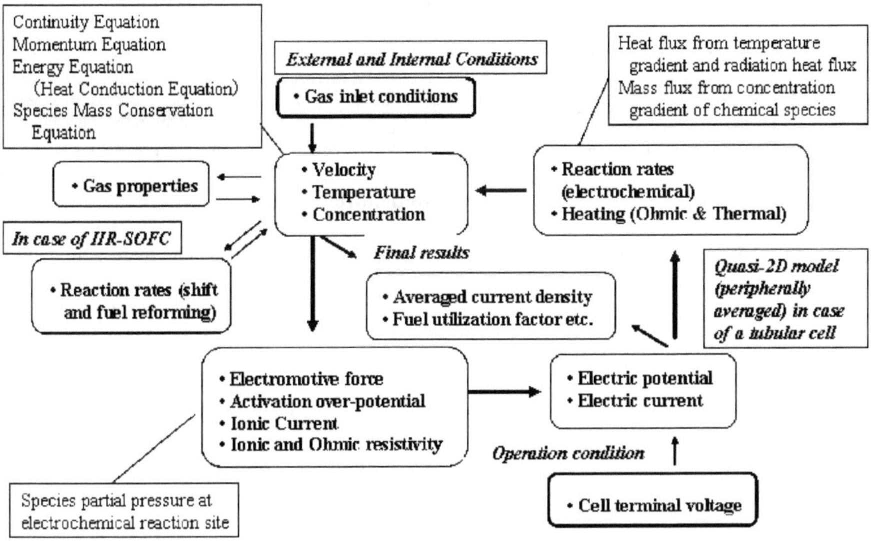

Fig. 7.5 Outlines of the numerical scheme.

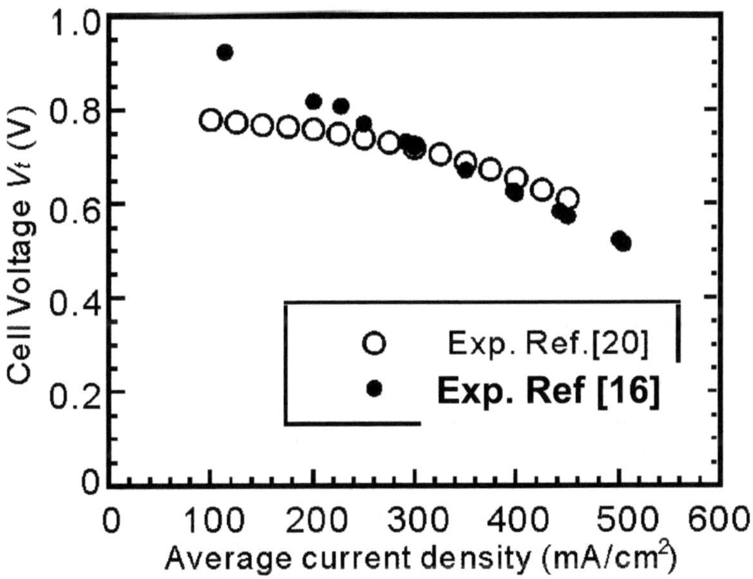

Fig. 7.6 V–I diagram of a tubular SOFC.

Figure 7.6 compares the calculated V–I diagram for a hydrogen-fuelled Siemens–Westinghouse tubular cell with the experiments [16]. Figure 7.7 illustrates the distributions of temperature and species concentrations of hydrogen and oxygen obtained at an average current density. It is seen that the temperature distribution is almost

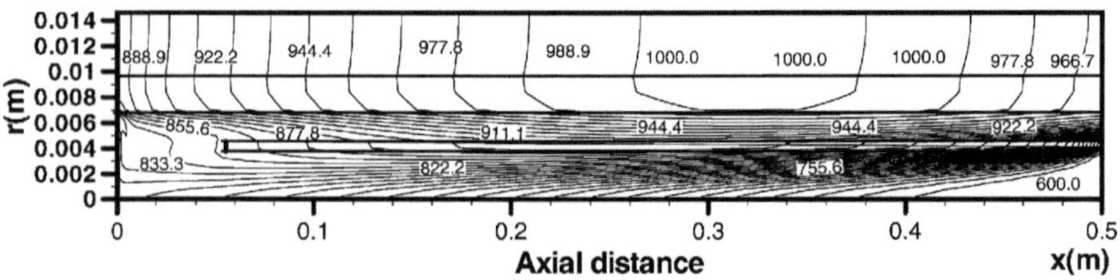

Fig. 7.7 Distributions of temperature and concentrations of oxygen and hydrogen.

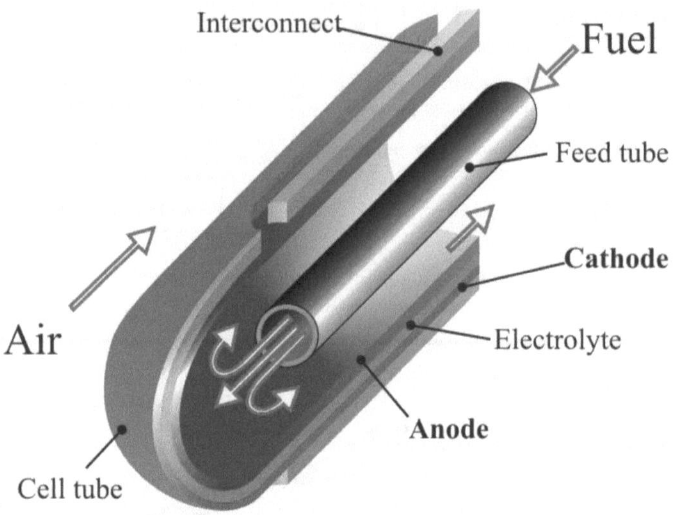

Fig. 7.8 Internal Indirect Reforming Tubular SOFC.

uniform and that the peak temperature is lower than the critical hot spot temperature 1050°C. This is considered to be the background for a good stability of the cell, as demonstrated by Hagiwara et al. [16] in their experiments.

Figure 7.8 shows the schematics of another type of tubular SOFC, an Internal Indirect Reforming Tubular SOFC, for which fuel is supplied inside the cell and air flows around the cell. Fuel reforming proceeds internally within the fuel feed tube in which a fuel reforming catalyst is embedded as a porous plug. This type is desirable because every necessary process including fuel reforming can be completed within a single cell. However, a large temperature gradient can be caused by the strong endothermic nature of the fuel reforming process and thermal cracking is likely to occur. Figure 7.9 shows the longitudinal distribution of the cell temperature for three conditions of different densities of the catalyst filling. In this computation, fuel is assumed to be methane; details can be found in [12, 14, 17]. It is seen that distribution of the cell temperature can be controlled by changing the filling density of the fuel reforming catalyst. A similar model has been developed for a disk-type

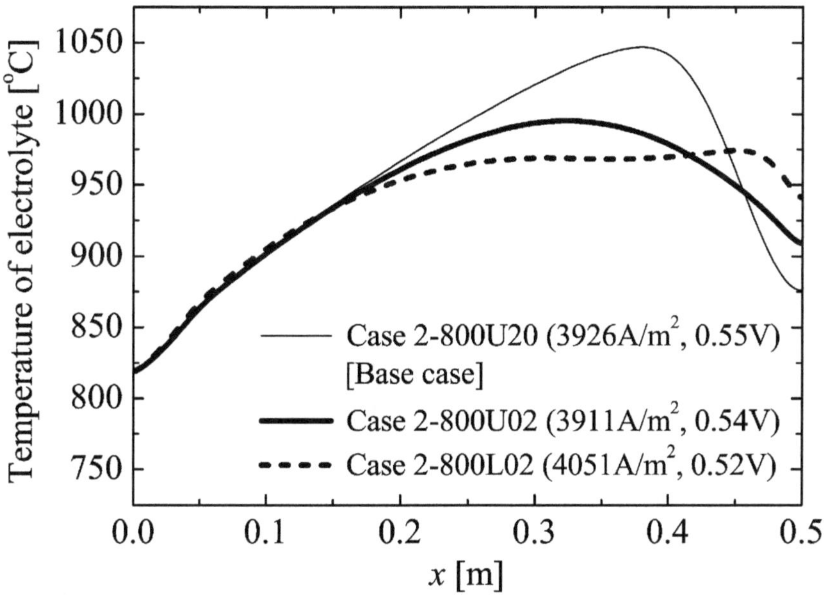

Fig. 7.9 Control of cell temperature by changing the filling rate of reforming catalyst.

SOFC [18] and simulation has been performed recently using a similar model to study the dynamic response of a tubular SOFC [19].

As an example, Figure 7.10 schematically illustrates the Siemens–Westinghouse internal reforming SOFC stack module. As seen in this figure, tubular fuel cells are packed into a stack in the module. Electricity is generated in the stack region of the modules but heat is also generated there. Generated heat is partially transferred to a reformer which provides the reformed fuel to the stack region. How effectively heat is supplied from the stack region to the reformer is an important point of the module design. In this sense, development of a numerical simulator is needed also for an SOFC stack module. Here are discussed some results obtained for an SOFC module with a numerical simulator. The simulator has been developed by the Kyoto University group lead by the present authors as one of the activities of the aforementioned CREST projects [4, 20–24]; it is based on a model for the stack module called a multi-stage model or quasi-2D model. This model uses in principle a scheme similar to the one illustrated in Figure 7.5 for a single cell unit but in a slightly simplified manner.

Details of the scheme are found in the references [14, 20, 21, 24]. Figure 7.11 shows some results compared with the experimental data presented by Singhal [25]. The calculated results agree fairly well with the experimental counterparts. The developed model can provide all necessary details of the local thermal condition in the stack module [4, 20].

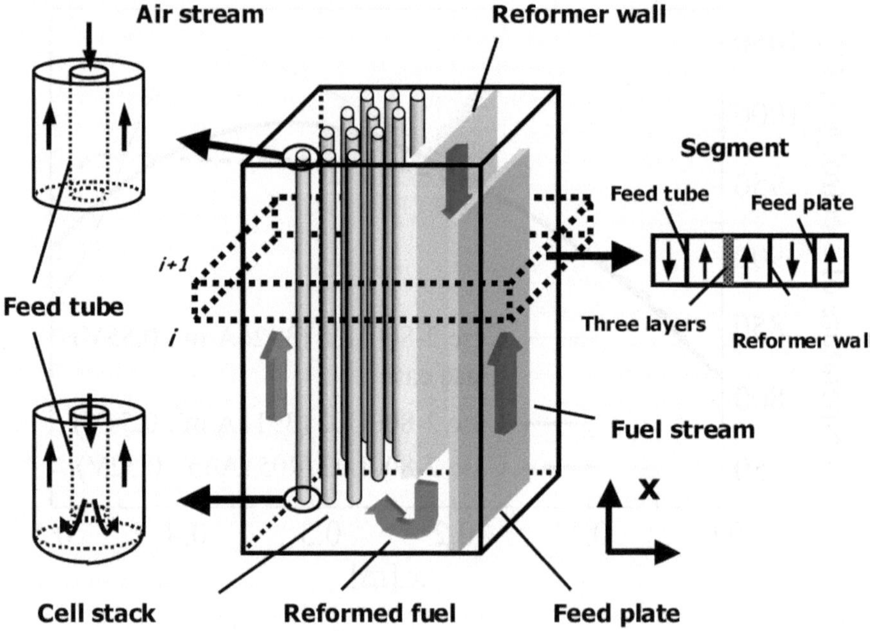

Fig. 7.10 Internal Indirect Reforming SOFC module.

As demonstrated in the three examples presented above, a numerical simulator can be used as a tool for thermal management and a tool for optimum design both of the SOFC single unit and the SOFC stack module. Therefore, it can work as a tool for the optimum design of SOFC and SOFC-based distributed energy generation units.

7.7 Concluding Remarks

It will take a long time before the hydrogen produced by a method with zero or low carbon dioxide emission rate becomes the major or significant fuel in energy conversion worldwide. In the meantime, it is necessary to develop other energy conversion systems which will satisfy the growing demand for electricity, heat and propulsion, and yet keep emission rate of carbon dioxide within sustainable limits. One of the candidates for this substitution is the Solid Oxide Fuel Cell (SOFC). Experiments and simulations indicated that the SOFC and especially its combination with a micro gas turbine offer interesting prospects for small self-contained units of high efficiency, suitable for distributed electricity supply. Cost reduction and durability are the two pertaining major problems of SOFC technology awaiting still to be resolved. The durability is related to the local thermal condition in a single SOFC unit

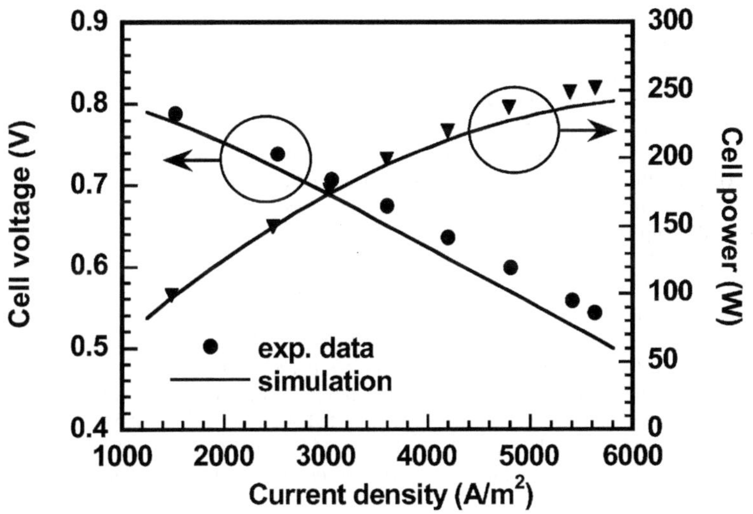

Fig. 7.11 Comparison of V–I diagram and cell power of the module.

and in an SOFC stack module. A numerical simulator has been developed which can be used as a tool both for thermal management and for optimum design of the SOFC.

References

1. Nishimura, Y., R&D hydrogen production technology, *Energy and Resources* **24**(6), 2003, 398–403 [in Japanese]
2. Ishihara, T., Sammes, N.M. and Yamamoto, O., Electrolytes, in *High Temperature Solid Oxide Fuel Cells – Fundamentals, Design and Applications*, S.C. Singhal and K. Kendall (eds.), 83, Elsevier, 2003, Chapter 4.
3. Suzuki, K., *Proceedings of Open Forum on CREST Project "SOFC and MGT Hybrid System for Distributed Energy Generation"*, Kyoto, Japan, 2002.
4. Suzuki, K., *Proceedings of Open Forum on CREST Project "SOFC and MGT Hybrid System for Distributed Energy Generation"*, Kyoto, Japan, 2003.
5. Suzuki, K., *Proceedings of Open Forum on CREST Project "SOFC and MGT Hybrid System for Distributed Energy Generation"*, Tokyo, Japan, 2005.
6. Suzuki, K., *Final Report of Crest Project on "SOFC and MGT Hybrid System for Distributed Energy Generation"*, Japan Science and Technology Corporation, Tokyo, Japan, 2005.
7. Suzuki, K., Teshima, K. and Kim, J-H., Solid oxide fuel cell and micro gas turbine hybrid cycle for a distributed energy generation system, in *Proceedings of 4th JSME-KSME Thermal Engineering Conference*, Vol. 3, 2000, pp. 1–8.
8. Suzuki, K., Iwai, H. and Kim, J.H., Micro gas turbines and fuel cells, in *Distributed Energy Generation and Fuel Cells*, M. Hirata (ed.), CMC, 2001, pp. 258–273 [in Japanese].
9. Suzuki, K., Song, T.W. and Iwai, H., Micro gas turbine – Solid oxide fuel cell hybrid system for distributed energy generation, in *International Symposium on Distributed Energy Systems in the 21st Century*, Tokyo, Japan, 2002.

10. Suzuki, K., Iwai, H., Kim, J.H., Li, P.W. and Teshima, K., Solid oxide fuel cell and micro gas turbine hybrid cycle and related fluid flow and heat transfer, in *12th International Heat Transfer Conference*, Grenoble, France, 2002.
11. Suzuki, K., Song, T.W. and Iwai, H., Fuel cell and gas turbine hybrid system, *Science of Machine* **56**(5), 2004, 543–550 [in Japanese].
12. Nishino, T., Komori, H., Iwai, H. and Suzuki, K., Numerical investigation on the strategies for reducing the cell temperature gradient of an indirect internal reforming tubular cell, in *2nd International Symposium on Fuel Cell Science, Engineering and Technology*, Rochester, USA, 2004, pp. 353–360.
13. Suzuki, K., Research on SOFC at Shibaura Institute of Technology, in *Proceedings of 1st International Workshop on Fuel Cell and Fuel Cell Hybrid*, Tokyo, Japan, 2004.
14. Suzuki, K., Iwai, H. and Nishino, T., Electrochemical and thermo-fluid modeling of an internal reforming type tubular solid oxide fuel cell, in *Transport Phenomena in Fuel Cells*, B. Sunden and M. Faghri (eds.), WIT Press, 2005, Chapter 3, pp. 83–131.
15. Li, P.W. and Suzuki, K., Numerical modeling and performance study of a tubular solid oxide fuel cell, *Journal of the Electrochemical Society* **151**(4), 2004, A548–A557.
16. Hagiwara, A., Michibata H., Kimura A., Jaszcar M.P., Tomlins G.W. and Veyo S.E., Tubular solid oxide fuel cell life tests, in *Proceedings of the 3rd International Fuel Cell Conference*, 1999, pp. 365–368.
17. Nishino, T., Komori, H., Iwai, H. and Suzuki, K., Development of a comprehensive numerical model for analyzing a tubular-type indirect internal reforming SOFC, in *Proceedings of 1st International Symposium on Fuel Cell Science, Engineering and Technology*, Rochester, USA, 2003, pp. 521–528.
18. Dang, Z., Iwai, H. and Suzuki, K., Numerical modelling of a disk shape planar SOFC, in *Proceedings of 2nd International Symposium on Fuel Cell Science, Engineering and Technology*, Rochester, USA, 2004, pp. 361–368.
19. Shiomi, D., Iwai, H., Suzuki, K. and Yoshida, H., Numerical simulation on transient characteristics of a tubular SOFC cell, in *Proceedings of 3rd International Symposium on Fuel Cell Science, Engineering and Technology*, Ypsilanti, USA, 2005, FUELLCELL2005-74171.
20. Song, T.W., Kim, J.H., Ro, S.T. and Suzuki, K., Quasi-2D model of IIR-SOFC stack for SOFC/MGT hybrid system, in *Proceedings of ICOPE-3*, Kobe, Japan, 2003.
21. Song, T.W., Sohn, J.L., Kim, J.H., Kim, T.S., Ro, S.T. and Suzuki, K., Parametric studies for a performance analysis of a SOFC/MGT hybrid power system based on a quasi-2D model, in *Proceedings of ASME Turbo Expo 2004*, Wien, Austria, 2004.
22. Song, T.W., Sohn, J.L., Kim, J.H., Kim, T.S., Ro, S.T. and Suzuki, K., Detailed performance analysis of a solid oxide fuel cell – Micro gas turbine hybrid power system, in *Proceedings of ASME Heat Transfer/Fluid Engineering Summer Conference*, North Carolina, 2004.
23. Song, T.W., Sohn, J.L., Kim, J.H., Kim, T.S., Ro, S.T. and Suzuki, K., Performance analysis of a tubular SOFC/MGT hybrid power system based on a quasi-2D model, *J. Power Sources* **142**(1/2), 2005, 30–42.
24. Suzuki, K., Song, T.W., Kim, J.H. and Ro, S.T., 2004, Multi-stage model of methane-fueled SOFC and performance of SOFC-GT hybrid system – Effects of heat loss from the stack, in *Proceedings of ATI2004*, Genova, Italy, 2004.
25. Singhal, S.C., Progress in tubular solid oxide fuel cell technology, *Electrochemical Society Proceedings* **99**(19), 1997, 39–51.
26. Suzuki, K., Song, T.W. and Li, P.W., SOFC-MGT hybrid system for distributed energy generation, *Energy and Resources* **26**(4), 2005, 246–250 [in Japanese].

Chapter 8
Current International Initiatives for Sustainable Nuclear Energy

Dan Gabriel Cacuci

Abstract. The currently operating light and heavy water nuclear reactors have reached a high level of safety, reliability and cost-effectiveness, demonstrated over thousands of years of cumulated operation. Mature industrial solutions are available for all stages of nuclear fuel cycles, including spent-fuel treatment, conditioning of medium-level waste, vitrification of high-level waste, and various technologies for interim storage. Considerable progress has been achieved in several countries to develop practical solutions for geological disposal of high-level, long-lived nuclear waste. To establish itself as a viable option for long-term energy production, accepted by public opinion, nuclear fission energy generation must meet the following criteria for sustainable development: (i) life-cycle cost advantage over other energy sources, together with a level of financial risk comparable to other energy projects; (ii) high-level safety and security, eliminating the need for offsite emergency response; (iii) protection of workers, population and environment against radiological hazards; (iv) minimal generation of long-lived, high-level radiotoxic waste; (v) safe disposal of all remaining waste products; (vi) optimised use of natural resources; (vii) closed fuel cycle, in harmony with the goal of "sustainability"; (viii) proliferation resistant, by fabricating the fuel so that it would be very difficult to extract fissile material from spent fuel, while minimising material accounting difficulties within the international safeguards system. This chapter aims at summarising the current status of nuclear fission energy generation vis-à-vis the above criteria.

Key words: Sustainable nuclear energy, light-water reactors, fast reactors, high-temperature reactors.

8.1 Current Status of Nuclear Fission Energy Generation

Reactor designs can be conveniently categorised according to technological "generations". The reactors of "Generation I" are defined as the prototype commercial

K. Hanjalić et al., Sustainable Energy Technologies: Options and Prospects, 159–176.

reactors of the 1950s and 1960s ("*Atoms-for-Peace*" era plants) in the USA, in the Russian Federation, in France and in the UK. The reactors of "Generations II" are the commercial reactors deployed in the 1970s and 1980s, as a result of the first fossil energy crisis (OPEC oil crisis of 1974). Most of these reactors are still in operation currently, worldwide.

The reactors of "Generation III" are "evolutionary" boiling (BWR) and pressurised water reactors (PWR) designed in the 1990s; some are in operation in East-Asian countries and others are under construction (Finland). These reactors represent developments of "Generation II"-reactors, with simpler design, passive safety systems, and enhanced performance characteristics. Since their designs emphasise improvements based on proven technology and experience, no prototype is needed for their industrial deployment. The increased use of passive safety features reduces considerably the core damage frequency, and their strengthened containment function limits the off-site consequences of a hypothetical severe accident. "Generation IV"-reactors are called "revolutionary" or "innovative" since their designs emphasise the use of new features, such that prototypes will be needed for their industrial deployment. From a safety point of view, the main aim of these reactors is to "practically eliminate severe accidents" (e.g., by increasing the safety margins regarding core damages and by relying extensively on passive processes). Gen-IV reactors aim at generating electricity and/or heat economically, with increased safety and performance while optimising the waste management process in conjunction with innovative sustainable fuel cycles; they will also be resistant to proliferation. In the following, the essential characteristics of the four generation of nuclear reactors will be depicted succinctly.

8.1.1 Generation I ("Atoms-for-Peace" – Era Plants)

The Magnox reactors in the UK, still in operation currently but scheduled to be closed by 2010, are representative Gen-I reactors. These reactors are fuelled by natural uranium, moderated by graphite or heavy water (to avoid the need for enrichment used), and cooled by CO_2. Experimental sodium cooled fast breeder reactors (FBR), aiming at a better utilisation of natural resources by converting fertile U^{238} into fissile Pu^{239}, were also developed during this pioneering period (the *Enrico FERMI* reactor in 1963 in the US, the *Rapsodie* in 1967 in France, the *BOR-60* in 1968 in the former Soviet Union, the *Joyo* in 1978 in Japan). Such fast-spectrum reactors can also improve the nuclear waste management process.

8.1.2 Generation II (Currently in Operation)

These are commercial reactors deployed in the 1970s and 1980s, as a consequence of the OPEC oil crisis of 1974, and are still currently operating world-wide. Most of these reactors are light-water reactors (LWR), either boiling water reactors (BWR) or pressurised water reactors (PWR), fuelled with ceramic uranium dioxide UO_2, typically encased in zirconium alloy tubes; the U235 is enriched to 3.5 to 5.0%. Other types of operating Gen-II reactors include advanced gas-cooled reactors (AGRs) in the UK, CANDU reactors (*CAN*adian *D*euterium *U*ranium heavy-water moderated and natural uranium fuelled) in Canada and Romania, and VVERs and RBMKs in the Russian Federation.

At the end of 2004, 148 reactors were in operation in 13 of the 25 European Community's Member States. Considering the electricity generation per primary energy type (gross production in "millions tonnes of oil equivalent" or Mtoe), nuclear energy is the Community's largest single energy type (32%), ahead of coal (21%), natural gas (20%), oil (15%), and hydro (9%) and other renewable energy sources (3%). With most nuclear plants operating continuously with an average capacity factor of 90% (i.e., approximately 8000 hours per year), the nuclear energy sector provides an essential source of reliable and secure base-load power. Note that the average capacity factor (i.e., the ratio of energy produced to the amount of energy that could have been generated at continuous full-power operation) of 90% indicates that nuclear energy is much more efficient than all other means of electricity generation.

The current concerns regarding the operation of Gen-II reactors are related to the security of energy supply (availability, diversification, independence), and to environmental compatibility (sustainability, low-carbon economy. The overall CO_2 emissions per unit of energy quantity emitted during nuclear fuel cycle processes and plant construction are extremely low, about 20 times less compared to fossil fuels. This is equivalent or inferior to the CO_2 emissions associated with the construction of wind power stations. In operation, nuclear power stations do not emit CO_2 and therefore play a major role in a low-carbon economy, just like renewable energies. Without nuclear energy, the CO_2 emissions in the EU would increase by 30%. It is also noteworthy nuclear power produce no particles (or aerosols) and no acid rain gases (or smog-forming NO_x).

8.1.3 Generation III ("Evolutionary" Reactors)

The Gen-III reactors were designed in the 1990s, taking into account the increased public and political concern following the two severe reactor accidents at Three Mile Island (Pennsylvania, USA, in 1979: a loss-of-coolant accident with substantial core melting and some hydrogen burning, but no large escape of radioactive material from the containment) and Chernobyl (Ukraine, in 1986: a criticality accident due to positive voiding reactivity feedback in a reactor of demonstrably inferior design without containment building and with poor operational standards). A dozen Gen-III reactors (PWR, BWR and CANDU designs) are being developed currently; among these designs, the ABWRs of Hitachi-Toshiba-General Electric (advanced boiling water reactors of 1350 MWe) are already operating in Japan. As formulated by the Near-Term Deployment Group (NTDG), organised by the US Department of Energy as part of the "Nuclear Power 2010" programme,[1] the performance criteria agreed upon for Gen-III NPPs (horizon 2010) are: (i) standardised designs to expedite licensing, reduce capital cost and construction time; (ii) more robust designs, making the plants easier to operate; (iii) higher availability and longer operating life (typically 60 years); (iv) enhanced security and safety (in particular, reduced probability of severe accidents); (v) minimal effect on the environment.

Four designs are being developed in the European Union to meet the "European Utility Requirements" (EUR), as follows:

1. The European Pressurised Water Reactor (EPR), which is being developed by AREVA NP. The EPR is a large reactor (1550–1750 MWe) based on the very successful Konvoi series in Germany and N4 series in France, and has the highest rated thermal efficiency (36%) among all LWRs. The EPR core is loaded with mixed uranium/plutonium oxide (MOX) fuel, with an expected burn-up of 65 GWd/t; its availability is expected to be 92% over a 60-year service life. The staff dosimetry-level is also significantly reduced. The EPR confirmed in 1997 as the new standard design for France.
   The world's first EPR is currently under construction by the AREVA-Siemens consortium for its Finnish customer, TVO, and is currently the most advanced reactor, worldwide, regarding economic competitiveness, safety and environmental protection. The pre-construction process included the following major steps: (i) TVO and Fortum submitted in 1999 the environmental impact assessment reports (Olkiluoto and Loviisa sites) to Ministry of Trade and Industry (KTM); (ii) TVO submitted its application for a Decision in Principle (DIP) to KTM on 15 November 2000; (iii) Finland's Parliament ratified in May 2001 with clear majority (with votes 157 *versus* 3) a DiP for a spent fuel disposal and encapsulation facility to Olkiluoto; this ratification was one of the cornerstones for the positive DiP one year later on the fifth reactor; (iv) Finland's Government made a DiP for FIN5 on 17 January 2002; (v) Finland's Parliament voted on that DiP in its plenary session on 24 May 2002 and ratified it with votes 107 *versus* 92. Following the good experiences from the operation of existing reactors, the main arguments presented in 2000 in the application for the DiP for a new NPP unit (FIN5) included: FIN5 covers partly the additional electricity demand and replaces old power plants; it enables the fulfilment of the Kyoto commitments; it secures stable and predictable electricity prices; it reduces the dependence on electricity import. On July 12, 2006, the AREVA-Siemens consortium has forecast that the new reactor will be commissioned during the second quarter of 2010, with electricity being generated for the first time at the end of 2009. TVO and AREVA have confirmed that the exceptionally stringent safety requirements (e.g., a first-of-the-kind and very detailed investigation into the nuclear island's ability to withstand the impact of a large commercial aircraft) are being fully satisfied. Furthermore, despite some minor delays (due to the time required by the pool's subcontractors to achieve the level of performance required by the

[1] See http://www.ne.doe.gov/NucPwr2010/NucPwr2010.html.

new generation of nuclear power plants), the profit targets for 2006 remained unchanged.
Recent discussions on the new National Energy and Climate Strategy in Finland will provide the basis for national actions to constrain emissions of greenhouse-gases (GHG). Even though the Government's proposal for a new energy and climate policy is not specifically proposing the expansion of nuclear capacity, the role of nuclear power is mentioned as an effective means to implement GHG reduction measures planned already during the previous strategy of 2001. In addition, it is stated that the GHG emissions are dependent on the amount of carbon-free/carbon-neutral or low-carbon energy sources, such as renewables, nuclear power and electricity imports. Currently, the Finnish Parliament is discussing the governmental report on the proposed national energy and climate strategy for implementation of the Kyoto protocol. Concurrently, TVO has announced that it has prepared a preliminary schedule for the sixth reactor unit (FIN6). If planning of the project would be started in 2007, and a DiP was granted by the Parliament, the commissioning of FIN6 could occur already before the year 2016.
2. The SWR 1000 is a 1290 MWe BWR, designed by AREVA NP, which has been precertified in the USA. The SWR1000 includes new passive safety systems (e.g., emergency condensers, as have been successfully operating in the plants at Gundremmingen A in Germany, and Dodewaard in the Netherlands), building condensers, passive pressure pulse transmitters and gravity-driven core flood lines. The SWR 1000 is overall simpler than the older BWRs, and uses high-burnup fuels that allow refueling intervals of up to 24 months.
3. The European/Economic Simplified Boiling Water Reactor (ESBWR), 1390 MWe with passive safety systems, developed by General Electric Energy. The ESBWR is under certification in the US.
4. The BWR 90+, an evolutionary reactor of 1500 MWe designed by BNFL Westinghouse (Sweden), which features improved safety and performance systems that improves the control of the plant, eases the burden on the operator and facilitates operation and maintenance.

The Euratom Research and Technological Development (RTD) programme within the 6th Framework Programme (FP6) contains projects, in the areas "safety of existing installations" and "innovative concepts", aimed at medium-term (2010) evolutionary improvements that will enable the next Gen-III of nuclear power plants (NPPs) to be inherently safer and more cost effective. The most important projects with such goals are the following: *Severe Accident Research Network of Excellence* (SARNET); *Prediction of Irradiation Damage Effects on Reactor Components* (PERFECT); and the *European Platform for Nuclear Reactor Simulations* (NURESIM).

8.2 Generation IV ("Revolutionary") Nuclear Power Plants

Studies regarding energy consumption, needs, and supply sources generally indicate that the worldwide energy consumption per habitant is very uneven, ranging from as low as 0.18 tonne-oil-equivalents (toe) in Bangladesh to as high as 8.2 toe in the US. Even assuming a modest population growth (from today's level of circa 6 billion) to circa 10 billion people by 2050, the total world energy consumption would be expected to reach circa 19.7 Gtoe. The future of energy, and especially the substitution of carbon-based economies by hydrogen economies, has come to the forefront of many international debates. For example: (i) in his State of the Union address (28 January 2003), President Bush announced a massive RTD effort on hydrogen in the USA; (ii) the initiative "Hydrogen Energy and Fuel Cells – A vision of our future", EC 2003 was launched in the EU to prepare a roadmap for the transition towards a hydrogen economy; (iii) the "International Partnership for the Hydrogen Economy" was established, involving academic and industrial organisations from the USA, the EU and 14 other countries worldwide. Seen in isolation, hydrogen produced from renewable energies or from nuclear fission is practically CO_2-free. However, there are only two truly "clean", hydrocarbon-free, processes for producing large quantities of H2 from the decomposition of H_2O using an external heat source, namely: (i) thermo-chemical water splitting cycles (TCWSC), and (ii) high-temperature electrolysis of steam (HTES). Using nuclear fission systems to produce process heat would greatly accelerate the growth rate of nuclear energy, but such a scenario would also lead to an accelerated depletion of the uranium resources in a few decades, if the once-through fuel cycle currently in use in most countries would not be discontinued. Furthermore, the fast neutron reactors would become essential for recycling and sustainability.

The global price of primary energy sources is currently characterised by the following developments: (i) the price of crude oil has increased from the level of less than 45 US$/barrel early 2005 up to over 70 US$/barrel in August 2006; (ii) prices of natural gas and coal have increased recently as well, but with a clearly slower rate; (iii) the uranium price has also increased, but this affects less the total power cost; (iv) the price of electricity will remain high also in the future, as emphasised by the tightening conditions for obtaining emission allowances (e.g., the NordPool price of CO_2 allowances remained above 20 €/tonne CO_2 from the middle of 2005 until the end of April 2006, but dropped thereafter roughly by 50%).

According to the so-called Red Book, the world nuclear industry is expected to grow between 22% (450 GWe) and 44% (530 GWe) from the current capacity of 370 GWe. This will raise annual uranium consumption requirements to 80,000–100,000 t. The currently identified uranium resources that can be mined at circa 130$/kg (the current spot-price) is about 4.7 Mt; these reserves may last for circa 60 years.

The "Generation IV" nuclear systems will aim not only at producing electricity, but also at hydrogen production from water, synthetic fuel (synfuels) production for transportation, heat production for the (petrochemical) industry and for water desalination. The above challenges are discussed in a world-wide context, particularly within the *Generation IV International Forum* (GIF), launched by the US; the "*INternational PROject on innovative nuclear reactors and fuel cycles*" (INPRO), launched by the IAEA; and the "*Global Nuclear Energy Partnership*" (GNEP), recently launched by the US.

The GIF (http://gen-iv.ne.doe.gov/) consisted originally of a group of nine countries: Argentina, Brazil, Canada, France, Japan, South Africa, South Korea, the UK, and the USA. In February 2002, Switzerland became a member of the forum, and the European Atomic Energy Community (Euratom) joined in July 2003. The contractual structure of GIF comprises a Framework Agreement (signed in February 2005), six System Arrangements (the first one, regarding the Sodium Fast Reactor, was signed in February 2006), composed in turn of many Project Agreements (due to be signed in the coming months; the first one will be the project "Advanced Fuels" of the system SFR). The IAEA is a permanent observer, which enables the coordination between GIF and INPRO activities. The INPRO international initiative was launched by IAEA in 2000;[2] INPRO is complementary to GIF As of June 2006, INPRO comprises 26 members: Argentina, Brazil, Bulgaria, Canada, Chile, China, Czech Republic, France, Germany, India, Indonesia, Republic of South Korea, Netherlands, Pakistan, Russian Federation, South Africa, Spain, Switzerland, Turkey, Armenia, Morocco, Ukraine, US, Slovakia, Japan, and the European Commission (Euratom). The IAEA General Conference has invited all interested Member States to consider jointly innovative actions in nuclear reactors and fuel cycles that use economically competitive technology, based on systems with inherent safety features, while minimising the risk of proliferation and the impact on the environment.

Four technological "performance areas" for industry and society were identified within the GIF initiative,[3] as follows:

1. *Sustainability*: This performance criterion comprises two aspects, namely: (i) Gen-IV systems will provide sustainable energy generation that meets clean air objectives and promotes long-term availability of systems and effective fuel utilisation for worldwide energy production; and (ii) Gen-IV systems will also minimise and manage their nuclear waste and notably reduce the long-term maintenance burden in the future, thereby improving protection for the public health and the environment. Gen-IV systems could operate as breeder reactors (fertile U^{238} or Th^{232} is converted to fissile Pu^{239} or U^{233}, respectively, during reactor operation) or as "burners" of plutonium and higher actinides. A fully closed fuel cycle for minimising the amount of high-level waste for geological disposal can be developed either by fully recycling the actinides (which would not be separated during reprocessing, but would be returned to the fast-neutron spectrum reactors, where they are consumed), or by "partitioning and transmutation", which involves using MOX-fuel simultaneously with separating and recycling the minor actinides (MA) in subcritical or critical reactors.

---

[2] See http://www.iaea.org/OurWork/ST/NE/NENP/NPTDS/Downloads/Brochure/2005_INPRO_Brochure.pdf

[3] See http://gif.inel.gov/roadmap/pdfs/000_contents.pdf and http://www.gen-4.org/Technology/roadmap.htm

2. *Economics*: This performance criterion also comprises two aspects, namely: (i) Gen-IV nuclear energy systems will have a clear life-cycle cost advantage over other energy sources; and (ii) Gen-IV nuclear energy systems will also have a level of financial risk comparable to other energy projects. To meet this performance criterion, Gen-IV systems must penetrate new energy markets; examples include; (i) producing electricity and high temperature heat through increased core outlet temperatures; (ii) using advanced technologies ranging from hybrid approaches (deterministic and probabilistic) for reactor design to Brayton cycles for power generation; (iii) reducing containment-system structural requirements by using coolants with very high boiling points or providing the containment function in a non-conventional manner.
3. *Safety and Reliability*: Gen-IV nuclear energy systems will have a very low likelihood and degree of reactor core damage, and they will eliminate the need for offsite emergency response. Measures for attaining this performance criterion include: (i) "practical elimination" (by design) of situations and phenomena which could lead to unacceptable core damage and/or uncontrolled large early releases of fission products; (ii) using passive decay heat removal (as a step beyond the last defense against fuel melting), and using advanced and compatible materials (e.g., incore materials with increased thermal capacity and coolants that do not react chemically with air or water); (iii) designing reactors with negative temperature and void reactivity coefficients, thereby causing the reactor power to decrease if core temperatures increase and/or coolant is lost from the core.
4. *Proliferation Resistance and Physical Protection*: Gen-IV nuclear energy systems will increase the assurance that they are a very unattractive and the least desirable route for diversion or theft of weapons-usable materials and provide increased physical protection against acts of terrorism. This performance criterion would be attained by (i) developing a closed fuel cycle, in harmony with the goal of "sustainability"; (ii) operating at a breeding ratio of near one, which would alleviate the need for a uranium blanket; (iii) fabricating the fuel so that it would be very difficult to extract fissile material from spent fuel, while minimising material accounting difficulties within the international safeguards system.

The GIF has selected six reactor systems, as follows:

1. Sodium-Cooled Fast Reactor System (SFR); leaders: Japan and USA; purpose: electricity production and actinide management; reference power: modules of 150 MWe or large plant of 1500 MWe; demonstration planned for 2020;
2. Very-High-Temperature Reactor System (VHTR); leaders: Japan and France; purpose: cogeneration for synfuel or hydrogen production; reference power: modules of 600 MWth (275 MWe); demonstration planned for 2020;
3. Gas-Cooled Fast Reactor System (GFR); leaders: France and USA; purpose: cogeneration for hydrogen production and actinide management; reference power: 600 MWth (275 MWe); demonstration planned for 2025;
4. Supercritical-Water-Cooled Reactor System (SCWR); leaders: Canada and USA; purpose: efficient electricity production with an option for actinide management; reference power: 1700 MWe; demonstration planned for 2025;

Table 8.1 Selected Generation IV reactor systems.

	Neutron spectrum	Coolant	Temp. (°C)	Fuel	Fuel cycle	Size(s) (MWe)
Gas-Cooled Fast Reactors (GFR)	fast	helium	850	U^{238}	closed	275
Lead-Cooled Fast Reactors (LFR)	fast	Pb-Bi	480–800	U^{238}	closed	10–100 600
Molten Salt Reactors (MSR)	epithermal	fluoride salts	700–800	UF in salt	closed	1000
Sodium-Cooled Fast Reactors (SFR)	fast	sodium	550	U^{238} and MOX	closed	150 1500
Supercritical Watercooled Reactors (SCWR)	thermal or fast	water	500	UO_2 or MOX	open (thermal) or closed (fast)	1700
Very High Temperature Gas Reactors (VHTR)	thermal	helium	1000	UO_2 prism or pebbles	open	275

5. Lead-Cooled Fast Reactor System (LFR) demonstration planned in 2025; leaders: USA and Euratom (Italy); purpose: cogeneration for hydrogen production and actinide management; reference power: 600 MWe;
6. Molten Salt Reactor System (MSR); leaders: France and USA; purpose: electricity production and waste burndown; reference power: 1000 MWe; demonstration planned for 2030.

Waste management as a key priority; the overwhelming scientific consensus is that geological disposal is safe and technically feasible, albeit with further refinement of models and techniques, though political and societal acceptance has been less forthcoming. Increasing emphasis is being focused on fully closing the fuel cycle so that the only high-level wastes, remaining from losses in recycling and re-fabrication, will comprise a relatively small quantity of short-lived nuclides. Ongoing EU-research on fully closed fuel cycles aims at: (i) substantial reduction of long-term radioactivity, radiotoxicity and fissile materials inventories that might contribute to improving the public acceptance of the geological repositories; and (ii) the minimisation of short- and medium-term heat sources to enable a reduction of the volume required by the high-level wastes in repositories, thus increasing the effective capacity and reducing the number of repositories. The most important ongoing Euratom FP6 projects addressing Gen-IV systems, and the partitioning and transmutation aspects are: (i) RAPHAEL: "*ReActor for Process heat, Hydrogen And ELectricity generation*" (GIF system VHTR); (ii) GCFR: "*Gas Cooled Fast Reactor*" (GIF system GFR); (iii) EUROPART: "*EUROpean Research Programme for PARTitioning of Minor Actinides within High Active Wastes Issuing from the Reprocessing of Spent Nuclear Fuels*"; (iv) EUROTRANS: "*EUROpean Research*

Programme for the TRANSmutation of High-Level Nuclear Waste in an Accelerator Driven System (ADS)" (v) ACTINET-6: "*ACTInide sciences NETwork*".

Important roles in the process of European integration of nuclear research and innovation are played by the "European Utility Requirement" /EUR/ association, which fixes the technical specifications for Generation III (http://www.europeanutilityrequirements.org/), the European Atomic Forum FORATOM (http://www.foratom.org), the "Western European Nuclear Regulators' Association" (WENRA http://www.wenra.org/extra/pod/), and the EC DG Transport and Energy (http://ec.europa.eu/comm/energy/nuclear/safety/index_en.htm). Collaboration with nuclear institutions outside the EU is also important, particularly with the International Science and Technology Centre in the Russian Federation (ISTC, http://www.istc.ru/), and the Science and Technology Centre in Ukraine (STCU, http://www.stcu.int/). Keeping the nuclear energy-option open can be achieved only by ensuring adequately qualified personnel in research organisations and in nuclear installations, thereby ensuring that the nuclear competences will remain at a high level in the long term. This concern is shared worldwide.[4] Preparing the next generation of experts to ensure continuous improvement of current nuclear fission technologies is based on attaining three basic objectives, namely: (i) modular approach and common qualification criteria in order to offer a coherent training and education framework comprising a wide variety of modules; (ii) uniform mutual recognition system across the European Union by using the European Credit Transfer and Accumulation System of "Bologna" /ects/; and (iii) mobility for teachers and students. In order to achieve the above objectives, the "European Nuclear Education Network" (ENEN) non-profit association was formed in September 2003, comprising 42 institutions for nuclear education and training (www.enen-assoc.org and enen.seccea.fr). Furthermore, the "*Nuclear European Platform of Training and University Organisations*" (NEPTUNO) Coordination Action was initiated in January 2004 within the FP6 Euratom framework. In addition, the World Nuclear University (WNU), with a secretariat in London, was launched in September 2003 by the World Nuclear Association under the umbrella of IAEA.

8.3 The Euratom Coordination Action "Sustainable Nuclear Fission Technology Platform (Snf-Tp)", Coordinated by CEA

A Euratom Coordination Action entitled "Sustainable Nuclear Fission Technology Platform" (SNF-TP) will commence in October 2006. The goal of the SNF-TP is to establish the long-term *sustainability* of nuclear energy, including economic competitiveness, safety, waste minimisation and disposal, radical technological change, environment, energy production, and non-proliferation. The SNF-TP is structured into two sub-projects and seven work-packages, which will be described briefly in the following.

[4] See, e.g., http://www.nea.fr/html/ndd/reports/2000/nea2428-education.pdf, www.iaea.org/km/

8.3.1 Sub-Project 1 (SP1): Strategic Research Agenda and Platform Organisation & Operation

The Strategic Research Agenda (SRA) is a non-prescriptive, non-authoritative, collaborative, informed, and dynamic strategy for developing technologies to achieve a long-term vision. The SRA is driven by the challenges (industrial and policy) faced by the nuclear sector, and harmonises ambitious goals and milestones across the short-, medium- and long-term planning ranges.

Materials & Fuel Development (WP1.1). In nuclear power reactors, materials undergo aging processes due to neutron irradiation. The current databases of material properties (yield stress, fracture toughness, corrosion crack propagation rate, etc.), which is needed to account for such degradations in the design and safe operation of installations, were established during long-term irradiation programmes in test reactors, and during mechanical and corrosion testing in specialised hot cells. Numerous experimental programmes have been performed for the ferritic steels used for manufacturing Reactor Pressure Vessels (RPV) for Light Water Reactor (LWR), and for the stainless steels used for manufacturing their internal structures. However, the predominantly empirical methods currently in use cannot entirely cover this need, particularly since certain in-service conditions are impossible to reproduce within the existing test facilities. Moreover, the number of test reactors is steadily decreasing, while costs and regulatory constraints are steadily increasing. In view of these complex circumstances, the experimental approaches can be complemented by post-irradiation research programmes for decommissioned reactors, and by developing physical and numerical modeling tools to simulate the effects of irradiation on the mechanical and corrosion behaviour of nuclear materials. To accomplish this goal, numerical modeling would need to have a "multi-scale" character, bridging atomic (nanometres), mesoscopic (micrometres), and macroscopic (centimetres) scales, aiming at reducing uncertainties and establishing safety margins by comparison to empirical component analyses. Ongoing developments aim at predicting the tensile and fracture toughness properties of the irradiated RPV steel by accounting for the size and constraint of the cracked body (from lab specimens to shells of RPV), and by simulating the irradiation-induced evolution of the microstructure (i.e., irradiation-induced damage). The irradiation-induced evolution of the microstructure (irradiation-induced damage) must also be considered in order to provide the tensile and swelling properties of irradiated stainless steel. To implement the transition to Generation IV reactors, a high priority must be placed on developing new fuels (ceramic fuel capable of high fission products confinement, compatible with high-temperature coolant, and self-sustainable), and structural and component materials resistant to both fast neutrons and high temperatures, for example, mastering the behaviour of Oxide Dispersion Strengthened (ODS) ferritic/martensitic steels and developing advanced composite ceramics materials for Gen IV fast reactors.

Simulation Tools for Reactor Design and Safety (WP1.2). Design and safety activities rely on numerical simulation tools that must be validated against experiments. These activities have a dual character, in that they comprise both generic issues (e.g., development of generic numerical solvers for mathematical problems that are common to all types of reactors) and issues that are specific for each type of reactor system. An example of a generic issue of importance to all reactor types concerns the development of methods for global sensitivity and uncertainty analysis in order to reduce the uncertainty margins. Since the deterministic and, respectively, statistical methods for sensitivity and uncertainty analysis have been, historically, developed separately, each class of methods has limitations that cannot be overcome by further developments within the current paradigms. A breakthrough could be achieved by developing a new theoretical and computational framework, in which the strengths of the deterministic and statistical methods would be combined to eliminate (most of) their current limitations. Most of the research topics mentioned in the foregoing are currently being addressed within the FP6-Integrated Project NURESIM, which aims at providing the initial step towards a Common European Standard Software Platform for modeling, recording, and recovering computer data for the next- generation nuclear reactors simulations. Such a common platform would also facilitate the exchange of data across sites, application codes and computing platforms. The tasks of WP1.2 would be to: (i) coordinate the activities of key organisations participating in Euratom FP6 programmes (including NURESIM and SARNET) involving simulation tools for reactor design and safety, in order to elaborate the technical agenda and Roadmap for future activities, commencing with FP7, for RTD in simulation tools for reactor design and safety; and (ii) commence Platform Organisation & Operation for RTD in simulation tools for reactor design and safety.

System Integration: Thermal and Fast Reactors, Fuel Cycles with Partitioning & Transmutation, and Waste Processes (WP1.3). Reactor system integration is envisaged at two levels, namely: (i) integration of each reactor-type with its own specific fuel cycle, and (ii) integration of the various reactor systems (thermal and fast) together with a fuel cycle (multi-disciplinary, transverse activity) that would lead to sustainability of nuclear energy. The commercial deployment of fast neutron reactors, with closed fuel cycles, is envisaged to commence by 2040. Among the fast reactor systems, the technological basis for the Sodium-Cooled Fast Reactor (SFR) is currently the most comprehensive, based on experience gained internationally from operating research, demonstration and commercial-size reactors.

The technological basis gained from these reactors includes key elements of the overall reactor design (pool- or loop-type), large-core design, fuel types (metal, carbide, nitride, oxide), safety, and fuel cycling. The main issues to be addressed in order to reduce the costs and improve the safety of SFRs include: design simplification (e.g., suppression of the intermediate sodium circuit, replacement of the water-cycle by innovative high-efficiency cycles), improvement of in- service inspection and repair, mitigation of the hypothetical recriticality severe accident, use of innovative fuel allowing grouped-actinide separation for minimising the high-level waste production and enhancing non-proliferation within a closed fuel cycle. None of these remaining issues appear unsolvable, so that the SFR is currently considered as a promising fast reactor system for achieving sustainability of nuclear energy production in the medium term.

The Gas-Cooled Fast Reactor (GFR) appears to have the potential of outstripping the SFR, by reaching higher operating temperatures and efficiency, thus opening the way to further industrial applications. Advantages of the GFR would be neutronic and optical transparency, as well as chemical inertness of Helium coolant. However, the GFR technology is considerably less developed than the SFR technology. The most important technological challenges to be resolved include the development of self-sustainable cores using innovative fuel, safety and system development for managing the decay heat removal under accident conditions, and establishing a sustainable fuel cycle. Innovations are needed for the development of GFR fuel, and structural materials resistant both to fast neutrons and high temperatures.

The Lead-Cooled Fast Reactor (LFR) has a large thermal inertia of the heavy liquid metal, and its non-reactivity with air and water could allow an integrated design with heat exchanger in the reactor vessel. Furthermore lead is inexpensive and can operate at low pressure and low pressure losses. These advantages, along with the elimination of the secondary loop, permit a compact lay-out and, in principle, a competitive electric energy generation and long-term investment protection. However, the LFR technology is less developed than the SFR technology; in particular, the corrosion/erosion at high speed and temperature and the choice of a suitable fuel cladding material have not been sufficiently established. Other issues to be addressed for this system are similar to the ones of the SFR, e.g., improvement of in-service inspection and repair, use of innovative fuel allowing grouped-actinide separation for minimising the high-level waste production and enhancing non-proliferation within a closed fuel cycle, etc. Other reactor systems that could be designed as fast reactors are appropriate variants of the Supercritical Water Cooled Reactor (SCWR), and the Molten Salt Reactor (MSR).

The fuel cycle and waste management strategies are integral parts of any reactor system. It is not possible to consider advanced reactor technology without factoring in fuel manufacture, spent fuel treatment and disposal. For Generation IV systems, the fuel cycle must demonstrate full sustainability by maximising resource utilisation through recycle, while minimising waste generation and remaining economically competitive. Furthermore, the safety and proliferation resistance constraints on reactor fuel cycles will likely become even more stringent in the future. Meeting these constraints will require closed fuel cycles with the separation of radioactive species (fission products from minor actinides), refabrication and recycling of fissile material and minor actinides. Since the recycling of minor actinides is a key issue to be addressed for closed fuel cycles within the Fission Platform, the future prospects and research direction for partitioning and transmutation need to be properly evaluated. This work package will need to assess the various fuel processing technologies, such as fuel fabrication (active and inactive), spent fuel handling and separation (aqueous and non-aqueous) and waste management options (encapsulation and sentencing). The reactor systems to be considered will include thermal and fast reactors with either open or closed fuel cycles. As the transmutation of actinides in fast systems constitutes a key component of the closed fuel cycle, the relative economical and technological advantages or disadvantages of using critical systems *versus* sub-critical systems, such as Accelerator Driven Systems (ADS), must be assessed in detail, in order to make optimal choices for the sustainable development of nuclear energy. Thus, the work performed in this work package will be based on results to be delivered by the CA-PATEROS.

Training and R&D Infrastructures (WP1.4). The objective of WP1.4 is to develop a European approach and Roadmap to infrastructures, covering both the creation of new installations, the functioning of existing ones and access to them. Implementing up-to-date research infrastructures, operated as a European user-facility, is an effective way for advancing the ERA concept by networking and giving impetus to existing facilities, and for attracting a new generation of scientists and engineers to contribute to challenging new programmes. The SNF-TP will also co-opt the NEPTUNO network of European universities. This work package comprises two tasks, namely: (i) to coordinate the activities of key organisations participating in Euratom FP6 programmes in order to elaborate the technical agenda and Roadmap for the future activities, commencing with FP7, for RTD in training and R&D infrastructures; and (ii) to commence Platform Organisation & Operation for training and R&D infrastructures.

Light Water Reactors (WP1.5). For Gen-II, III, and III+ LWRs, innovative research continues to play a major role in improving PSA methods, life time management, passive safety features, mitigation of severe accidents, the performance of fuels, improving the Pu-conversion capabilities to reduce uranium consumption (increasing the flexibility of the fuel cycle), optimising the economic performance, and developing a set of modern, standardised simulation tools based on multi-scale and multi-physics capabilities for reducing uncertainties and improving operation margins. Further research is needed to evaluate the deployment of Generation III and III+ technologies, including, in particular, evaluation of their performance and safe operation.

This work package comprises two tasks, namely: (i) to coordinate the activities of key organisations participating in Euratom FP6 programmes in order to elaborate the technical agenda and Roadmap for the future activities, commencing with FP7, for LWR-RTD; and (ii) to commence Platform Organisation & Operation for LWR-RTD.

High Temperature Reactors and Processes (WP1.6). The gas-cooled Very High Temperature Reactor (VHTR) is envisaged to produce hydrogen and process heat on an industrial scale. The current scientific and technological challenges for this reactor system include the development of dedicated fuel and structure materials resistant at very high temperatures (>950°C), helium loops and intermediate heat exchangers capable of operating at such high temperatures, and efficient procedures for high temperature electrolysis or thermo-chemical decomposition of water for producing hydrogen at industrial scales. This work package comprises two tasks, as follows: (i) to coordinate the activities of key organisations participating in Euratom FP6 programmes in order to elaborate the technical agenda and Roadmap for future activities, commencing with FP7, for RTD regarding high temperature reactors and processes; and (ii) to commence Platform Organisation & Operation for RTD regarding high temperature reactors and processes.

SNF-TP Strategic Research Agenda and Platform Organisation & Operation (WP1.7). The Strategic Research Agenda (SRA) is a non-prescriptive, non-authoritative, collaborative, informed, and dynamic strategy for developing technologies to achieve a long-term vision. The SRA is driven by the challenges (industrial and policy) faced by the nuclear sector, and harmonises ambitious goals and milestones across the short-, medium- and long-term planning ranges. It is also important to recall that the SRA includes two intertwined components, namely (i) a "Technical Agenda", and (ii) a strategy comprising essential enabling factors that enable the implementation ("The Enablers") of the "Technical Agenda".

The enabling factors span a wide spectrum, from educational policies to research infrastructures, from certification and qualification aspects to coordination of programmes, etc. The SRA must be revised every 2–3 years to remain always adapted to changing circumstances or requirements. The "Technical Agenda" of the SRA comprises, in turn, the research and technological development (RTD) plans of each of the six previous work packages WP1.1 through WP1.6. The tasks of this work package are to (i) elaborate the Strategic Research Agenda for the SNF-TP by merging the Roadmaps developed in WP1.1 through 1.6, and the Roadmap for P&T elaborated in other EU Programmes; and (ii) to commence Platform Organisation & Operation by merging the results from WP1.1 through 1.6.

8.3.2 Sub-Project 2 (SP2): Platform Deployment Strategy

The Deployment Strategy Group will include end users such as utilities, oil companies (for process heat and hydrogen production) together with public bodies. Furthermore, several additional types of organisations (e.g., regulators, oil companies, socio-economic groups) are expected to join the CA partners in the Platform. The participants in this CA consider that the SNF-TP is the proper body for consolidating the European position, and for decisions regarding future joint undertakings within the Euratom treaty. For example, a Joint Undertaking could emerge, in some years down the road, depending on the industrial possibilities that might develop for some specific goal within the SNF-TP. Recent analyses have failed to propose 50-year scenarios, based on sustainable development principles, in which nuclear fission would not provide large-scale amounts of highly intensive energy. Of course, renewable energy sources could also contribute for meeting small-scale (and especially dispersed), low- intensity needs. All the reputable energy scenarios show the main load being carried by coal, gas, and nuclear, with the balance among them depending on economic factors in the context of various levels of greenhouse constraints.

The notion of sustainability needs to be supported politically, in order to factor in all external costs, so that sustainability could become an important factor driving the economic choices among fuels for electricity production. The proposed SNF- TP is envisaged to provide the mechanisms for consolidating and deciding future joint undertakings within the Euratom Treaty. The SNF-TP would also consolidate the European and Euratom positions within the GIF- initiative, including waste management related to closed fuel cycles involving fast neutron systems. The CA-participants consider that the SNF-TP is the proper body for consolidating the European position, and for decisions regarding future joint undertaking within the Euratom treaty. For example, a Joint Undertaking could emerge, in some years down the road, depending on the industrial possibilities that might develop for some specific goal within the SNF-TP.

8.4 Conclusions

At this time, existing nuclear technologies are mature and proved; in particular, the currently operating (second-generation) nuclear power plants, light and heavy water reactors, have reached a high level of safety, reliability and cost- effectiveness, demonstrated over thousands of years of cumulated operation. Mature industrial solutions are available for all stages of nuclear fuel cycle, including spent fuel treatment, conditioning of medium-level waste, vitrification of high-level waste, and various technologies for interim storage. Considerable progress, aiming at industrial implementation in the early 2020s, is being made in several countries to develop practical solutions for geological disposal of high-level, long-lived nuclear waste. To establish itself as a viable option for long-term energy production, accepted by public opinion, nuclear fission energy generation must meet the following criteria for sustainable development: (i) economic and reliable, (ii) high-level safety and security; (iii) protection of workers, population and environment, against radiological hazards; (iv) minimal generation of long-lived, high-level radiotoxic waste; (v) safe disposal of remaining waste products; (vi) optimised use of natural resources. Existing (second generation) reactors already provide a strong basis for meeting these criteria, and R&D activities for further improving fuel performance (higher burn-up, less radiotoxic waste) and safe waste disposal are currently in progress. Building on this basis, third generation reactors are expected to further improve reactor safety (lower risk of accident, no need for evacuation and relocation outside the plant site), reduce radioactive waste, and lower natural uranium consumption per kWh (an expected gain of 200%). The advent of fourth generation innovative reactors will fulfill the criteria for sustainable development, through additional reduction of volume, thermal load and toxicity of long-lived radioactive waste (due to advanced fuel technology and fuel cycle technology), lowest possible uranium consumption (conversion yield close to 100%, through combining fast neutron spectrum and closed uranium-plutonium cycles), and production of high temperature (>1000°C) process heat for applications (especially for hydrogen production). Several RTD areas are especially important for achieving the criteria for sustainable development, and are therefore likely to receive priority in the upcoming decades. These RTD areas include: (a) fuels and materials (advanced fuels; materials resistant to high temperature and/or fast neutron fluence); (b) fuel cycles (advanced spent fuel treatment, including partitioning of actinides; recycling processes for actinides; transmutation; geological waste disposal technology); (c) reactor design (innovative concepts, for power generation and/or heat generation; advanced safety concepts); (d) radiation protection (refined assessment of the impact of radiation and radioactive materials on man and the environment). Furthermore, research infrastructures for nuclear fission technology are large and costly. Several aging, but currently operating facilities will need to be replaced in the short- or mid-term. Therefore, existing facilities should be pooled as much as practicable, while future research infrastructures (e.g., the planned Jules Horowitz Materials Testing Reactor; the MYRRHA fast spectrum irradiation facility) should be designed and operated as European facilities. A large potential global market exists in Europe and Asia for European nuclear technology, and the European industry is well-structured, well-positioned on this market, and should strive to keep its competitiveness. To ensure that the European Union retains its leading position in the field of civil nuclear technology, it is paramount to mobilise representatives from all interested stakeholders, including: (i) research institutions, both public and private; (ii) utilities, as the end-users of nuclear power plants, as well as oil companies interested in process heat and the hydrogen market; (iii) nuclear industry, embracing reactor construction, fuel cycle and waste disposal; (iv) public authorities, both in their roles of regulators and policy makers and as promoters of technology development. Specific actions that would need to be pursued include: (i) creating mechanisms for supporting trans-European synergy, emphasising, in particular, the creation of a coordinated training and educational system for maintaining nuclear competence in Europe; (ii) strengthening the European science base and expertise; (iii) optimally integrating research teams and RTD tools, optimising the use of existing research infrastructures, and creating new infrastructures when needed; (iv) integrating scientific and technological advancements dedicated to Generation III+ and IV systems; (v) using the basis created by the above-mentioned integration process in order to prepare a Technology Platform dedicated to specific Generation IV technologies; (vi) coordinating RTD in support of geological waste disposal technologies; (vii) disseminating the results of the above activities to appropriate policymaking bodies, to ensure a common vision.

References

1. FP6 Coordination Action "Sustainable Nuclear Fission Technology Platform", Specific Programme for Research and Training on Nuclear Energy Proposal/Contract No. FP6-036410, 2006.
2. AREVA Press Release, Paris, July 1, 2006.

3. *Uranium 2005: Resources, Production and Demand*, OECD/NEA and IAEA, June 2006.
4. Van Goethem, G., Hugon, M., Bhatnagar, V., Manolatos, P., Casalta, S. and Deffrennes, M., Euratom research and training in reactor systems and fuel cycles: EU response to some of the main S/T challenges in nuclear fission, in *Proceedings of FISA-2006 Conference on EU Research and Training in Reactor Systems*, Luxembourg, 13–16 March 2006.

Chapter 9
Safety in Nuclear Power: A Proposal

Maurizio Cumo

Abstract. MARS is a fission nuclear reactor developed at the Department of Nuclear Engineering and Energy Conversion of the University of Rome "La Sapienza". The design was aimed at a multipurpose, modular reactor to be used in high population density areas not only for electricity generation, but also for industrial heat production and, in particular, for water desalination or district heating. This objective, not common for nuclear power plants, required the adoption of very high safety standards not achievable by the use of conventional systems; the only viable approach was the extensive use of "passive" safety. The whole plant is assembled with special flanged connections and even major components of the primary loop, such as the reactor vessel and the steam generator, may be easily substituted or repaired. The cost of KWh is competitive with that of large reactors.

Key words: Nuclear power plant, nuclear safety, generation III+, inherent safety, passive safety, decommissioning.

9.1 Introduction

Modern nuclear power plants, in order to be accepted and widely used, must be based on new safety targets and design criteria, with the principle rule that the overall safety is not impaired by component failures or malfunctioning, nor by human errors: the most reliable approach is to utilise incontrovertible natural laws to perform any protection action needed. This philosophy leads to the identification of plant and component solutions relying on "inherent safety" and generally utilising passive components instead of active ones and static components instead of moving ones. Inherent safety concepts must be applied not only to the functional design of circuits, but also to the structural design of components, structures and buildings.

The MARS (Multipurpose Advanced Reactor, inherently Safe) nuclear reactor, developed at the Department of Nuclear Engineering and Energy Conversions of the

K. Hanjalić et al., Sustainable Energy Technologies: Options and Prospects, 177–193.

University of Rome “La Sapienza?, belongs to the generation of “inherently safe” nuclear reactors and may be classified as a “Generation III+” reactor. Its concept emphasises the “prevention approach” up to extreme consequences, making any component relevant to safety practically free from primary stresses, thus highly reducing the probability of their failure. These results are achieved thanks to the new design of the primary loop, of its auxiliaries and of the whole primary building.

The MARS design was originally conceived with the purpose to adopt only passive safety systems, so as to reach the condition of an inherently safe reactor. However, the removal of residual decay heat with natural circulation of water, together with the economic competitiveness of the plant, imposes a limit on the size of the plant. In this sense, 600 thermal MW power represents an optimal compromise. So, looking for a proper market niche, the small-medium size is particularly fit for countries which have a limited electric network for transmission. Some of these countries have also a shortage of desalinated water and the MARS plant has also been studied in a cogenerative version, electricity plus heat for desalination of water.

Other characteristics have been progressively developed along the design line: the modularity of the plant to facilitate a chain production in factories, which was possible only with the introduction of flanged connections by means of an external steel envelope which practically annuls the pressure difference in the primary loop confinement. The possibility to assemble the reactor from pieces on the site, while all the construction work is done in factories ensuring all the necessary quality controls in clean environment, thus reducing construction times and costs. The reactor life is prolonged to over a century through easy substitutions and, all the plant structures being made of steel, at the end the decommissioning costs are greatly reduced. A small part of the total steel used that is radio-activate, can be melted and frozen directly in the form of final waste, while its largest part can be recycled.

The huge increase in oil (and gas) prices has recently drawn attention to the competitiveness of nuclear energy and a greater interest, worldwide, has been focussed on small-to-medium size plants. For instance, a recent initiative of the US DOE (Department of Energy), called Global Nuclear Energy Partnership (GNEP), which received attention in France, Russia, Japan, and other countries, is applicable also to countries that are interested in building small-medium size plants. These countries will have the possibility to employ such reactors, even in co-generation of electricity-heat, receiving fuel from countries that have enrichment and reprocessing plants at a pre-defined, convenient cost equal to that of the selling countries. The only condition to subscribe is that, to avoid risks of proliferation, all the irradiated fuel must be given back when receiving new, fresh fuel. In this case, for instance, the MARS reactor might be employed. Recent developments in the MARS nuclear design, devised jointly with CEA researchers, have held out the possibility of raising enrichment and burn-up, thus prolonging fuel life up to five years, and in the meantime neutralising the boric acid content (and the complication of the related systems) through its substitution with burnable poisons.

9.2 Design Objectives

MARS is a 600 MWth, single loop, pressurised light water reactor (PWR). The design was originally focused on a multipurpose reactor to be used for private consumption in high population density areas as well as for industrial heat production, particularly for water desalination, and for district heating.

Using the well-proven technology and the operation experience of Pressurised Water Reactors (the most diffused nuclear reactors), the project introduces a lot of innovative features hugely improving the safety performance while keeping the cost of KWh competitive with traditional large power plants (this is a fundamental aspect to make the plant appealing from the marketing point of view). Extensive use of passive safety, in-depth plant simplification, new and decommissioning oriented design, were the principal guidelines in the design development. The MARS design mainly refers to "inherently" safe operation, relying on natural laws to assure protective actions, instead of active systems and components. In particular, physical laws are used both to detect protection requirements and to perform the safety action itself.

The main safety-related objectives set for the MARS design were:

- extremely low probability of severe accidents causing core melting;
- capability of the plant to manage even extremely severe accidents with core melting;
- "insensibility" to human errors;
- no requirement for an external emergency plan, even in the case of a severe accident;
- very low levels of doses to operation and maintenance personnel; and
- easy operation.

In addition, the following non-safety-related objectives were set:

- use of proven technologies;
- fast and easy assembling and disassembling of the whole plant, including components of the primary loop;
- plant modularity, allowing the assembling of several reactor modules;
- extensive plant simplification;
- suitability for co-generation;
- economic competitiveness with large plants, both nuclear and fossil-fueled; and
- a very long operative life, of more than a century.

9.3 Reactor Description

The MARS reactor is moderated and cooled by pressurised light water; the nominal core thermal power is 600 MW. Well-proven technological features of most experienced PWRs (geometry and materials of fuel rods and of fuel assemblies; layout of

Table 9.1 MARS reactor characteristic data.

Rated power	600	MWth
Inlet core coolant temperature	214	°C
Outlet core coolant temperature	254	°C
Rated pressure	75	bar
Fuel bundles	89	
Fuel rod array	17×17	
Fuel rods per fuel bundle	264	
Fuel rod external diameter	0.95	cm
Fuel rod active length	260	cm
Fuel rod pitch	1.26	cm
Control rod diameter	0.978	cm
Core heat transfer surface	1823	m^2
Core average linear power density	98.2	W/cm
Core average heat flux	32.9	W/cm^2
Core average power density	56.5	kW/litre

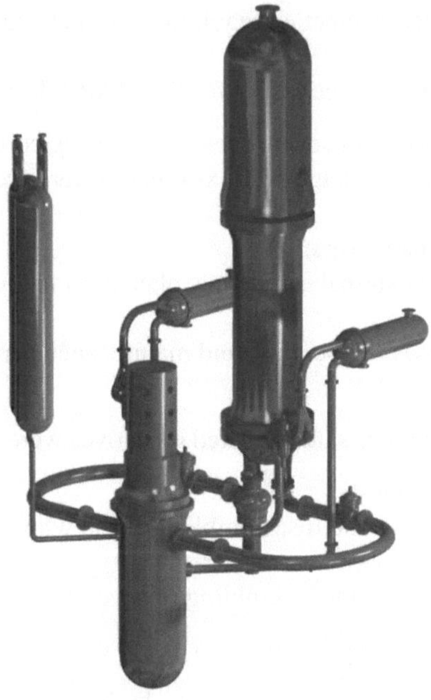

Fig. 9.1 MARS Primary coolant system.

the primary cooling system; geometry and materials of control rods and of control rod drive mechanisms; core internals; etc.) have been chosen as components of a basis for the design of the primary coolant system. Table 9.1 contains the main data of the MARS reactor.

The primary cooling system (Figure 9.1) includes one loop only with one vertical-axis U-tube steam generator and one canned pump. The pressure inside the primary cooling system is controlled by a large-volume vapor-bubble pressuriser. On/off valves (primary loop Main Isolation System, MIS) are installed in the primary cooling loop, in order to isolate, if necessary, the Steam Generator and the primary pump (e.g., in the event of an SG tube rupture). The pressure in the primary system has been chosen equal to 75 bar.

This choice, which leads to a loss in thermodynamic efficiency of the plant, has nevertheless allowed the adoption of systems which substantially eliminate the possibility of loss-of-coolant accidents (see the description of the CPP system below).

The low operating temperature (254°C max) allows one to reduce damage probability in materials; for example, corrosion crack and intergranular stress corrosion have never been experienced in Inconel 600 steam generator tubes operating at temperatures lower than 300°C, with the practical elimination of Steam Generator Tube Rupture accidents. Large safety margins in thermalhydraulic design (low thermal flux, low power density, etc.) have been adopted in order to make a thermal crisis impossible (the minimum DNBR is higher than 5).

The primary coolant inventory is significantly greater than in traditional PWRs (relevant to core power), thus the primary loop has a large thermal inertia; this makes all thermal transients (even accidental transients) very slow.

9.4 Main Innovative Safety Systems

9.4.1 Safety Core Cooling System

Connected to the reactor vessel is the Safety Core Cooling System (SCCS, see Figure 9.2) that provides the cooling of the core in emergency. It is designed to transfer the core decay heat directly from the reactor pressure vessel to the external environment, without the intervention of any energised (active) system or component.

The system operating principle relies on fluid natural circulation and its intervention is passively actuated by a reduction in the primary coolant flow; a cascade of three circuits provides redundant barriers between activated reactor coolant and external environment. The final heat sink of the SCCS is an atmospheric-pressure condenser cooled by external air in natural convection, avoiding any pressure control system and any energised system in the final heat sink.

During the normal operation, the flow in the SCCS primary loop is inhibited by a special check-valve. This valve is kept closed by the pressure difference existing between the reactor vessel inlet and outlet (proportional to the square of coolant flowrate inside the reactor vessel) that pushes the valve shutter in the upper position; when the coolant flow through the vessel decreases (whatever is the cause), the pressure difference decreases faster and becomes insufficient to keep the shutter in the upper position, so it falls, driven by gravity, leaving the valve completely opened (with a very low pressure drop).

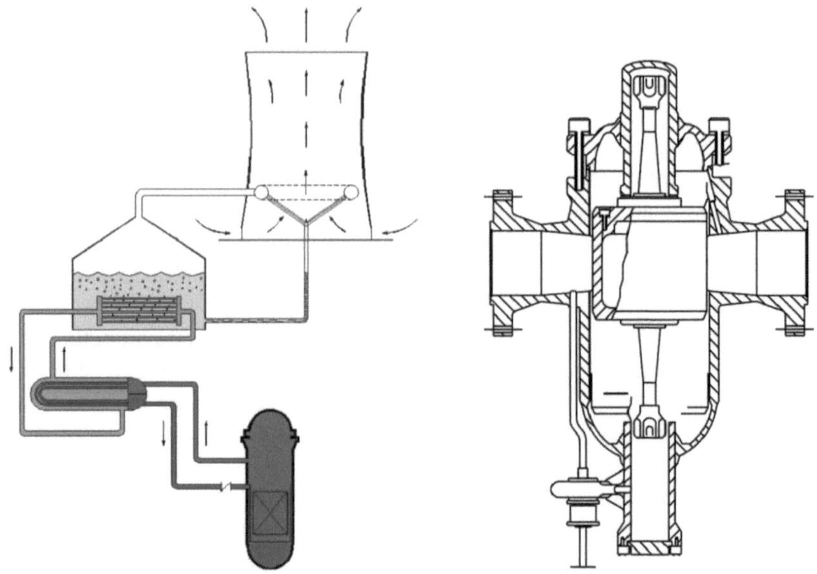

Fig. 9.2 SCCS scheme and innovative check valve.

9.4.2 Additional Temperature actuated Scram System (ATSS)

A passive scram system (called ATSS, see Figure 9.3) has been introduced in addition to the traditional scram system to provide an automatic and safe shutdown of the reactor as soon as the fluid temperature in the core rises above a selected setpoint.

The ATSS, whose working principle is based on differential thermal expansion of a bi-metallic sensor, is directly actuated by any excessive temperature rise inside the reactor core and eliminates the consequences of Accelerated Transients Without Scrams (ATWS).

A bimetallic rod is located directly in the core (in the place of a fuel rod); any increase in the coolant (and rod) temperature causes different elongations of the two elements of the rod. The differential displacement of the two ends is directly used to actuate (through a leverage) the releasing of a scram cluster.

9.4.3 Pressurised Containment for Primary Loop Protection

The primary cooling system and the primary loop of the safety core cooling system are enveloped by a steel-made pressurised containment, filled with water at the same pressure as the primary coolant, but at a lower temperature (about 70°C), called CPP (pressurised Containment for Primary loop Protection, Figure 9.4), which eliminates

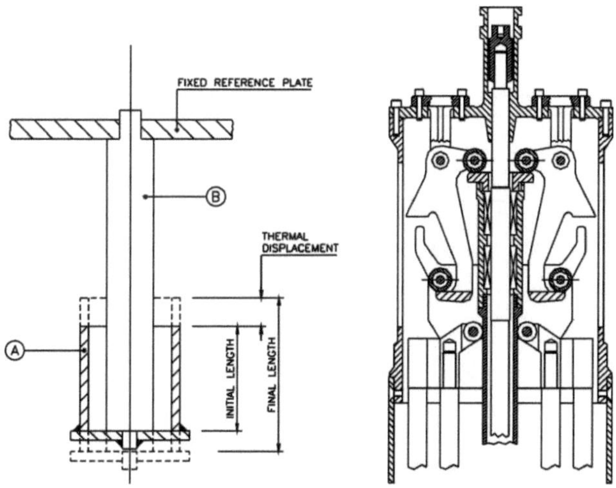

Fig. 9.3 ATSS scheme and mechanical detail.

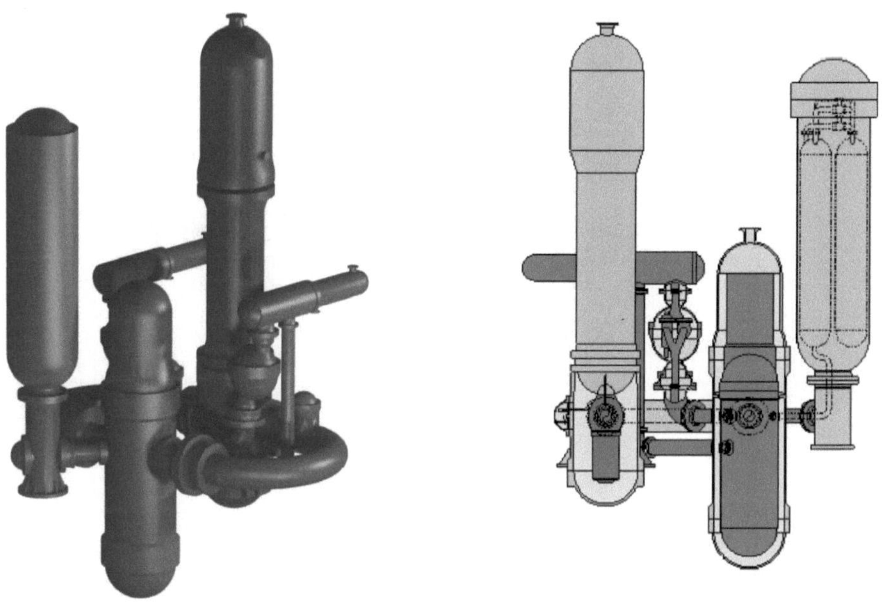

Fig. 9.4 CPP view and scheme.

the primary stresses on the primary coolant boundary. The CPP practically eliminates the possibility of Loss Of Coolant Accidents and Control Rod Ejection accidents.

The inclusion of the primary coolant system inside the CPP requires thermal insulation to reduce heat losses from the primary coolant system. An insulating sys-

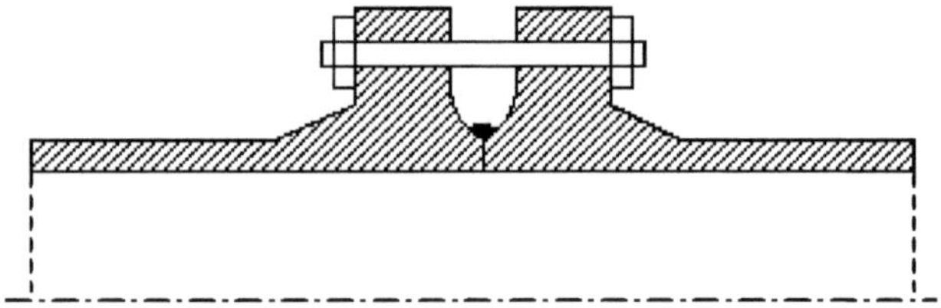

Fig. 9.5 Flanged connection.

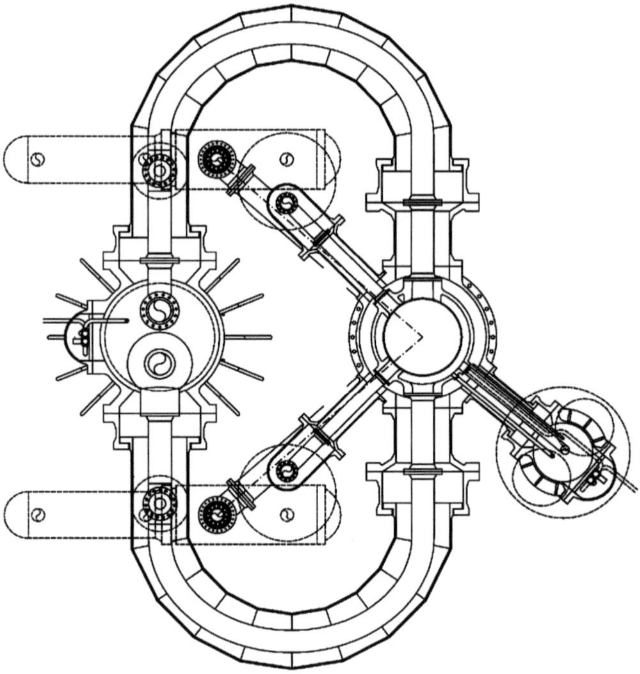

Fig. 9.6 Primary loop.

tem has been designed on the external side of the whole primary coolant boundary (excluding the pressure vessel lower head), taking into account the presence of the pressurised water. The pressure vessel lower head is, instead, internally insulated using stagnant water devices in a honeycomb system, to allow the external coolability in case of core melting.

The presence of low temperature water in the CPP, even with the insulating system, causes losses of about 0.3% of the reactor thermal power. On the other hand, the absence of primary stress due to pressure differences allows the adoption of flanged, leak-free, connections between components and piping, instead of welding techniques traditionally used in primary loops. This solution makes assembling

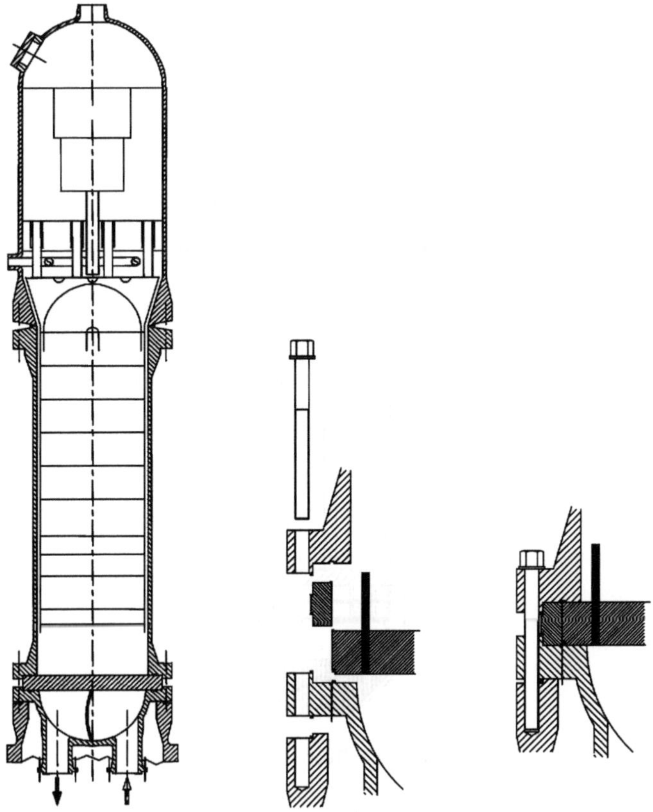

Fig. 9.7 Steam generator.

and disassembling operations undoubtedly much easier and faster than in traditional PWRs.

The presence of cold water outside the reactor vessel also allows one to deal with severe accidents guaranteeing the in-vessel cooling of molten corium.

9.5 New Approach to Construction

The enclosure of the primary coolant boundary in the CPP allows one to use flanged connections between components and piping. Special flanged connections, derived from solutions already used in chemical and petrochemical industries (Figure 9.5), have been designed to allow the transmission (between components and piping) of loads coming, for example, from thermal expansion or earthquakes, while tightness is assured through a welded gasket in the joining sections.

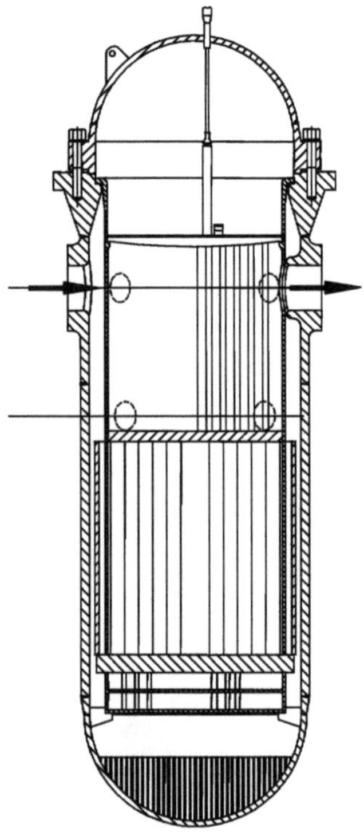

Fig. 9.8 Reactor vessel.

Any component in the primary loop, including the reactor vessel and the steam generator, can be easily substituted or repaired, with obvious benefits for plant availability, maintenance costs, life extension, easy and rapid decommissioning. Also, for the main components of the primary loop (Figure 9.6) an innovative design has been introduced aimed at an easy construction, testing and final assembling on site. In particular the steam generator, the largest component in the primary loop, has been completely redesigned to allow its complete dismounting, including the tube bundle (Figure 9.7).

These solutions allow the in-shop prefabrication of all components and eliminate the need of on-site full width welding, thus improving the quality control. Moreover, by adopting such a design, the plant erection time is also reduced.

9.6 Containment Building

The design of the reactor building has been completely innovated (Figure 9.9). In particular, an extensive use of bolted steel structures has been introduced for components' support and for working floors, instead of the more traditional concrete

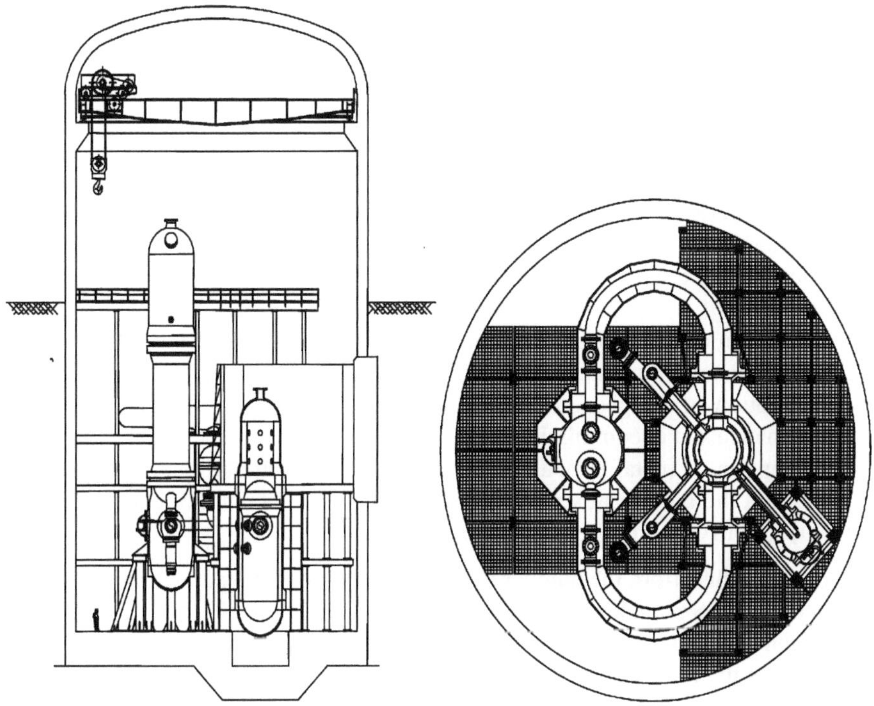

Fig. 9.9 Containment building section and plan.

structures (that require longer erection time and make decommissioning more complex).

A new design has also been developed for the biological shield, made of steel boxes filled with paraffin. This choice allows reduction of plant erection cost and time and makes the final plant dismantling easier.

9.7 Plant Simplification

The whole design of the MARS plant has been aimed at strongly simplifying the plant layout, the component construction and assembling on site in order to reduce construction time and costs (to make the plant competitive with larger, traditional ones) and to make plant decommissioning easy, fast and inexpensive. In particular, the following aspects were relevant for circuit simplification:

- the selected plant characteristics (specific power, temperature, pressure, thermal inertia, etc.) allowed simplification of all non-safety-related auxiliary systems, thanks to the reduced performance requirement;

- the extensive use of passive systems to assure plant safety allowed elimination of some traditional safety-related auxiliary systems (e.g., injection systems) or to strongly simplify other ones, reducing the number of redundant components (e.g., boron emergency shutdown);
- the selected plant power allowed reduction in the size of main components, making them easy to construct, to assemble and to disassemble;
- the strong reduction of radioactive streams and of their contamination allowed simplification in waste treatment systems;
- the low boron content in the primary coolant allowed simplification of boron systems;
- the adoption of an innovative design for all big components allowed their easy and fast construction, assembling and disassembling;
- the adoption of flanged connections allowed simplification of plant erection, maintenance and decommissioning;
- the adoption of metal structures to support components, to build the working floors and the biological screen, allowed to limit the amount of concrete in the plant, simplifying and speeding the construction phase.

As a result of this effort, the number of the main components (pumps, valves, tanks, etc.) in the MARS plant is reduced to about 50% of traditional PWRs, and the buildings volume is reduced by a similar factor. Also the plant construction time is reduced to less than four years.

9.7.1 Decommissioning Oriented Design

The major effort invested in simplifying the plant construction and erection has also resulted in a parallel, but by no means secondary, huge simplification of all decommissioning activities. The most important feature of the MARS plant relevant to decommissioning concerns the adoption of flanged connections in the primary loop. This, besides allowing the easy dismounting of each component (even SG and Pressure Vessel) for maintenance or substitution, makes all disassembly operations precisely the reverse of the assembling operations.

In the MARS plant, decommissioning the adoption of innovative or dedicated techniques or of special equipment is not necessary; in particular, the cutting of large thickness components is not required, thus hugely simplifying the dismounting operations and reducing the wastes. In addition, the reduced size of all components (the biggest ones may be also disassembled in transportable sub-components) makes them easily removable.

The general plant simplification allows a huge reduction, up to 50% in comparison with an equally powered traditional plant, of the number of contaminated or activated components, with a corresponding reduction in the amount of radioactive materials. In addition, the selected structural materials, together with the use of a cleaner primary coolant allows a reduction in total and specific activation of contaminated materials.

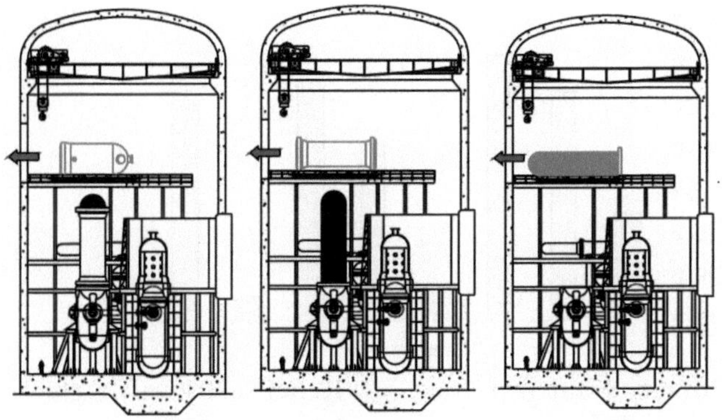

Fig. 9.10 Steam generator disassembling.

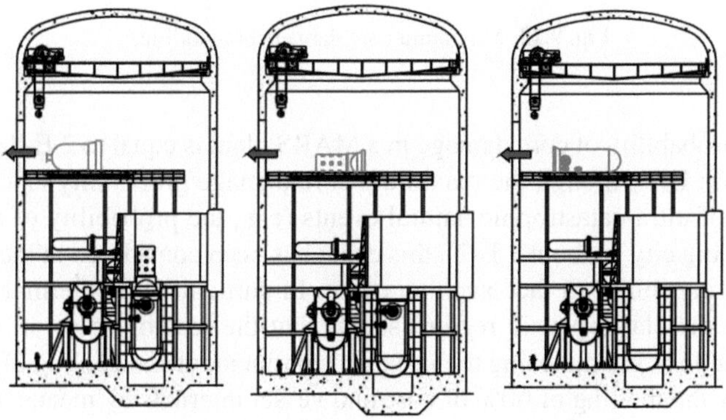

Fig. 9.11 Pressure vessel disassembling.

Last, but not least, the presence, inside the MARS containment building, of only steel structures (working floors, shields, etc. are made all of steel frames) allows complete dismantling of the plant, leaving a “clean” building to be demolished or reused for other purposes. In Figure 9.10 the disassembling of the steam generator is shown; one can note the possibility of a sub-disassembling of the SG; first the upper shell is removed and transferred outside the reactor building, then the lower shell and the tube bundle can be removed. Figure 9.11 shows the dismounting phases for the pressure vessel.

9.7.2 Probabilistic Safety Assessment

The complete safety analysis of the MARS nuclear plant, performed using analytical methodologies (such as HAZOP, Fault Tree analysis and Event Tree analysis) together with events listed in regulations (such as NRC-RG 1.70) shows that the

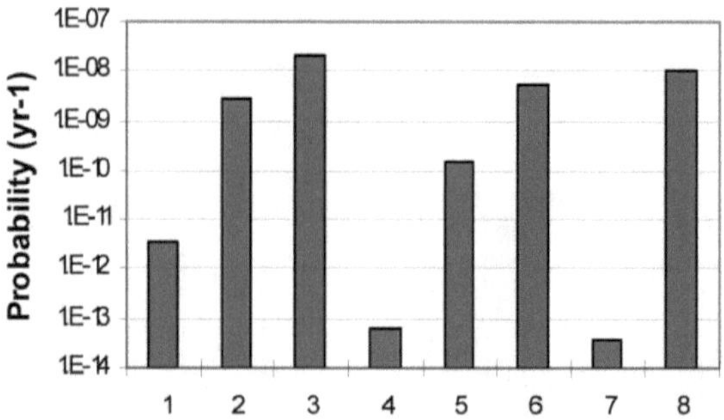

1 - Primary pump trip 2 - Relief/safety valves stuck open
3 - S.G. failure 4 - Loss of on/off site power
5 - Loss of coolant from aux. 6 - Steam generator tube rupture
7 - Primary pump trip 8 - Steam line break

Fig. 9.12 Maximum core damage probabilities.

maximum probability of core damage in a MARS plant is equal to 2.E-8 events/year (Figure 9.12). Even though the maximum core damage probability is less than the probability of ultra-catastrophic natural events (e.g., the probability of a meteorite impact on a big city is about 1.E-7), this event has been considered in the evaluation of the plant capability to face such an event. In particular, the thermal-mechanics analysis (in the elasto-plastic region) shows that the reactor vessel of the MARS plant is capable of guaranteeing the in-vessel confinement and cooling of the corium produced by the melting of 60% of core and vessel internals by means of the water contained in the CPP.

9.7.3 Economic Competitiveness

To make the MARS plant appealing from the market point of view, it was necessary to make its construction, operation and maintenance costs comparable with competitors' plants (both nuclear and fossil-fueled). This was achieved thanks to the described plant simplification and also to simplification in plant operation and maintenance.

A detailed analysis of electric energy costs production in a 450 MWe nuclear power station equipped with 3 PWR MARS reactors demonstrated the competitiveness of this type of plant for the production of electric energy. Table 9.2 lists the major direct construction costs of the plant (referred to year 2000). A specific direct cost of about 1,840 Euro/kWe was obtained, quite similar to the direct construction cost of large traditional PWRs, in spite of the unfavorable scale factor.

The specific indirect construction costs have been assessed to about 330 Euro/kWe.

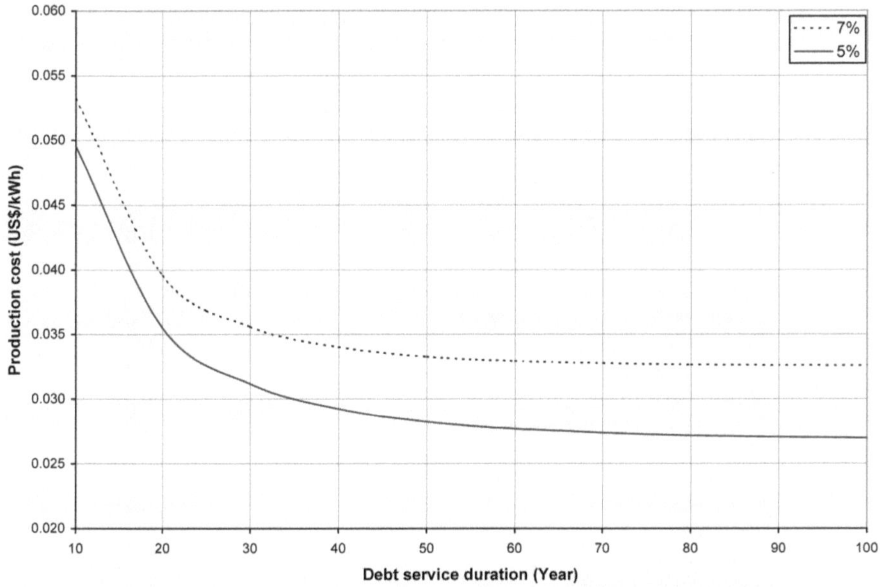

Fig. 9.13 Average cost of kWh during debt service period *versus* debt service period duration.

Figure 9.13 shows the cost of produced energy in the case of only electric production for two interest rates (5% and 7%) as a function of the period of debt repayment for a load factor of 85% (even if a load factor up to 95% is possible for a MARS plant) during the debt repayment period. Today (2006) the two used interest rates appear pessimistic for this kind of investment, because these were typical for institutional investors just a few years ago (around year 2000). By using, for instance, an interest rate equal to 2.5%, a 15–20% reduction in the kWh cost is achieved, depending on the mortgage period. However, such an estimate based on actual data are considered as overoptimistic and not recommendable, since the capital repayment is the most relevant portion of kWh cost (more than 60%). For an interest rate of 5%, the production cost of a kWh comes to 0.039 Euro under the assumption of debt service over a 20-year period, and to 0.034 Euro for debt service over 30 years.

It should be pointed out that the technical life of a MARS plant is much longer than the debt repayment period since every component, when reaching the end of its technical life, may be easily and cheaply replaced. During this life-extension period, the produced energy cost is due to operation and maintenance costs only. Even taking into account the cost of the largest components substitution (i.e., pressure vessel and steam generator), the produced energy cost is reduced to about 35–45% of that during the mortgage period.

The above estimates have been evaluated taking into consideration, besides normal maintenance costs, an investment cost for extraordinary maintenance equal to 15% of the initial investment every 30 years, to be repaid within the next 30 years.

Table 9.2 Direct construction cost of a 450 MWe power station (3 MARS units).

System	Cost (k Euro)
Buildings	68,638
HVAC systems	11,528
Closed circuit water cooling system and main condenser	40,826
Control rod systems	9,144
Fuel handling and storage system	8,869
Reactor coolant system (RCS)	125,502
Pressurised containment for primary loop protection (CPP)	66,232
Safety core cooling system (SCCS)	13,216
Main RCS auxiliaries	10,272
CPP auxiliaries	4,808
Reactor auxiliaries	9,889
Containment building safeguards	929
Radwaste system	2,653
Turbine	110,313
Condensate system	30,820
Feedwater system	14,389
Main steam system	13,837
Electric power station	53,527
Protection systems	1,962
Control systems	9,636
Plant supervising system	10,221
Environmental monitoring system	522
Plant monitoring system	2,047
Electric boards and panels	7,176
Neutron monitoring system	2,278
Demineralised water system	359
Auxiliary steam system	286
Instrumentation air system	889
BOP fire protection system	137
Elevators and lighting system	6,956
Common services common buildings	14,273
Condenser cooling water system	48,461
Minor RCS auxiliaries	15,912
Turbo-alternator lubricating system	162
Plant electrical system	20,252
Other plant auxiliaries	16,510
Total	753,429

9.8 Conclusions

A modular, fission nuclear reactor "MARS", developed at the Department of Nuclear Engineering and Energy Conversion of the University of Rome "La Sapienza" for multipurpose applications, offers a number of novel key features, paramount for future safe use of nuclear energy: simple design, modularity, moderate operating parameters (temperatures and pressure), and, above all, completely passive safety systems. At a cost comparable with the current conventional reactors (1840 Euros/kWe), MARS offers decisive advantages in the class of small-to-medium (up to 600 MW) plants for future safe use of nuclear energy not only for electricity generation, but also for industrial heat production, desalination or district heating. Because of its light construction (made totally of steel), simple design using proven technologies for auxiliary systems from chemical and petrochemical industries, the time for erection of such a reactor is much shorter (up to four years) than for conventional reactors. The same features make the reactor dismantling and decommissioning much simpler, much less expensive and faster. Some of the safety features have already been proven by laboratory testing of their components. Recent studies have shown the possibility of prolonging the core life up to five years and of substituting the boron acid control system with burnable poisons.

In view of the increasing world-wide concerns about green gas emissions and consequent climate changes, as well as continuous increase in prices of fossil fuels, it is believed that the MARS reactor will in the near future find its niche as the optimum option in electricity generating and other industries.

References

1. *600 MWth MARS Nuclear Power Plant*, Design Progress Report, University of Rome "La Sapienza", Department of Nuclear Engineering and Energy Conversion, Rome 2003.
2. Weinberg, A.M., *The First Nuclear Era*, American Institute of Physics Press, New York, 1994.
3. IAEA-TECDOC-1485, Status of innovative small and medium sized reactor designs 2005 (reactors with conventional refueling schemes), IAEA, March 2006.
4. Weinberg, A.M., On immortal nuclear power plants, November 19, 2002.
5. Naviglio, A., Improved nuclear technologies under safety profile for more international customers, in *The Pacific Use of Nuclear Energy Since Geneve 1995 till Today: The Italian Case*, Rome, 8–9 March, 2006.
6. Cumo, M., Naviglio, A. and Sorabella, L., MARS, 600 MWt nuclear power plant, in *Americas Nuclear Energy Symposium (ANES, 2004)*, Miami, October 3–6, 2004.
7. Golfier, H., Caterino, S., Poinot, C., Delpech, M., Mignot, G., Naviglio, A. and Gandini, A., Multipurpose Advanced "inherently" Safe Reactor (MARS): Core design studies, in *PHYSOR 2006, American Nuclear Society's Topical Meeting on Reactor Physics*, September 10–14, Vancouver, BC, Canada, 2006.
8. Caruso, G., Cumo, M., Minervini, G. and Naviglio, A., Experimental investigation of a passive system for heat removal in emergency conditions, in *NURETH-11*, Avignon, France, October 2–6, 2005, Paper No. 124.

Chapter 10
CO_2 Emissions Mitigation from Power Generation Using Capture Technologies

Philippe Mathieu

Abstract. Technologies for capturing CO_2 from power generation, that is separating CO_2 from other components in a gaseous or liquid mixture, are examined from a triple point of view: of impacts of capture on performance, of costs and of pollutant emission. It is shown that there is currently no winner among the main three capture options, namely post-combustion removal of CO_2 from flue gas, pre-combustion carbon removal from fuel and from oxy-combustion power systems. Although post- and pre-combustion capture is an available technology in industrial applications, it must still be scaled up to the size of large power plants and from that point of view the three options are not at the same stage of development. Pilot plants are expected to be in operation in the period 2010–2015, especially in European funded projects, Europe being in the forefront of reducing strongly CO_2 emissions from power plants.

Key words: CO_2 emission reduction, capture technologies, zero emission systems.

10.1 Introduction

It is suspected that human activities are altering the global climate by increasing the concentration of heat trapping gases in the atmosphere. Alongside water vapour, the largest contributor to the global warming amongst these gases is CO_2 (namely 60% of the six identified anthropogenic greenhouse gases: CO_2, methane, NO_2, 2 ChloroFluoroCarbons and SF_6), which is mainly produced by the combustion of fossil fuels. The UN Framework Convention on Climate Change's ultimate objective, approved by 188 nations including the USA and Russia, is to achieve "a stabilisation of greenhouse gases (GHG) concentrations in the atmosphere at a level that would prevent dangerous anthropogenic interference with the climate system". However the UNFCCC did not indicate a specific level of stabilisation. This is an essential point in the following discussion since what has to be done strongly depends on that

K. Hanjalić et al., Sustainable Energy Technologies: Options and Prospects, 195–205.

level. In Kyoto, in 1997, a protocol was adopted as a first step towards the reduction of greenhouse gases emissions by around 5% compared to its 1990 emissions taken as the baseline by 2008–2012. In that framework, Europe committed itself to reduce its emissions by 8% shared amongst the 15 countries, Belgium having to reduce its emissions by 7.5%. The Kyoto objective could be achieved by implementation of the following technological options:

1. increase of the efficiencies of conversion of primary energy both on the supply and demand sides while setting up a reduction of the demand growth and of the consumption of end-users;
2. switch to low carbon-intensive fuels (replacement of coal by natural gas and by biomass);
3. highest possible use of renewable energies (nuclear energy is not mentioned);
4. increase of natural sinks (plants, soils, forests, ocean, etc.).

Today and in the near future, however, the world is not on the way to GHG concentration stabilisation and certainly not to reduction of GHG emissions. Europe will not be able to comply with the Kyoto target by 2010. As an illustration, if the current emissions are projected up to 2010, Europe-15 will be 7.5% pts below the target (reduction by 0.5% below the 1990 level instead of 8%) and Belgium will be at +23% instead of −7.5% [1]. In addition, the real environmental target is asking for a seriously deeper reduction than that considered in the Kyoto protocol. If a specific level of stabilisation of CO_2 concentration is set at 550 ppmv [2], a reduction in the CO_2 global emissions by 75 to 90% compared to the 1990 baseline by 2100 is required and the emissions should start to decrease by 2030.

For substantial reduction of worldwide CO_2 emissions by 50% and more from 2030 on, capture and storage technologies such as Carbon Capture and Seque- stration (CCS) offer a particularly well suited option for power generation and other heavily polluting industrial sectors in order to fill the gap between the Kyoto scenario and a 50% and more emissions reduction.

The idea of CCS technology is to separate CO_2 from other gases (e.g. nitrogen, hydrogen) and to store it away from the atmosphere. Power plants are the best candidates for CCS since they are large centralised sources contributing to 35% of the world emissions in 2001. The purpose of CO_2 capture is to produce a concentrated stream from which CO_2 is extracted and then, if needed, compressed above the supercritical pressure for its transportation to a storage site. Many of the technologies for CO_2 capture require either a removal of air pollutants such as SO_2, NOx, heavy metals (mercury) and fine particulates before the capture unit, or simultaneously remove them in the CO_2 capture process, and thus contribute to clean air emission from fossil fuel use at the same time. As fossil fuels are predicted to remain the dominant form of primary energy used worldwide (about 85%) and are responsible for about 75% of anthropogenic CO_2 emissions for the next decades [3], the CCS technology allows us to continue using them in a clean way.

Note that storage of the total flue gas is not cost-effective (because of the high amount of N_2 to compress and store). The removal of CO_2 from flue gases and fuel gas for storage is hence unavoidable. Considering storage, the capacity of CO_2

storage sites is very large in the deep ocean or in geological reservoirs such as deep saline aquifers, depleted gas and oil reservoirs, unminable coal beds, with possible enhancement of oil and methane recovery from oil fields and coal seams respectively using CO_2 injection instead of water. The single existing storage at commercial scale at present is a deep saline aquifer under the North Sea, operated by Statoil (Norway) since 1996 with a CO_2 injection rate of 1 Mtonne per year [4]). From this experiment, the permanence, the leakage rate and safety of the storage should be demonstrated. It should be noted that transport and storage are comparatively less expensive than capture (typically 20% of capture and compression cost, but this is very much site dependent).

CO_2 is routinely separated today at some large industrial plants such as natural gas processing hydrogen and ammonia production facilities, although these plants remove CO_2 to meet process demands and not for storage. Today CO_2 capture has been applied to several small power plants also. However, there are no current applications at large-scale power plants that are the major source of CO_2 emissions.

For power plant applications, the main options for CO_2 capture may be divided into three categories [5, 6]:

- de-carbonisation of the flue gas from energy-intensive industry : power generation, refinery, chemistry, cement;
- de-carbonisation of the fuel gas; and
- oxy-fuel combustion or near-zero emission technologies.

The first two options are based on a de-carbonisation process, that is on the removal of a large amount of carbon atoms injected via the fuel in the system, either from the flue gas or from the fuel itself, generally in the form of CO_2. The third option is based on the de-nitrification of air, that is on the separation of oxygen from nitrogen in air.

Separation of components from mixtures, either CO_2 from N_2 in the flue gas de-carbonisation, or CO_2 from hydrogen in fuel de-carbonisation or be it O_2 from N_2 in an air separation unit, is of course at a cost. From thermodynamics, these separation processes induce serious penalties on the performance and significant additional capital and operation and maintenance costs.

10.2 Flue Gas De-Carbonisation

In post-combustion capture or flue gas decarbonisation (Figure 10.1), a large fraction of the CO_2 generated in the combustion process is removed from the flue gas (air + water). CO_2 is at a low partial pressure of CO_2 in the flue gas, namely 3–5% volume for natural gas and 10–15% volume for coal, in a mixture with 85–97% N_2 so that the separation will be costly both in energy and money. For CO_2 separation from the stream, one of the four following processes can be used:

- chemical and physical absorption with suitable solvents,
- cryogenics,

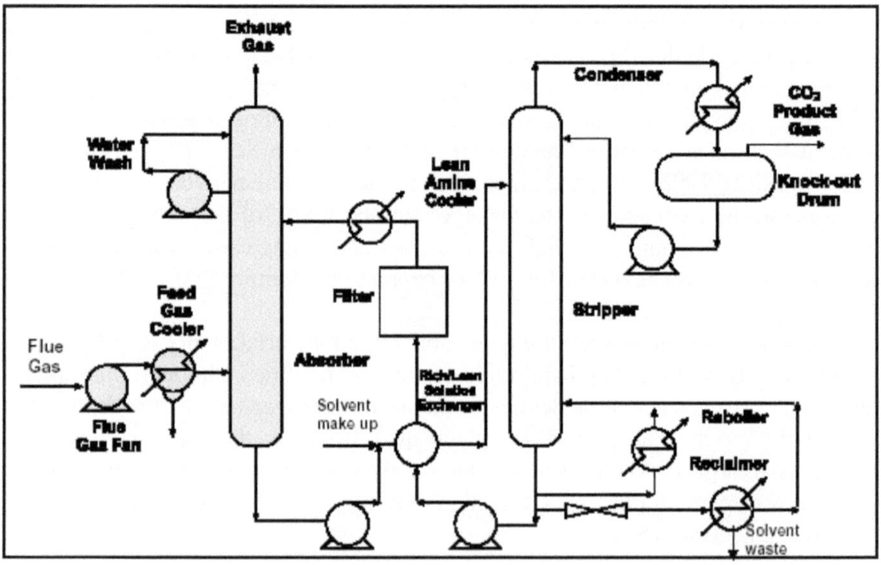

Fig. 10.1 Absorber/stripper for CO_2 removal from flue gas using a solvent. Heat is provided at the reboiler for CO_2 desorption from the rich solution. The CO_2 product is compressed above 80 bar.

- membranes, and
- selective adsorption solids.

The CO_2 chemical absorption technique is best adapted to small CO_2 concentrations at atmospheric pressure and large flue gas mass flow rates. It will be available in the near future (possibly within this decade) for large centralised sources. The other three techniques are not well suited to large power plants because they require higher CO_2 partial pressures and smaller throughputs. CO_2 scrubbing is just like flue gas de-sulfurisation (FGD) and, as such, can be used in retrofit of existing (coal-fired) boilers, at low capital cost. This technique "saves" the boiler (no change of the plant itself). Currently, the CO_2 retention rate is typically 80–85% but, with large scrubbers, may reach 90% of the amount of CO_2 generated in combustion. Most of the acid gases NOx and SO_2 (FGD) have to be removed from the flue gas before entering the absorber to minimise consumption of solvent.

After water condensation, the separated CO_2 being at atmospheric pressure has still to be compressed up to 135 bar where CO_2 is in supercritical or liquid state (critical point: 73 bar; 30°C). The energy requirements in the form of steam extracted from the steam cycle (low-pressure turbine) are high and typically around 4 GJ/t CO_2 captured so that the performance is heavily penalised, namely 20–25% decrease of efficiency and power output, which means about 10% efficiency points drop for coal and gas fired power plants. Chemical scrubbing is commonly used for CO_2 removal from natural gas (separation from methane) and from water shifted syngas (CO_2 + water), but despite existing in a few small units, it is *not* available *yet* for removal from flue gas at the scale of large power plants.

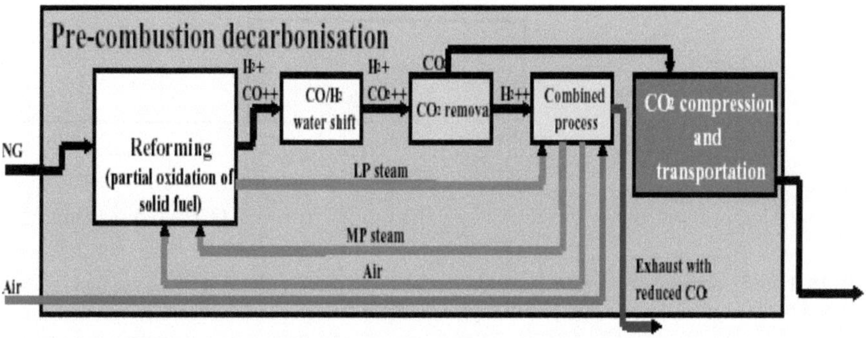

Fig. 10.2 Pre-combustion capture with natural gas reforming or coal partial oxidation, shift of the syngas CO + H_2 into CO_2 + H_2 using water and separation of CO_2 and H_2 in a scrubber. H_2 in mixture with N_2 is used as the fuel of the gas turbine in a CC. Courtesy: IEA GHG.

10.3 Fuel Gas De-Carbonisation

In pre-combustion capture or fuel de-carbonisation (Figure 10.2), a large fraction (90 to 95%) of carbon is extracted from the fuel before its combustion. This technique is based on the gasification of fossil fuels, as well as biomass, and their conversion in a synthetic gas, a mixture of mainly CO and hydrogen. CO may be converted with water into CO_2 and H_2. The mixture may then be efficiently separated, leaving H_2 as a fuel and CO_2 for possible re-use and storage. In a multi-generation plant, H_2 and CO_2 are chemicals co-generated with electricity and heat. H_2 may be used in fuel cells both for transport and power generation and CO_2 for enhanced oil and natural gas recovery from oil, gas and coal reservoirs.

Although this technology is also available from chemical engineering, its adaptation to solid and liquid fossil fuels partial oxidation and natural gas reforming at the power generation scale still requires further R&D efforts. On the other hand, the best suited power plants (integrated gasification combined cycle or IGCC) are not commercially available yet so that IGCC with pre-combustion capture is not expected before 2020.

Many commercial H_2 and ammonia plants based on gasification are in operation and produce pure H_2 and CO_2. Gas turbines firing H_2-rich fuel gas (H_2 + N_2) are currently under development and near to commercial availability for advanced heavy duty gas turbines (General Electric's experience is however mostly on older low-power gas turbines.) CO_2 removal from the fuel is less penalising than in flue gas de-carbonisation: 10–15% capacity decrease, which means 5–8% efficiency points drop. Gasification based chemical plants and IGCC, even without CO_2 capture, are complex processes and thus expensive and not suitable to off-design and transients. They also require much space on the ground and are still at the demonstration level (Buggenum, NL and Puertollano, SP, in Europe). A significant asset is the opening of the pathway towards an H_2 economy.

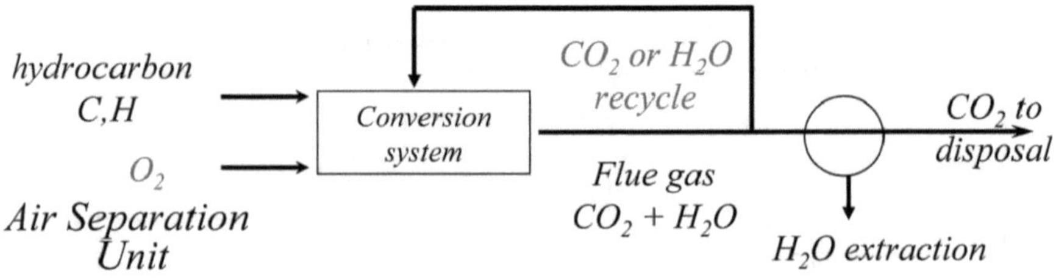

Fig. 10.3 Basic principle of a zero emission power cycle using pure oxygen as the fuel oxidiser and one of the combustion products as the cycle working fluid.

10.4 Oxy-Fuel Combustion

The basic idea behind oxy-fuel (implying a mixture of fuel and O_2) based combustion processes and cycles is quite simple (Figure 10.3): use as pure an O_2 stream as possible as the fuel oxidiser in stoichiometric conditions to generate mainly CO_2 and water (H_2O) from the combustion process. If either CO_2 or water generated from the combustion process is used as the working fluid in a thermodynamic cycle (e.g., in a Rankine or Brayton cycle), then CO_2 can be more readily separated from the exhausted working fluid of the power cycle. However, because fuel combustion in pure O_2 generates very high temperatures, the combustion or exhaust products are partly recycled back to the upstream combustion process in order to control the flame temperature and to meet temperature limitations of materials used in the construction of process equipment. Because CO_2 and/or H_2O replaces the nitrogen in air as a diluent, the production of thermal NOx is practically zero. Thus, in oxy-fuel combustion, fuel is burnt in a mixture of nearly pure O_2 and partially recycled flue gas.

Although oxy-fuel combustion is commonly used in the glass and steel-making industries, modern boilers, process heaters and gas turbines cannot work with a mixture CO_2/H_2O as the working fluid without a partial or complete redesign. Technical issues linked to the new working fluid composition impose a need for the adaptation of existing equipment or the development of new combustors, boilers and turbines. There is now a need for the use of an oxygen production plant which induces energy and cost penalties.

Novel gas turbine conversion systems based on this concept are called "Zero Emission Power Plants" (ZEP) and are technically feasible but are not available yet since they require new developments, in particular of combustors and turbomachines. They are expected in the long term (beyond 20 years) and are known as "water or CES cycle", Graz cycle, AZEP cycle and Matiant cycle. This latter cycle has been designed and developed at the University of Liège (Belgium) for 15 years. It consists of a gas turbine operating in a semi-closed and combined cycle using O_2 as the fuel oxidiser and CO_2 as the working fluid. The efficiency of all these cycles is in the range 45 to 50% depending on the configuration.

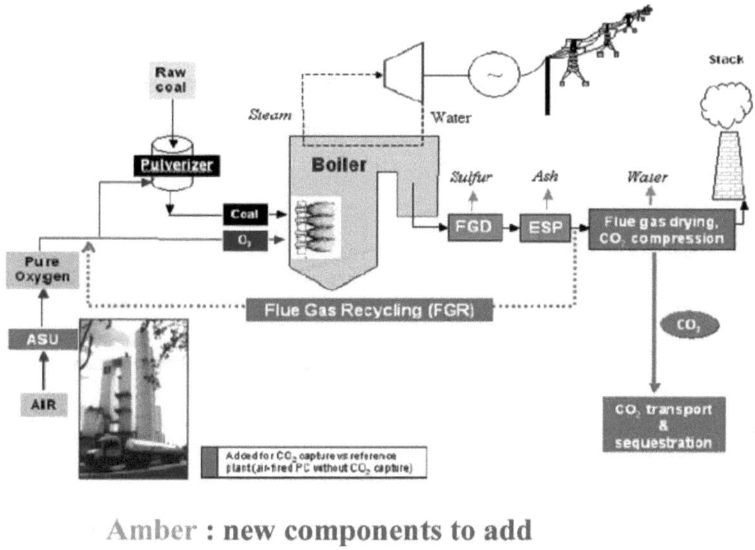

Fig. 10.4 Oxy-fuel combustion steam boiler. The new components to add to an existing "air boiler" are: ASU, flue gas recycling, flue gas drying and CO_2 compressor.

At present, only tests on pilot plant boilers and retrofit of existing boilers and heaters have been carried out (Figure 10.4). This option "saves" the boiler but needs compression of CO_2 to the supercritical state (typically at or above 110'bar). In oxy-fuel combustion, while nearly 100% of CO_2 from combustion can theoretically be removed, the other acid oxides (NOx, SO_2) may be left in the "raw" CO_2 together with other impurities (Ar; N_2; O_2; H_2O). If pure CO_2 is needed for enhanced oil recovery (EOR) or storage, purification of the raw CO_2 gas, its compression and liquefaction and stripping of acid oxides and inerts require more power and capital. Consequently, NOx and SOx could be removed by the purification of the raw CO_2 flow instead of by removal from the flue gas. Here, a high amount of O_2 (more than three times per kWh that in IGCC) is required and despite the avoidance of costly chemical absorption in the flue gas, there is also a high energy penalty due to O_2 production and compression. In other words, separation of CO_2 from N_2 after the combustion, or separation of O_2 from N_2 before combustion, induce roughly the same efficiency drop. As a result, efficiencies of new oxy-fuel combustion systems based on boilers typically drop by 7–8% efficiency points compared to the performance of the same boilers operating on air while releasing nearly no air pollutants.

10.5 Costs of Capture for the Different Options

For all applications considered here, recent studies provide data on the performance and cost of commercial or near-commercial technologies, as well as that of

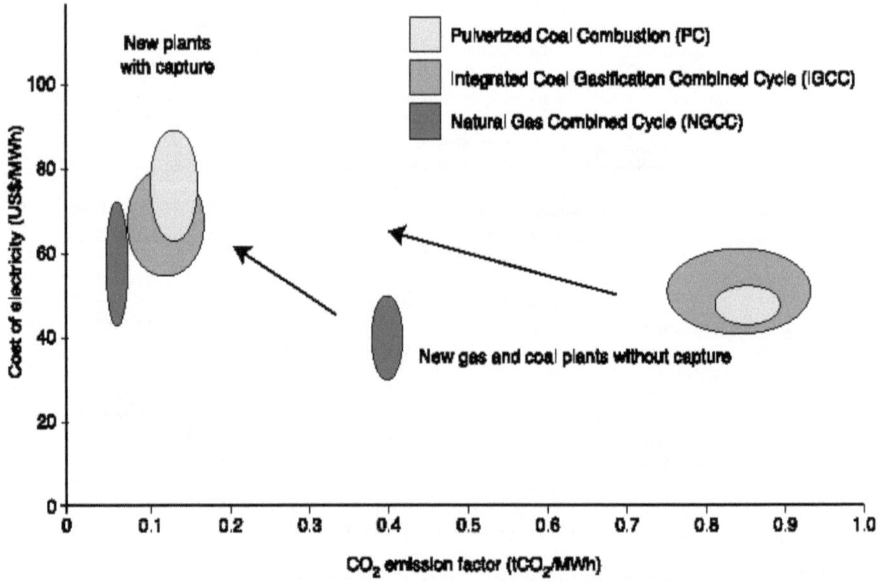

Fig. 10.5 COE *versus* specific emissions g CO_2/kWh for coal (PC and IGCC) and gas (NGCC) power plants without and with capture. Ellipses show the dispersion of data and uncertainties for each technology. The slope of the lines joining ellipses on the right to ellipses on the left provide the capture cost in US$/tonne CO_2 captured.

newer CO_2 capture concepts that are the subject of intense R&D efforts worldwide. For power plants, current commercial CO_2 capture systems can reduce CO_2 emissions per kilowatt-hour (kWh) by 85–90%. The cost of electricity production (COE) increases on average by 70% (60–85%) for a new pulverised coal (PC) plant, by 50% (30–70%) for a new natural gas combined cycle (NGCC) plants, and by 40% (20–55%) for a new integrated gasification combined cycle (IGCC) plant using coal (Figure 10.5). Overall, the COE for fossil fuel plants with capture ranges from USc$4.1 to 8.7 per kWh. The production cost from windmills being in the range 4 to 8 c$/kWh, the production cost of capture plants is currently in the same range as that of the most cost-effective renewable energies. In terms of the cost of avoidance of one tonne of CO_2 , namely the ratio of increase in cost of electricity and of emission reduction due to capture, capture cost amounts typically to 40 to 60 $/tonne CO_2 avoided, depending on technological factors (efficiency, capacity factor, capture process, site conditions) and economic factors (capital, O&M and fuel costs; discount rate).

10.6 Conclusions

Capture is a proven and mature technology but has now to be adapted to large installations such as in power generation. It has a rather well-known cost and can penetrate the market in a near term only with financial incentives. R&D show much room for cost reductions. In Europe, which is at the forefront, some 10 pilot plants have now been approved and expected to come into operation in 2010–2015. Pilot plants on post-combustion capture (Ejberg in Denmark) and on oxyfuel combustion boilers (Schwarze Pumpe, Berlin) are already in operation for the first or in construction for the second (operation expected by 2008). Oxy-fuel combustion allows the use of fossil fuels with nearly zero emission in the atmosphere at a cost similar to that of post-combustion capture for boilers, while for gas turbine systems it will appear later than de-carbonisation techniques since it requires completely new designs of turbo-machines. Retrofit of existing boilers and process heaters with oxy-fuel combustion is technically feasible. CO_2 capture could also advantageously be used in oil refining, in gas and chemical industries, and that at low cost as the CO_2 concentrations are much higher than in flue gas of power plants. Transportation of CO_2 is currently commercially available as it is already made in pipelines and tanker ships and trucks. Injection of CO_2 underground is well mastered by oil and gas producers, for instance for enhancement of oil recovery (Canada, USA). The cost of sequestration (transport and storage) depends of course on the option, but it is generally much cheaper than capture.

Capture technologies also open the way for large-scale production of low-carbon or carbon-free fuels, such as hydrogen in IGCC plants, for transportation and stationary small-scale or distributed applications, in fuel cells and gas turbines. At the same time, however, the energy requirements of current CCS systems significantly reduce the overall efficiency of power generation and other processes, leading to increased fuel requirements, solid wastes, and environmental impacts related to primary fuels production and distribution. Reductions in energy requirements and in costs is now the future objective. The target of EU framework programmes is a cost reduction by 50%, i.e., a capture cost of 20 to 30 $/tonne CO_2 avoided.

In conclusion, efficiency improvements and CO_2 capture and storage together should make it possible to reduce drastically CO_2 emissions in a cost-effective way.

References

1. EEA, European Environment Agency, Greenhouse Gas Emissions Trends and Projections in Europe, Environmental Issue Report 36, Copenhagen, 2003.
2. IPCC, *Climate Change 2001 – Mitigation. The Third Assessment Report of the Intergovernmental Panel on Climate Change*, B. Metz, O. Davidson, R. Swart, J. Pan (Eds.), Cambridge University Press, Cambridge, UK, 2001, 752 pp.
3. IPCC, *Climate Change 2001 – The Scientific Basis. The Third Assessment Report of the Intergovernmental Panel on Climate Change*, J.T. Houghton, Y. Ding, D.J. Griggs, M. Nogouer,

P.J. van der Linden, X. Dai, K. Maskell, C.A. Johnson (Eds.), Cambridge University Press, Cambridge, UK, 2001.

4. Gale, J.J., Christensen, N.P., Cutler, A. and Torp, T.A., 2001, Demonstrating the potential of geological storage of CO_2: The Sleipner and GESTCO projects, *Environmental Geosciences* **8**(3), 2001, 160–165.
5. Hendriks, C.A., Bok, K. and Turkenburg, W.C., The recovery of CO_2 from power plants, *Energy* **16**(11/12), 1989 1277–1293.
6. IEA GHG, *Leading Options for the Capture of CO_2 Emissions at Power Stations*, Report No. PH3/14, IEA Greenhouse Gas R&D Programme, Cheltenham UK, February 2000.
7. IEA GHG R&D, *Decarbonization of Fossil Fuels*, Report No. PH2/2 IEA GHG, Putting Carbon back in the Ground. IEA Greenhouse Gas R&D Programme, Cheltenham, UK, 2001.
8. IEA GHG, *Opportunities for Early Application of CO_2 Sequestration Technology*, Report Ph4/10. IEA Greenhouse Gas R&D Programme, Cheltenham, UK, 2002.
9. Bolland, O. and Undrum, H., Removal of CO_2 from gas turbine power plants: Evaluation of pre-and post combustion methods, in *Proceedings 4th International Conference on Greenhouse Gas Control Technologies*, Interlaken, Switzerland, B. Eliasson, P. Riemer, A. Wokaun (Eds.), Elsevier, 1998.
10. Chapel, D.G., Mariz, C.L. and Ernest, J., Recovery of CO_2 from flue gases: Commercial trends, in *Annual Meeting of the Canadian Society of Chemical Engineering*, Saskatoon, Canada, October 1999.
11. Mimura, T., Nojo, T., Iijima, M., Yoshiyama, T. and Tanaka, H., Recent developments in flue gas CO_2 recovery technology, in *Proceedings 6th International Conference on Greenhouse Gas Control Technologies*, Kyoto, Japan, October, Elsevier Publishers, 2002.
12. Rao, A.B. and Rubin, E.S., A technical, economic, and environmental assessment of amine-based CO_2 capture technology for power plant greenhouse gas control, *Environmental Science and Technology* **36**, 2002, 4467–4475.
13. Sander, M.T. and Mariz, C.L., The Fluor Daniel® Econamine™ FG process: Past experience and present day focus, *Energy Conversion and Management* **33**(5/8), 1992, 341–348.
14. Simbeck, D., A portfolio selection approach for power plant CO_2 capture, separation and R&D options, in *Proceedings 4th International Conference on Greenhouse Gas Control Technologies*, 45, Interlaken, Switzerland, B. Eliasson, P. Riemer, A. Wokaun (Eds.), Elsevier, 1998.
15. Feron, P.H.M., *Carbon Dioxide Capture: The Characterisation of Gas Separation/Removal Membrane Systems Applied to the Treatment of Flue Gases Arising from Power Plant Generation Using Fossil Fuel*, IEA/92/08, IEA Greenhouse Gas R&D Programme, Cheltenham, UK, 1992.
16. Chiesa, P., Consonni, S., Kreutz, T. and Williams, R., Co-production of hydrogen, electricity and CO_2 from coal with commercially ready technology. Part A: Performance and emissions, *International Journal of Hydrogen Energy* **30**, 2003, 747–767.
17. IEA GHG, *Potential for Improvement in IGCC with CO_2 Capture*, Report PH4/19, May 2003, IEA Greenhouse Gas R&D Programme, Cheltenham, UK, 2003, 788 pp.
18. Jordal, K., Bredesen, R., Kvamsdal, H.M. and Bolland, O., 2002 Integration of H_2-separating membrane technology in gas turbine processes for CO_2 sequestration, *Energy* **30**(9/10), 2002, 1269–1278.
19. Todd, D.M. and Battista, R.A., Demonstrated applicability of hydrogen fuel for gas turbines, in *Proceedings of 4th European Gasification Conference*, 11–13 April 2001, Noordwijk, the Netherlands, 2001.
20. Mathieu, P., Mitigation of CO_2 emissions using low and near zero CO_2 emission power plants, *Clean Air, International Journal on Energy for a Clean Environment* **4**, 2003, 1–16.
21. Bolland, O., Boden, J. and Kvamsdal, H.M., A thermodynamic comparison of the oxy-fuel power cycles: Water-cycle, Graz-cycle and Matiant-cycle, *Proceedings of International Conference on Power Generation and Sustainable Development*, Liège, Belgium, 8–9 October, Association des Ingénieurs sortis de Liège, 2001.
22. Wilkinson, M.B., Simmonds, M., Allam, R.J. and White, V., Oxy-fuel conversion of heaters and boilers for CO_2 capture, in *Proceedings of 2nd Annual Conference on Carbon Sequestration*, Virginia, USA, 2003.

23. Chatel-Pelage, F., Ovidiu, M., Carty, R., Philo, G., Farzan, H. and Vecci, S., A pilot scale demonstration of oxy-fuel combustion with flue gas recirculation in a pulverised coal-fired boiler, in *Proceedings 28th International Technical Conference on Coal Utilization & Fuel Systems*, Clearwater, FL, March 10–13, 2003.
24. Stromberg, L., Options for CO_2 free coal based power generation – Timing, technology and economics, in *Proceedings of Euro-CASE Workshop: CO_2 Management in Europe* The Norwegian Academy of Technological Sciences, Oslo, May 16.
25. Anderson, R., Brandt, H., Doyle, S., Pronske, K. and Viteri, F., Power generation with 100% carbon capture and sequestration, in *Proceedings of 2nd Annual Conference on Carbon Sequestration*, Alexandria, VA, 2003.
26. Nsakala, N., Liljedahl, G., Marion, J., Bozzuto, C., Andrus, H. and Chamberland, R., Greenhouse gas emissions control by oxygen firing in circulating fluidized bed boilers, in *Proceedings 2nd National Conference on Carbon Sequestration*, Alexandria, VA, May 5–8, 2003.
27. Mathieu, P., The Matiant cycle: An innovative zero-emission technology, in *PowerGen Europe98*, Milan, Italy, 1998.
28. Heitmer, F. and Jericha, H., Graz cycle – An optimised power plant concept for CO_2 retention, in *Proceedings of First International Conference on Industrial GT Technologies*, RTD Framework Programme of the European Union, Brussels, July, 2003.
29. Panesar, R., Gibbins, J., Allam, R.J., White, V. and Wall, A., Oxy-combustion processes for CO_2 capture from power plant, in *Proceedings of 7th International Greenhouse Gas Control Technologies Conference*, Vancouver, 2004.
30. Bill, A., Span, R., Griffin, T., Kelsall, G. and Sundkvist, S.G., Technology options for 'Zero Emissions' gas turbine power generation, *Proceedings of International Conference Power Generation and Sustainable Development*, Liège, Belgium, 8–9 October, 2001.
31. Rubin, E.S. and Rao, A.B., Uncertainties in CO_2 capture and sequestration costs, in *Proceedings of 6th International Conference on Greenhouse Gas Control Technologies (GHGT-6)*, Kyoto, Japan, October, Elsevier Publishers, 2002.
32. Herzog, H.J., The economics of CO_2 capture, in *Proceedings of the 4th International Conference on Greenhouse Gas Control Technologies*, Interlaken, Switzerland, B. Eliasson, P. Riemer, A. Wokaun (Eds.), Elsevier, 1998, pp. 101–106.
33. Rao, A.B., Rubin, E.S. and Morgan, M., Evaluation of potential cost reductions from improved CO_2 capture systems, in *Proceedings of 2nd Annual Conference on Carbon Sequestration*, Alexandria, VA, May 5–8, 2003.
34. Freund, P. and Davison, J.E., General overview of costs, in *Proceedings of IPCC Workshop*, Regina, Canada, 2002.

Chapter 11
Clean and Efficient Coal Technology Integrated with CO_2 Sequestration and Hydrogen Energy Systems

Ken Okazaki

Abstract. In view of the threatening global environmental issues and depletion of fossil fuel resources, a hydrogen energy-oriented society is expected to be realised around 2030–2050 supported by the introduction of a large amount of renewable energy. To achieve this, however, it will take a long time and a realistic intermediate scenario is needed which is practical and socially acceptable. For this purpose, a new system integration for CO_2-free coal utilisation combined with hydrogen energy and CO_2 sequestration is proposed and discussed. The method is based on the present status and future prospect of various related technologies in Japan including O_2/CO_2 (oxy-firing) coal combustion with CO_2 recovery, hydrogen gas turbine power station, highly sophisticated utilisation of hydrogen and CO_2 sequestration.

Key words: Clean coal technology, oxy-firing, carbon capture and sequestration (CCS), global warming, hydrogen energy, integrated energy system.

11.1 Introduction

CO_2-induced global warming, originating from the abnormally large and rapid consumption (combustion) of fossil fuel, has emerged only in the last few decades. To suppress the ever-growing global warming caused by CO_2, there is an urgent need to develop technologies which will reduce the quantity of CO_2 emissions on a global scale. In view of the depletion of energy resources such as oil and natural gas, technologies are needed that efficiently use the energy of coal, which has relatively long-lasting reserves and is reasonably evenly distributed throughout the world, while solving the problems of global warming and acid rain.

Japanese coal utilisation technologies are adopted by the most advanced pulverised coal-fired power plants and other power plants (power generated by coal-fired power plants is twice that by oil-fired power plants), emitting the world's smallest quantities of NOx, SOx, and fine particles (1/10th of NOx and one twentieth of SOx

K. Hanjalić et al., Sustainable Energy Technologies: Options and Prospects, 207–225.

compared with other developed countries). Thus, the clean coal technologies (CCT) in Japan have almost reached a mature stage. Nevertheless, there remains the serious issue of large quantities of CO_2 emissions. To address the issue, various technologies are being developed to actively recover and sequester CO_2 from combustion flue gas in large-scale coal-fired power stations. These technologies, however, consume energy for processing and so reduce net efficiency by about 10% (from 40% to 30%) and thus increase the depletion of fossil resources, even though CO_2 emissions are reduced. Several systems to solve the problem have been proposed. For example, CO_2 is separated and recovered in the course of Integrated Coal Gasification Combined Cycle Power Generation (IGCC) and Integrated Coal Gasification Fuel Cell Combined Cycle Power Generation (IGFC) which operate at high efficiency, through the shift reaction generating CO_2 and H2, thus achieving a total efficiency equivalent to or higher than that of the current pulverised coal-fired power plants while significantly reducing the CO_2 emissions. Those technologies are now at the stage of feasibility studies, and some are entering the technology development stage.

Furthermore, the CO_2-free high-efficiency coal utilisation technology which integrates into one system, the production of hydrogen from coal, the advanced utilisation of hydrogen, and CO_2 sequestration, has recently attracted attention as technologies for using hydrogen energy have rapidly progressed. The DOE of the US announced the FutureGen Project (1 billion dollars in 10 years) in February 2003 which has had a strong impact worldwide. The FutureGen Project is a clear step toward the above-described integrated technology linking with the Climate Change, Hydrogen Initiative (1.2 billion dollars in 5 years), which was announced by President Bush in January 2003, although there are political motives characteristic of the US which has large coal reserves. Since, however, Japan has long led the world in individual technologies relating to coal, hydrogen, and CO_2 sequestration, Japan should take the initiative in developing the integrated system technology.

This chapter describes the key issues and countermeasures for global warming, significance of CO_2 recovery and sequestration and technology trends, and outlines the significance of introducing hydrogen energy, their current state and future prospects. The chapter then discusses integrated technologies to prevent global warming by combining core technologies of coal, hydrogen, and CO_2 sequestration. Until the use of fossil fuels can be eliminated around 2030–2050 by introducing a large amount of renewable energy combined with technologies for using hydrogen as an energy carrier, this system integration technology could play a most important role in suppressing CO_2 emissions, which cannot be dealt with solely by efficiency improvements, in order to achieve the interim scenario for global environmental conservation while depending on fossil fuel.

11.2 Viewpoint of Energy and Environmental Countermeasures

When discussing energy and environmental measures, the judgment and evaluation significantly differ depending on the viewpoint: either the large, stationary, and centralised type such as large thermal power plants, or the small, mobile, and distributed type such as vehicles (facility characteristics and size); either local clean or global clean (space scale); either short period, medium period, or long period (time scale); and either core technologies or system technology (independent, combined, integrated, or global system). These aspects must be fixed before conducting an evaluation and mutual comparison of global environmental measures.

Furthermore, comprehensive environmental conservation measures should combine various options and include a correct understanding of: the magnitude of contribution to the objective in terms of quantity, the magnitude of net contribution in terms of LCA, and the quality of obtained energy (high for electricity and low for heat). Care should be taken to avoid the illusion that an apparently fancy technology that cannot make a large contribution can solve global warming caused by large global CO_2 emissions. Countermeasures for global warming are substantially different from the local cleaning of islands, villages, and towns.

11.3 CO_2 Recovery and Sequestration Technologies

11.3.1 Comparison of CO_2 Mitigation Technologies and Significance of CO_2 Recovery and Sequestration

In general, measures to reduce CO_2 emissions include the following:

1. energy saving and improvement in energy conversion and use efficiency;
2. switching fuel to low carbon light fuels;
3. full-scale introduction of renewable energy;
4. increase in utilisation of nuclear power energy; and
5. intentional sequestration of CO_2 (separation, recovery, and storage).

Although measure (2) might have short-term effects, it significantly increases the depletion of oil and natural gas resources and unavoidably increases the dependence on coal in the medium term. Measure (3) offers only a small contribution compared with the large amount of fossil fuel consumption at present (natural energy in Japan accounts for less than 0.2% of primary energy). Measure (4) has various difficulties in view of recent world concern about nuclear power, though it reduces CO_2 and is technologically feasible. Measure (4), however, is at the investigation stage as nuclear power is expected to be integrated with hydrogen.

On the other hand, measure (1) involves the development of existing technologies and is the most realistic, and various national projects are underway. However, saving energy and improving efficiency alone cannot compensate for the increase in CO_2 emissions that accompanies the increase in energy consumption. Therefore, measure (5) of intentionally sequestrating CO_2 (separation, recovery, and sequestration of CO_2 derived from thermal power plants and other sources), will become necessary in the future, even though the measure is not a fancy technology and ap-

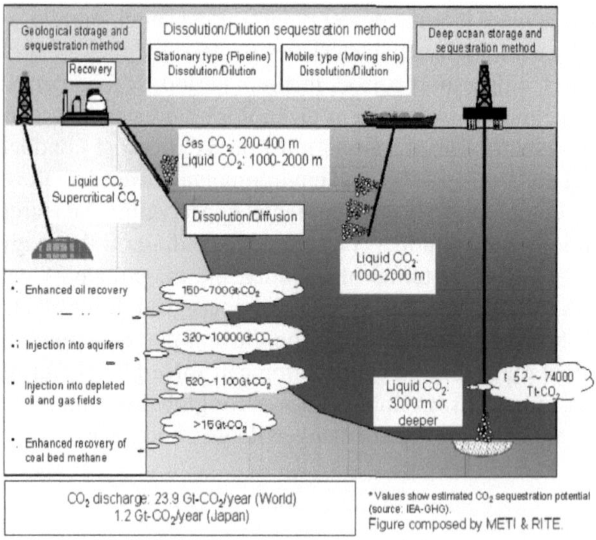

Fig. 11.1 Various CO_2 sequestration technologies and capacity for CO_2 storage.

pears to be avoiding the problem. Accordingly, measure (5) is investigated from various standpoints as a potential future option that can significantly reduce CO_2.

11.3.2 Sequestration of Huge Quantities of CO_2 from Large and Stationary Sources

Figure 11.1 shows typical examples of deliberately fixing large quantities of CO_2. One example is ocean storage using the oceans as a huge reservoir, and another example is geological sequestration by injecting CO_2 into underground aquifers and depleted oil and gas fields. The ocean storage method is classified into: the intermediate layer dissolving method in which the CO_2 emitted from combustion plants such as thermal power plants is recovered and liquefied to dissolve into an intermediate depth of about 1000 m; and the deep ocean storage method in which liquid CO_2 is stored for a long time in depressions on the bottom of the deep ocean at 3000 m or deeper where liquid CO_2 becomes heavier than seawater. In both cases, a crystalline membrane called a clathrate hydrate is formed at the interface between the liquid CO_2 and seawater, which significantly influences the behaviour of CO_2 dissolution.

However, there is very little quantitative data on mass transfer under specific conditions. Our laboratory is conducting studies on this issue, both experimentally and theoretically, and our quantitative data has revealed that the CO_2 emitted from a 1 million kW class coal-fired power plant over ten years can be stored in depressions

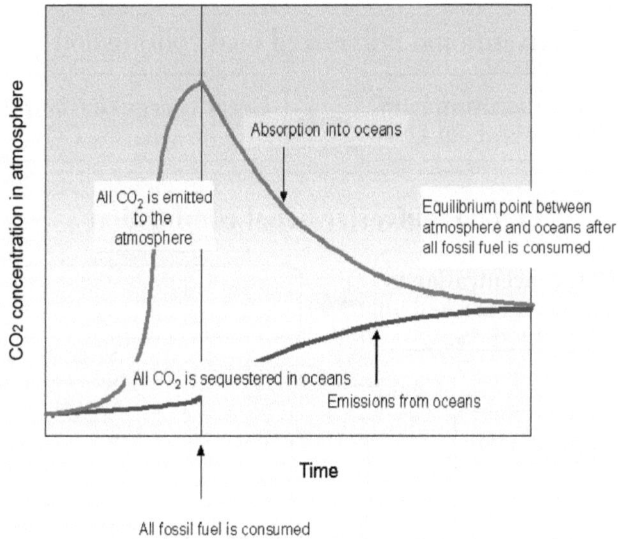

Fig. 11.2 Significance of ocean sequestration of CO_2 for the peak shaving in CO_2 concentration in the atmosphere.

on the bottom of the deep ocean at about 100 m in depth within an approximate area of 1 km × 1 km, and that the storage will last for 200 years or more owing to the clathrate hydrate membrane which suppresses the re-dissolution of CO_2 [1]. It is estimated that the redissolved CO_2 will take a further 1000 years or so until it is emitted to the atmosphere after being carried by ocean currents. Although the method has the drawback of increasing the depletion of fossil fuel resources because of the extra energy required during the treatment process, the method is at the development stage as a potential option that could be combined with other high-efficiency systems because the method can process huge quantities of CO_2.

However, before ocean storage technology can be applied, it is indispensable to prove that the method will not have a significant influence on the ocean ecosystem. According to the project of "Study of Environmental Assessment for CO_2 Ocean Sequestration for Mitigation of Climate Change (SEA-COSMIC)" which has been underway since fiscal 1997 led by the Ministry of Economy, Trade, and Industry (METI), and the Research Institute of Innovative Technology for the Earth (RITE), the development of intermediate layer dissolving technology is the main focus, and efforts are being made to correctly predict the new environmental influence on the ocean ecosystem and to develop technology to minimise the influence.

Regarding CO_2 ocean sequestration technology, there is the criticism that discarding waste generated from human activities into the ocean will contaminate the ocean. The criticism is unwarranted. The CO_2 concentration in the atmosphere depends on the balance of mass transfer between the ocean and the atmosphere. In this respect, as shown in Figure 11.2, CO_2 ocean sequestration technology can reduce the rise in sea level and climate changes by eliminating the high concentration

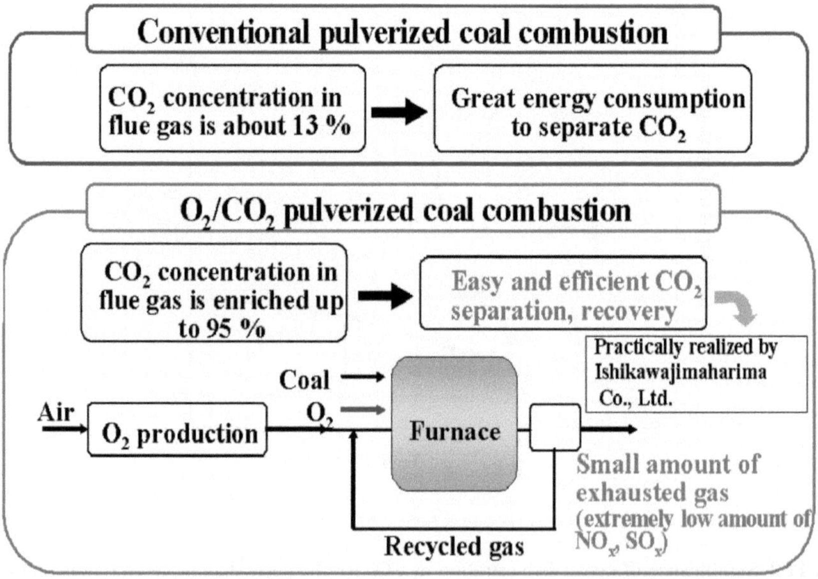

Fig. 11.3 O_2/CO_2 (or oxy-firing) coal combustion for easy CO_2 recovery and drastic reduction of NOx and SOx.

CO_2 peak in the atmosphere that occurs by emitting large quantities of CO_2, by intentionally increasing the inherent natural circulation [2].

11.3.3 CO_2 Recovery from Coal Combustion System

(1) *Easy CO_2 recovery by O_2/CO_2 coal combustion (oxy-firing).* Before CO_2 sequestration technology can enter practical application, a coal-fired power generation technology which efficiently and actively recovers CO_2 from flue gas and sequesters the CO_2 from the atmosphere will be required. The Center for Coal Utilisation, Japan (CCUJ) therefore investigated the technology under contract from NEDO between FY1996 and FY1999. For the conventional case of coal combustion with air, the concentration of CO_2 in flue gas is as low as about 13%, and high-efficiency separation and recovery of large quantities of CO_2 is difficult. For the CO_2 recirculation coal combustion (O_2/CO_2 or oxy-firing coal combustion) shown in Figure 11.3, a pilot plant test conducted at IHI proved that coal combustion in an atmosphere of oxygen and recycled gas gives a CO_2 concentration in the flue gas of about 95%, which allows the CO_2 to be readily and directly liquefied and recovered merely by compressing the gas without passing through the separation stage [3, 4].

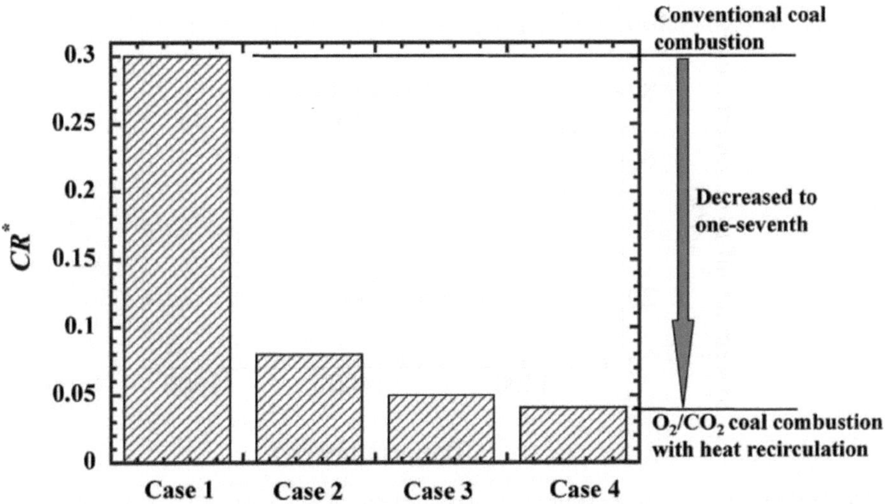

Fig. 11.4 Drastic reduction of conversion ratios, CR*, from fuel-N to NO for various cases in O_2/CO_2 pulverised coal combustion.

(2) *Drastic reduction of NOx in O_2/CO_2 coal combustion.* In addition, it was found that the CO_2 recirculation coal combustion has the unexpected advantage that the conversion ratio of Fuel-N in coal to exhausted NOx is automatically and significantly reduced to one quarter or one sixth of that by conventional air combustion (see cases 2 and 3 in Figure 11.4). We studied the mechanism in detail in our laboratory through both experiments [5] and theoretical analysis [6], and found that the significant reduction in NOx emissions to outside the combustion system is not influenced by the special reaction field of coal combustion in concentrated CO_2, and is mainly caused by the effective reduction of the NOx, enriched by self-recirculation together with the recycled CO_2, on re-entering the combustion flame zone. This is a typical example of significantly improving the characteristics of the entire system by integrating the core technologies rather than by controlling individual phenomena in core technologies.

(3) *Further reduction of NOx in O_2/CO_2 coal combustion with heat recirculation.* However, the combustion intensity of O_2/CO_2 coal combustion becomes lower than conventional combustion due to the high concentration of CO_2 in flue gas. To solve this problem, recently, a new scheme of further NOx reduction, i.e., O_2/CO_2 coal combustion with heat recirculation, was proposed.

According to the chemical stoichiometry of complete burn-out of coal and the overall mass balance, a low O_2 concentration led to a higher gas recirculation ratio, less exhausted flue gas, and accordingly a lower conversion ratio from fuel-N to exhausted NO, i.e., lower CR*. Based on the experimental results and the overall mass balance of the system, the conversion ratio from fuel-N to exhausted NO, CR*, was estimated as shown in Figure 11.4 for various cases [7]. The CR* of

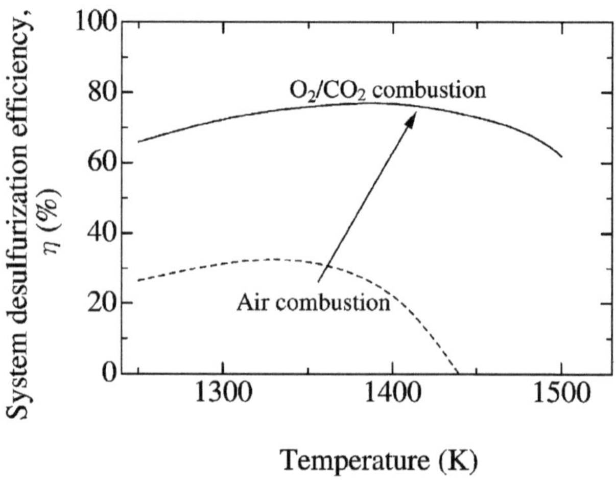

Fig. 11.5 Drastic increase of in-furnace desulphurisation efficiency in O_2/CO_2 coal combustion.

case 3 was decreased to about 1/6th of case 1 (conventional coal combustion), but the combustion intensity of case 3 was much lower than case 1. Cases 4 and 1 had approximately the same combustion intensities. In addition to that, the CR* of case 4, i.e., the new scheme, was decreased to about only one seventh of case 1. Therefore extremely low NO emission can be realised with this new scheme.

(4) *Enhancement of in-furnace desulphurisation in O_2/CO_2 coal combustion.* Moreover, our study through experiment and modeling revealed that the system desulphurisation efficiency in O_2/CO_2 pulverised coal combustion was increased to about four to six times as high as that of conventional pulverised coal combustion as shown in Figure 11.5, mainly attributed to: (1) practical residence time of SO_2 was extended and SO_2 was enriched inside the furnace due to flue gas recirculation; and (2) $CaSO_4$ decomposition was inhibited because of a high SO_2 concentration [8]. The contributions of various factors to the high in-furnace desulphurisation efficiency were quantitatively clarified.

11.3.4 Simultaneous CO_2 Recovery and High-Efficiency Power Generation by the CO_2 Recovery Type IGCC

The CO_2 recovery type coal-fired power generation involves problems: net efficiency is reduced from the 40% of normal pulverised coal-fired power generation to about 30%, and depletion of fossil fuel resources is increased, even if high-efficiency CO_2 recovery and reduction in NOx and SOx emissions are simultaneously attained. To solve these problems, a system of modifying the high-efficiency Integrated Coal Gasification Combined Cycle Power Generation (IGCC) to the CO_2 recovery type

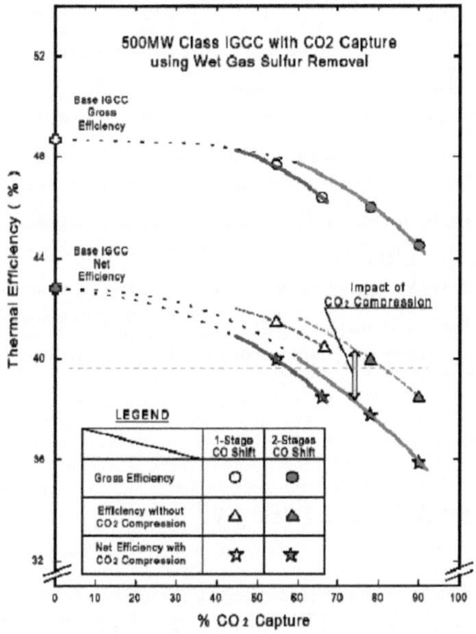

Fig. 11.6 Net efficiency of IGCC with active CO_2 recovery.

was studied, in order to simultaneously achieve a high CO_2 recovery percentage and power generation efficiency equal to or higher than that of the conventional process [9]. As shown in Figure 11.6, a 1300°C class gas turbine has a power generation efficiency of about 40% with a CO_2 recovery rate of 60–70% (90% CO_2 recovery rate for 1500°C class), which suggests that the CO_2 recovery type IGCC has good potential for global environment conservation.

11.4 Significance of Introducing Hydrogen Energy and Prospects for Technology Development

11.4.1 Expectations for the Introduction of Hydrogen Energy

The recent strong expectations for the introduction of hydrogen energy are the result of demands for global environmental conservation and for energy resources free from fossil fuel. Therefore, technologies for using hydrogen energy require a realistic scenario to meet these requirements in the long term. Since, however, hydrogen is a secondary energy source, and not a primary energy source like oil, natural gas, and coal, and since hydrogen is obtained through processing (which consumes energy), conservation of the global environment and fossil fuels can be attained only by drastically improving the total efficiency including the primary energy stage, which is higher than the present level; by applying an energy system based on hydrogen; or by using large quantities of renewable natural energy for producing hydrogen. The latter should be the ultimate mode of using hydrogen energy. At present, however, natural energy in Japan accounts for only 0.2% of total primary energy consumption (although the percentage has increased in recent years, photovoltaic power and wind power account for only around 0.01% each), and so the latter case is impossible to attain in the near future. Nevertheless, to solve the global environmental issues and fossil fuel depletion in the long term, a hydrogen energy-oriented society which links these types of natural energy in the future is essential. Thus, a strategy for practical application of hydrogen energy and an interim scenario for developing practical hydrogen technology including links with fossil fuel are needed.

11.4.2 Is Hydrogen a Clean Energy?

To the question of "Is hydrogen a clean energy?", many would answer that hydrogen is the ultimate clean energy because it burns to become water. This is true assuming that the process starts from hydrogen. But as described already, hydrogen is a secondary energy and so is not necessarily clean when the environmental influence is evaluated including the hydrogen production process. For example, assuming that hydrogen is industrially produced from natural gas as the primary energy, and is then used to drive a hydrogen internal combustion engine in a vehicle, about one-third of the primary energy is lost in the hydrogen production stage (including exploitation, liquefaction, and transportation of natural gas), and the net energy used for power compared with the primary energy is about 13% lower than that of existing gasoline-fueled internal combustion engines. As a result, although the exhaust gas itself is clean, the total quantity of CO_2 on a global basis actually increases. Specific cities and towns may be kept clean by producing the hydrogen at a distant place (local clean). Although the local clean scheme is important in itself, it has an adverse influence on global warming (global clean). The situation is the same with the methanol internal combustion engine for vehicles [10]. How about using natural energy such as solar energy and wind power for producing hydrogen? Such natural energy alone cannot produce the quantity required for global environmental conservation for the time being.

Recently, polymer electrolyte membrane type fuel cell vehicles (FCV) using pure hydrogen as the fuel have attracted much attention, and there is a fierce battle to develop FCVs. FCVs are of great significance for global environmental conservation because the fuel cell itself has an efficiency of almost 50% and because the net conversion efficiency to power ("Well to Wheel Efficiency", starting from the primary energy stage), is almost three times that of conventional gasoline internal engine vehicles. There are many other examples of overall systems that contribute to global environmental conservation and saving of fossil fuels, including the drastic improvement in performance of the energy conversion equipment, which can be achieved only by hydrogen, and with hydrogen utilisation as an energy carrier fully utilising the specific thermo-chemical features of hydrogen as described later. In this way, hydrogen will acquire the title of "clean energy".

11.4.3 Various Technology Developments in Hydrogen Utilisation

(1) *Hydrogen fuel cell vehicle.* Hydrogen-based polymer electrolyte membrane (PEM) fuel cell vehicles triggered the boom in hydrogen energy. Hydrogen-based PEM fuel cell vehicles reached a practical stage in the late 1990s when Ballard Co. of Canada successfully improved the power density per unit volume by a factor of 12 after six years of study. The power density exceeded that of gasoline engines, and is now almost 2.2 MW/m^3. Ballard first developed fuel cell buses because the infrastructure for supplying hydrogen can be easily built for buses, and operated three buses in Vancouver and in Chicago for two years from 1998. Other than the efficiency increase issue, there are many technological issues to be overcome before such vehicles can be mass-produced, including cost, durability, and a significant reduction in the consumption of noble metal catalysts such as platinum. Nevertheless, Toyota Motor Corp. and Honda Motor Co., Ltd. began selling fuel cell passenger cars at the end of 2002 for the first time in the world, though only on lease and in limited numbers. Since then, in Japan about 50 fuel cell vehicles started driving on public roads in June 2004, and in August 2003 Tokyo Metropolis Bus introduced some fuel cell buses.

(2) *Hydrogen refueling station.* In 1997, Munich Airport installed a 35 MPa high pressure hydrogen fuel station and a fully automated robot liquid hydrogen fuel station to supply hydrogen to some shuttle buses. The hydrogen is produced by pressurised alkali-water electrolysis, and the stations have metal hydride (MH) tanks as a buffer. The technology has reached the mature level. Other than Munich, various hydrogen fuel stations have already been installed in Vancouver, Chicago, Hamburg, and Palm Springs. Triggered by the California Fuel Cell Partnership (CaFCP) which started in FY2000 (a demonstration project combining vehicles with stations), infrastructure is now being rapidly built worldwide toward the practical use of hydrogen vehicles.

Also in Japan, Task 7 in Phase II of the World Energy Network (WE-NET) which began in FY1999 as a five-year project installed two types of hydrogen production and fuel station of a natural gas reforming type and a PEM water electrolysis type in February 2002 (1/10th of the commercial size, providing 30 Nm^3/h of hydrogen production capacity), at Osaka and Takamatsu for the first time in Japan. WE-NET was, however, abandoned in FY2002 (one year earlier than scheduled). As an alternative plan, the Japan Hydrogen and Fuel Cell Demonstration Project (JHFC) started integrating hydrogen fuel cell vehicles with hydrogen fuel stations at about 10 locations in the Tokyo Metropolitan area, which have different fuel sources and hydrogen production methods. The change of project reflects a change of national

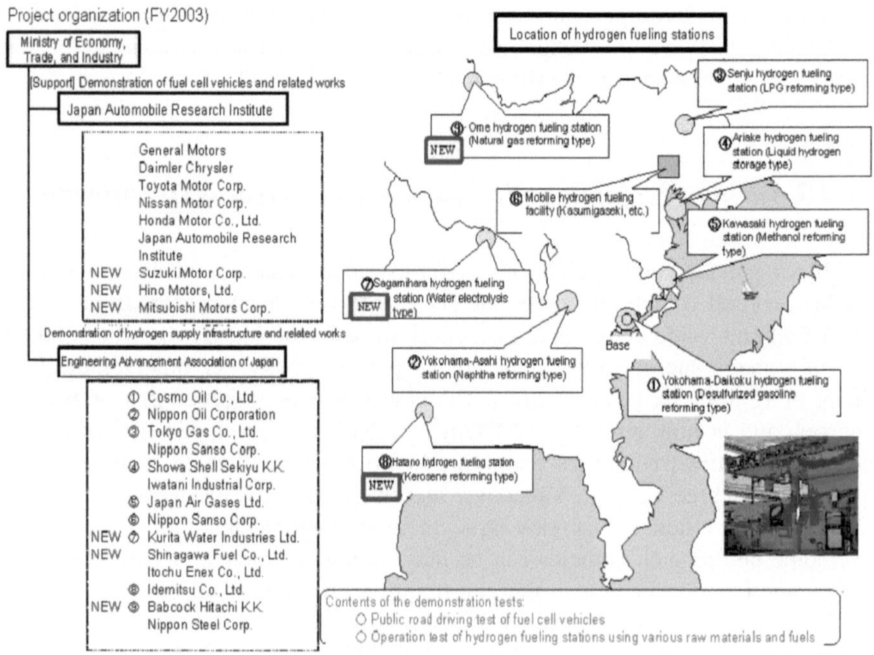

Fig. 11.7 Hydrogen refueling stations near the Tokyo area.

policy which aims to introduce fuel cell vehicles (50,000 cars by FY2010, 5 million cars by FY2020, and 15 million cars by FY2030). Figure 11.7 shows the installation of stations by the end of FY2003. The demonstration tests aim to acquire data on hydrogen production at the stations and driving the fuel cell vehicles on public roads, so that the total efficiency can be identified.

The efficiency of hydrogen fuel stations was disclosed as raw data at the recent JHFC Symposium in March 2005 and 2006 for the first time in the world [11]. In addition, not only individual technologies but also new technological issues as a system were identified through the demonstration tests, which will help to establish technological guidelines for developing more advanced systems for practical application.

To create a hydrogen-oriented society smoothly, a long-term approach to introducing hydrogen energy is essential. However, WE-NET, which is a long-term project including hydrogen production, storage, transportation, and conversion, and which includes system technologies relating to hydrogen utilisation for other than vehicles, was stopped, and hydrogen technology development is now short-term, focusing only on introducing hydrogen vehicles.

(3) *Hydrogen gas turbine power station with CCS (carbon capture and sequestration).* British Petroleum Co. recently presented a new project [12] to install two large-scale hydrogen gas turbine power stations combined with CO_2 sequestration

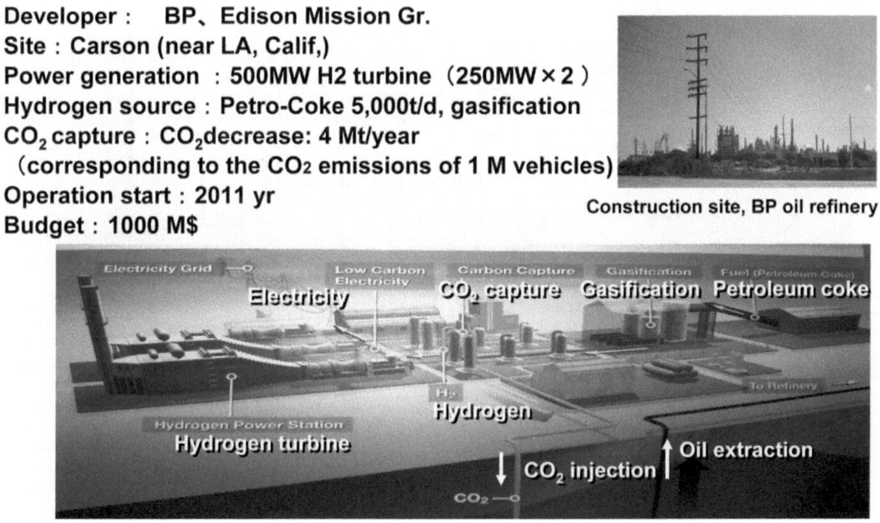

Fig. 11.8 Hydrogen gas turbine power station with CO_2 recovery and sequestration [12].

using EOR (enhanced oil recovery). One will be built in the US, with 500 MW output of electricity and CO_2 mitigation of 4 million tonnes per year, which corresponds to CO_2 emission from one million usual IC engine vehicles, as shown in Figure 11.8. The other one will be built in the Scotland, UK, with 350MW output of electricity and CO_2 mitigation of 1.2 million tonnes per year. They will start a commercialised operation in 2011 and 2010, respectively and the total budget size is 1600 million US dollars. This fact will have a strong impact on realising hydrogen economy earlier.

(4) *Decentralised energy system using hydrogen.* The development of a household cogeneration system based on PEM fuel cells using city gas reformed hydrogen is being aggressively promoted by gas companies. Tokyo Gas Co., Ltd. has developed a compact system for producing high-purity hydrogen (99.999%, 10 ppm or lower CO concentration) at low temperatures (about 550°C) using a membrane reformer. The decentralised energy system essentially needs an optimum total system design, with a community size, on the basis of the hierarchical structure of energy demand, not limited to fuel cells but also including micro gas turbines. The decentralised energy system for using hydrogen is expected to be combined with hydrogen production using natural energy and with the existing energy supply infrastructure in the future, and is expected to grow and mature to become a system that greatly contributes to global environmental conservation and energy sources free from fossil fuel. It is therefore important to effectively use natural energy which is available intermittently, level the power load, and eliminate the mismatch between supply and demand of energy. In this respect, hydrogen will have an important role not only as an energy source but also as an energy carrier. Thus, various technologies need

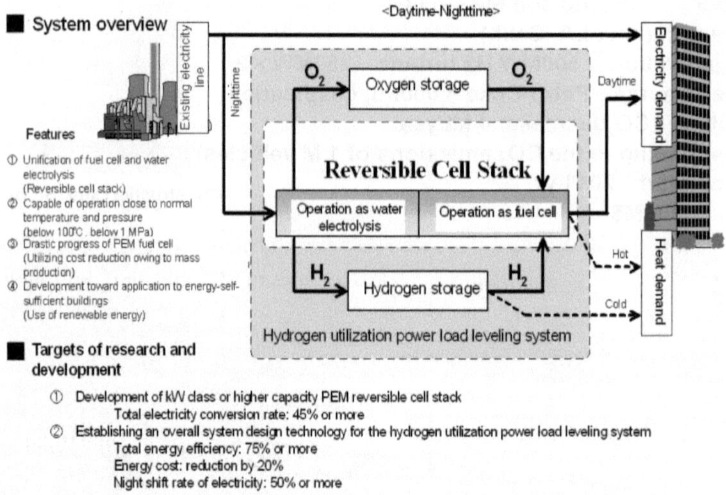

Fig. 11.9 Load levelling of electricity by reversible cell stack.

to be developed in succession to correctly understand the superior thermo-chemical features of hydrogen, which are described later, and to utilise the multi-functions of hydrogen.

Furthermore, a new energy system was proposed and investigated [13]. As shown in Figure 11.9, it combines hydrogen storage with a PEM reversible cell stack having the functions of both a fuel cell and water electrolysis, thus levelling the power load on the user side. A prototype of the reversible cell was fabricated in order to create an actual system, and the prototype showed excellent performance in both power generation and electrolysis.

11.5 Enhancement of Energy Quality Using Hydrogen as an Energy Carrier

Advanced effective utilisation of energy requires a system which enhances exergy to a higher energy quality in order to maximise useful work (electric power). That is, waste heat at low to medium temperatures (around 100–300°C) is processed to give high-quality energy for reuse, without using a conventional simple cogeneration system as a heat source of heating/cooling and of hot water. The exergy does not increase in quality by itself, but is increased (from low level to high level) only in combination with the downgrading of higher quality energy or substances. Hydrogen plays an essential role as a carrier in this type of system [14].

The quality of energy in substances (fuel, heat source, etc.) is evaluated by an index of exergy ratio (= exergy divided by enthalpy). In other words, the exergy ratio

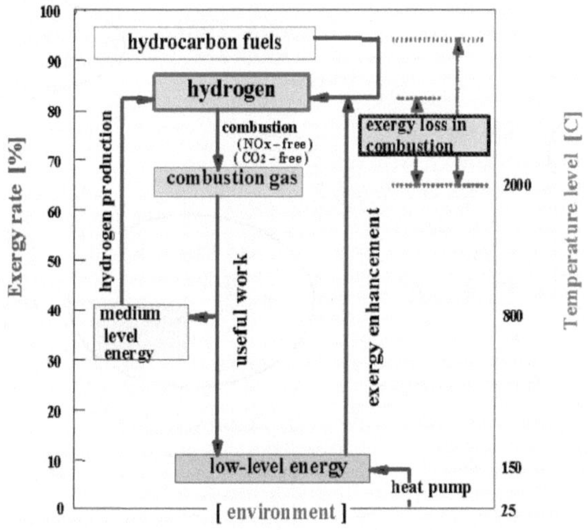

Fig. 11.10 Principles of exergy enhancement by fuel using hydrogen as an intermediate energy carrier.

is the derivable fraction of useful work (free energy, exergy) in the energy released (enthalpy, or heat value for fuel) until equilibrium with the surrounding system is attained. For example, the exergy ratio of methane is 91.9%, and that of methanol is as high as 96.8%. By comparison, the exergy ratio of hydrogen is 83.0%, which is the lowest among fuels. As shown in Figure 11.10, generally the exergy ratio of hydrocarbon-based fuels ranges from approximately 91 to 96%, which suggests theoretically that 90% or more of the chemical energy in fuel can be converted to useful work.

Nevertheless, interest has focused only on saving energy, in which fuel is combusted to convert into thermal energy having a low exergy ratio (about 50% at 800°C, about 60% at 1200°C, and about 70% at 2000°C), followed by a cascade (from high temperature to low temperature) or cogeneration. Although the combustion can attain very high temperatures, the combustion process generates much exergy loss because the thermal engines have a temperature limitation of materials and because the combustion process is highly irreversible. This is a wasteful activity. As for hydrocarbon-based fuels, conversion to hydrogen (enthalpy increases by endothermic reactions) rather than direct combustion to convert to thermal energy, raises the energy quality by converting low quality low- to medium-temperature waste heat into the chemical energy of hydrogen, the increased energy of which is then used by the above-described high-efficiency hydrogen utilisation system, thus significantly improving the energy system [10]. Even when the hydrogen is directly combusted, the exergy loss accompanied by combustion is halved. Hydrogen generation through steam reforming of methanol ($CH_3OH + H_2O \rightarrow CO_2 + 3H_2$) theoretically increases the quality significantly by converting a low quality thermal

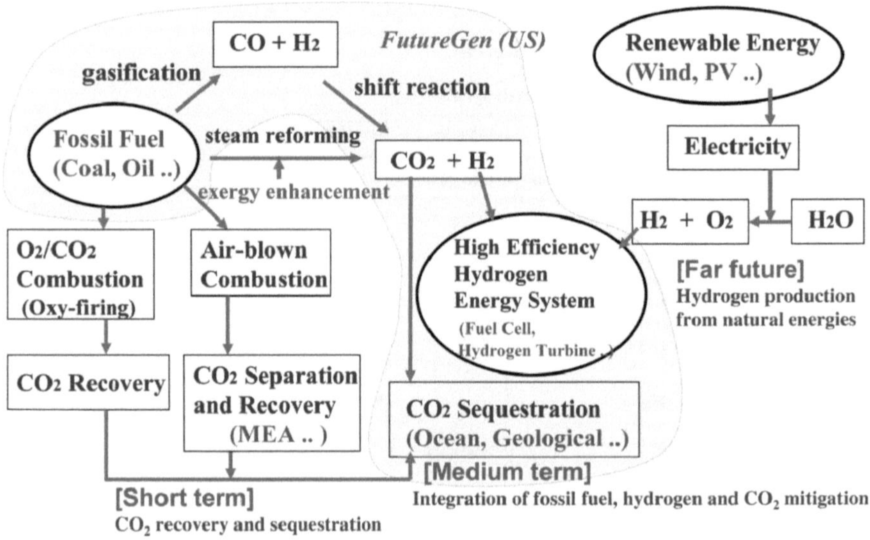

Fig. 11.11 System integration of coal, hydrogen and CO_2 sequestration.

energy at about 100°C having an exergy ratio of about 6%, into the chemical energy of hydrogen having an exergy ratio of 83%.

11.6 System Integration of Coal, Hydrogen and CO_2 Sequestration

Figure 11.11 summarises the integration of individual systems described above: CO_2 recovery coal combustion; coal gasification (including reforming); generation of H_2 and CO_2 by shift reaction; CO_2 sequestration; advanced hydrogen utilisation system including exergy enhancement; and integration of renewable energy and hydrogen utilisation expected in the long term [15]. The FutureGen Project announced by the DOE of the US in February 2003 as shown in Figure 11.12 is only a part of the drawing (shaded portion). The US share of proven coal reserves is 25% of the world total, while the share of oil and natural gas reserves is only 3%. Compared with Russia's share of natural gas reserves of 30%, it is a reasonable policy for the US to integrate coal, hydrogen, and CO_2 sequestration as a national policy to counter its delay in global warming countermeasures, while depending on coal. The CO_2 sequestration in the US, however, anticipates geological sequestration which has already been established by Enhanced Oil Recovery (EOR). This appears to be a smart political action by the US to take leadership of the world's activities.

On the other hand, Japan has invested heavily in developing related individual technologies, and has attained successful results, as shown in Figure 11.13, and

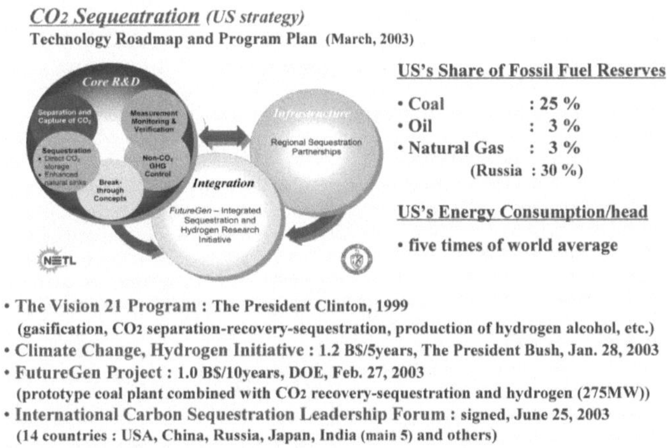

Fig. 11.12 US strategy for CO_2 mitigation – FutureGen.

Japanese Projects for Mitigation of Global Warming

- Coal Technologies for CO_2 Recovery
  - CO_2/O_2 Coal Combustion (NEDO-CCUJ• 1996-1999)
  - IGCC+CO_2 Recovery and Hydrogen-rich Gasification
  - HyPr-RING: New Hydrogen Production Process with CO_2 Recovery
- CO_2 Sequestration Technologies
  - SEA-COSMIC : Study of Environmental Assessment for CO_2 Ocean Sequestration for Mitigation of Climate Change (NEDO-RITE• Phase I: 1997-2001, Phase-II: 2002-)
  - Geological Sequestration
- Hydrogen related Technologies
  - WE-NET• World Energy Network Project) (NEDO-IAE• ENAA• Phase-I: 1993-1998, Phase-II: 1999-2002)
- Fuel Cell Vehicles and Hydrogen Refueling Stations
  - JHFC (Japan Hydrogen and Fuel Cell Development Project) (METI-JARI(JEVA)• 2002-)

Fig. 11.13 Japan's related national projects.

there have long been discussions among Japanese researchers. Unfortunately, however, Japan failed to lead the world in issuing policy proposals for building a system to attain significant synergies through the integration of these individual technologies. Full-scale collaboration among researchers in different research fields is crucial for establishing a large and long-term interim scenario that deals with complex problems such as global environmental conservation and that will smoothly lead to a hydrogen-oriented society free from dependence on fossil fuels.

Japan has just initiated discussions on collaboration, and several interdisciplinary projects have begun led by governmental organisations. Taking this opportunity, actions toward developing integrated complex system technologies as shown in Figure 11.14 should now be started.

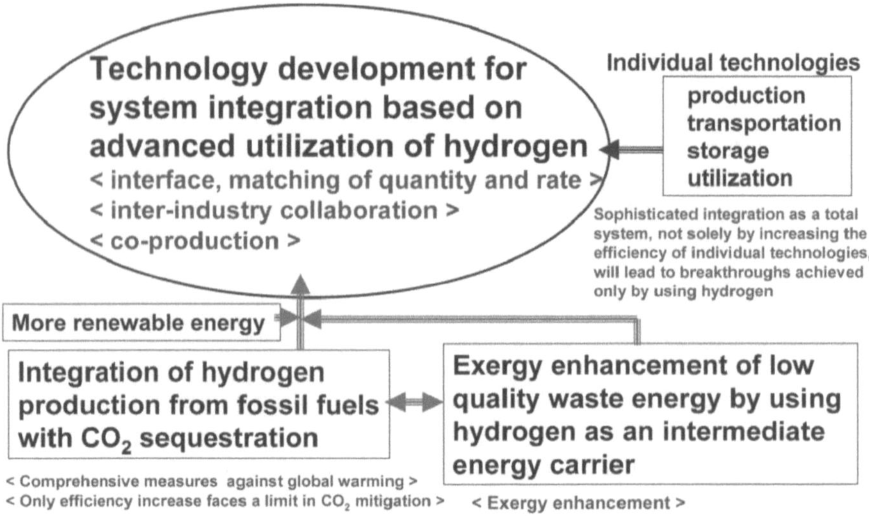

Fig. 11.14 System integration as a realistic intermediate scenario toward a hydrogen community.

11.7 Concluding Remarks

A realistic interim scenario before introducing a large amount of renewable energies by around 2030–2050 is necessary and must be practical and socially acceptable. System technologies suitable in time scale and space scale need to be developed and introduced while ensuring compatibility with other energy supply and demand systems including coal and other fossil fuels, in parallel with the development of various core technologies for hydrogen utilisation, including manufacture, transportation, storage, supply, and conversion. This chapter described the significance of clean coal technologies with CO_2 recovery and sequestration, and introducing hydrogen energy as measures to combat global environmental issues; outlined the comprehensive global warming countermeasures of integrating coal, hydrogen, and CO_2 sequestration, and discussed the project scheme, including the importance of collaboration among various researchers of different fields.

References

1. Hirai, S., Okazaki, K. and Hijikata, K., Mass transport phenomena of liquid CO_2 with clathrate, *Waste Management* **17**(5/6), 1997, 353–360.
2. Ishizaka, S. and Ohsumi, T., Development of technology to predict the influence of ocean sequestration of carbon dioxide on the environment: Study of prediction of the influence on organisms, *Oceanography in Japan (Umi No Kenkyuu)*, 1999, 213–218.
3. Wolsky, A.M., Daniel, E.J. and Jody, K.S., Recovering CO_2 from large and medium-size stationary combustors, *Journal of the Air & Water Management Association* **41**, 1991, 449–454.

4. Kimura, N., Omata, K., Kiga, T., Takano, S. and Shikishima, S., The characteristics of pulverized coal combustion in O_2/CO_2 mixtures for CO_2 recovery, *Energy Conversion & Management* **36**, 1995, 805–808.
5. Okazaki, K. and Ando, T., NOx reduction mechanism in coal combustion with recycled CO_2, *Energy, The International Journal* **22**(2/3), 1997, 207–215.
6. Feng, B., Ando, T. and Okazaki, K., NO destruction and regeneration in CO_2 enriched CH_4 flame, *JSME International Journal* **41**, 1988, 959–965.
7. Liu, H. and Okazaki, K., Simultaneous easy CO_2 recovery and drastic reduction of SOx and NOx in O_2/CO_2 coal combustion with heat recirculation, *Fuel* **82**, 2003, 1427–1436.
8. Liu, H., Katagiri, S. and Okazaki, K., Drastic SOx removal and influences of various factors in O_2/CO_2 pulverized coal combustion system, *Energy & Fuels, An American Chemical Society Journal* **15**(2), 2001, 403–412.
9. Amaike, S., Carbon dioxide capture in IGCC plant using CO shift converter and conventional acid gas removal, in *Proceedings of 5th ASME/JSME Thermal Engineering Joint Conference*, AJTE99-6410 (CD-ROM), 1999, p. 170.
10. Okazaki, K., Significance of hydrogen energy introduction and technology prospects, *PETROTECH, The Japan Petroleum Institute* **25**(8), 2002, 627–632.
11. Okazaki, K., Hydrogen fuel station, Test results, in *FY2005 JHFC Seminar*, JHFC (Japan Hydrogen and Fuel Cell Demonstration Project), 2006, pp. 49–70.
12. Okano, K., Report of NHA conference and world prospect of hydrogen, in *Proceedings of 118th HESS Meeting*, Tokyo, HESS (Hydrogen Energy Systems Society, Japan), 2006, pp. 1–20.
13. Kozawa, Y. and Okazaki, K., Total energy system based on fuel cells, *Chemical Engineering* **65**(10), 2001, 530–533.
14. Okazaki, K., Exergy enhancement and effective use of low-to-medium-temperature waste heat with an example of biomass hydrogenation, *Hydrogen Energy System, Hydrogen Energy System Society of Japan (HESS)* **29**(1), 2004, 18–25.
15. Okazaki, K. and Akai, M., Countermeasures to global warming by integration of coal, hydrogen and CO_2 sequestration, and system evaluation, in *Proceedings of the 36th Clean Coal Seminar*, CCUJ (Center for Coal Utilization of Japan), 2004, pp. 17–46.

Chapter 12
Advanced Steam Generator Concepts for Oxy-Fuel Processes

Hartmut Spliethoff

Abstract. A comparison of possible CO_2 separation methods for coal-based power plant processes is considered. The requirements concerning development and research of oxy-fuel processes are presented. Steam generator concepts suited for combustion with pure oxygen are introduced, focusing on the new controlled staging with rich/lean burners.

Key words: Coal, oxy-fuel combustion, CO_2 separation, flue gas circulation, staging.

12.1 Introduction

Along with the increase of efficiency and the use of low-carbon, carbon-neutral or carbon-free primary energy sources, CO_2 retention methods offer a highly promising possible method of CO_2 reduction in fossil fuel-fired power plants. Starting out from the actual generation system structure and the power generation costs of the various technologies, fossil energy sources will cover the bulk of the power demand also in the next decades.

Due to the high carbon-to-hydrogen ratio in hard (highly bituminous) and brown (low bituminous) coal, the combustion of these fuels produces a very high specific level of carbon dioxide emissions. Nevertheless, coal, compared with natural oil and gas, shows itself as the most important primary energy source for power production owing to its worldwide and long-term availability and price stability. Therefore, the reduction of CO_2 emissions from coal-fuelled power plants will be one of the most essential issues in research.

The objective of this study is to compare advanced steam generator concepts with oxy-fuel combustion and, in particular, to discuss the feasibility and need for research concerning the new scheme of controlled staging with lean/rich burners.

K. Hanjalić et al., Sustainable Energy Technologies: Options and Prospects, 227–236.

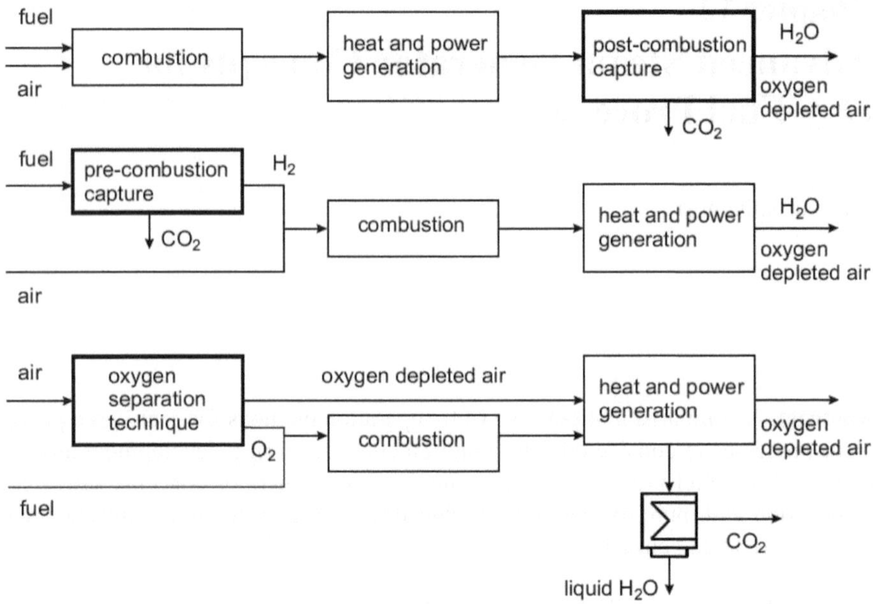

Fig. 12.1 Classification of power plant technologies of CO_2 sequestration.

12.2 Classification of CO_2 Capture Technologies

The superordinate characterising feature of CO_2-removing technologies is the process-integrated separation. Three groups are distinguished in this respect, as given in Figure 12.1.

12.2.1 Post-Combustion Capture/Flue Gas Decarbonisation

The combustion process remains unaffected; carbon dioxide is removed only at the end of the power plant process by separation from the flue gas. Flue gas decarbonisation is therefore especially suited for retrofitting power plants. The removal is carried out chiefly by chemical absorption. Adsorption techniques or membranes are theoretically conceivable but cannot compete with absorption. The costs and the loss in efficiency associated with flue gas decarbonisation are the highest, though [1, 2]. The individual components and processes are state of the art. However, the process as a whole still needs to be demonstrated successfully.

12.2.2 Pre-Combustion Capture/Fuel Gas Decarbonisation

This process requires gasifying coal and can therefore be employed only in Integrated Gasification Combined Cycle (IGCC) power plants. The synthetic gas produced by gasification first has to pass a shift reaction in order that hydrogen accumulates in the fuel gas. Membrane separation, ad- and absorption methods are capable of removing the carbon dioxide from the fuel gas. Efficient gas turbines which make the combustion of hydrogen possible have yet to be developed [3–6].

12.2.3 Oxy-Fuel Firing/Combustion with Pure Oxygen

The combustion reaction takes place using oxygen as pure as possible. Owing to the absence of nitrogen in the flue gas, a CO_2-rich gas can be achieved after the condensation of the water vapour. The provision of oxygen is carried out by means of a conventional cryogenic air fractionation plant or by ceramic oxygen membranes, yet in the stage of development, which will increase the total efficiency by about 10% as opposed to cryogenic air fractionation. The Graz Cycle [7], the Water Cycle [8], the Matiant Cycle [9] and the AZEP (Advanced Zero Emissions Power Plant) concept [10], which have been or are being developed for gaseous fuels, are special examples of the oxy-fuel method. The Chemical-Looping method [11] plays a special part among the oxy-fuel processes.

12.3 Oxy-Fuel Processes

The so-called oxy-fuel process was an issue for the first time in the early 1980s in the discussion about the reduction of CO_2 emissions, and was then advertised by Horn and Steinberg [12]. Vattenfall has already begun constructing an oxy-fuel power plant (30 MW pilot plant) at *Schwarze Pumpe*. Plans envisage the plant to go into service in 2008 [13].

Oxy-fuel firing is a combustion concept which burns the fossil energy carrier in oxygen-rich atmosphere. The oxygen is provided by an air fractionation plant. Nitrogen from air being absent in the furnace, a flue gas mixture is achieved with carbon dioxide and water vapour as the essential components. The water vapour can easily be separated by condensation. The level of impurities in the flue gas thus depends exclusively on the type of fossil fuel, provided a high purity level of the supplied oxygen and possible gas-tight operating conditions of the firing plant.

Besides the simple separation of the carbon dioxide, another advantage of *oxy-fuel firing* is reduced NOx formation: in the ideal case, the formation of prompt and thermal NOx becomes totally stopped. Basically, the nitrogen contained in the fossil energy carrier can alone form the so-called fuel NOx. In order to avoid a higher rate of conversion of the fuel nitrogen into NO, the application of combustion-

Table 12.1 Overview of the state of development of oxy-fuel units [14].

Components	Studies	Laboratory Tests	Pilot Plants	State of Development	Readiness for Marketing
Air fractionation plant	x	x	x	x	x
Oxygen heater	x	x			
Brown coal predrying	x	x	x	x	
Oxy-fuel steam generator	x	O			
Flue gas recirculation	x	x	x	x	O
Electrostatic precipitator	x	x	x	x	x
Desulphurisation plant	x	x	x	x	x
Carbon dioxide: cleaning/removal	x	O			
Water/steam unit: cycle system/ steam-water cycle	x	x	x	x	x

engineering measures such as staged oxygen and fuel supply are an absolute necessity.

Commercial and series-production air-fractionation plants today are based on the liquefaction of air and subsequent separation in rectification columns. This air-fractionation unit in particular causes a clear efficiency loss in the entire process due to its considerable demand for energy.

One of the big advantages of the oxy-fuel process is the simple adaptation of a conventional coal-fired power-generating process to the modified conditions. The already present components being largely kept, it is possible in general to retrofit the plant of an existing power station.

12.4 Oxy-Fuel Steam Generator Concepts

The combustion of coal in a pure oxygen atmosphere can lead to flame temperatures rising above 3000°C. At such high temperatures, the ash softening temperature is clearly exceeded. This will involve the volatilisation of alkalies but of silicon or other ash components, too, and may result in heavy fouling in the convective heating surfaces. Also, modern materials cannot stand these high temperatures. The actual combustion temperature should not exceed 1500–1800°C. In principle, there are two possible methods to limit the combustion temperature.

Table 12.1 lists the most important components of an oxy-fuel process with brown coal firing and flue gas recirculation taking into account their readiness for marketing [14]. It shows that the highest development requirements arise from the adaptation of the steam generator to the oxy-fuel firing system. Firing concepts suited for oxy-fuel processes will be discussed in the next section.

On the one hand, it is possible to control the temperature by suitable admixture of thermal ballast, in the form of recirculated flue gas, solids or water vapour for instance. On the other hand, the combustion process can be carried out in a proportion not corresponding to the stoichiometric mixing ratio of fuel to air. Both excess-air and air-deficient combustion results in a reduction of the adiabatic combustion temperature until the stability limit of the flame is reached. Commonly, hard coal-fired furnaces are operated with a slight excess of air in order to even out the non-ideal mixture and to guarantee complete combustion.

Examples from practice for the admixing of thermal ballast and for air-deficient or excess-air combustion are described in the following.

12.4.1 Flue Gas Recirculation

The upper temperature limit is met through the thermal ballast of the recirculated flue gas. A distinction is made between internal and external flue gas recirculation.

12.4.1.1 Internal Flue Gas Recirculation

Every conventional jet burner has a well-developed recirculation zone to homogenise combustion and to avoid hot spots which have increased NOX emissions in consequence. This is described as internal recirculation.

Another option for applying internal flue gas recirculation is the use of FLOX (FLameless OXidation) burners, such as are used in the steel industry at present. Flue gas from the environment of the flame is mixed with the flame by skilful flow control. The advantage of this technique is the low costs. In addition, it has already proved worthwhile in other applications. Concerning the duty range of power plants, applications of FLOX burners are not known so far. It remains open, however, whether it should be possible by this technique exclusively to recirculate sufficiently large flue gas quantities to achieve adequate cooling. Internal recirculation can contribute, though, to obtain local uniformity of the flue gas temperatures in combustion with oxygen.

12.4.1.2 External Flue Gas Recirculation

In the process of external flue gas recirculation, the cooled down flue gas is conducted into the furnace again. Depending on the cooling degree, differently high volumetric flow rates are needed to reduce the combustion temperature. About one third of the flue gas has to be recirculated to keep the temperature level below the ash melting point, the boiler material stress and the temperature triggering nitrogen formation of the fuel nitrogen. Which means, about 65% of the flue gas formed in the steam generator has to be recirculated when raw brown coal is fired, about 80% with dry brown coal, and about 78% with hard/bituminous coal.

Due to the high volumetric flow that needs to be transported, the dimensions of the plant and the auxiliary power requirement for the recirculation fan increase. External recirculation is established technology which, however, entails a number of problems (e.g., construction size, wear and tear, corrosion when temperatures fall below the dew point, distribution of the individual flows).

Owing to the retrofittability of existing power stations and the application of conventional methods, recirculation provides a solution standing at the logical beginning of the realisation of CO_2-emission-free power plants.

12.4.2 Water/Steam Spraying

Applying this method, the flame is cooled by injecting water and/or steam. For gas turbines in particular, water-cooled burners are a cost-effective and well-tried option. In steel furnaces, these burners belong to the first generation of oxy-fuel burners in practical application. However, the heat loss in the steam generator increases because the vapourisation enthalpy in the steam cycle cannot be used completely.

12.4.3 Staged Combustion

Staged combustion technologies such as air or fuel staging are employed in conventional steam generators operated using air to reduce nitrogen oxide emissions. Staged supply of the oxidant or of the fuel is applied in the following concepts to control the temperature in the flue gas during combustion with oxygen. The flue gas heated up in the single stages through step-by-step burning becomes cooled between stages by radiative heat exchange with the furnace walls. Different concepts for staged combustion are conceivable.

The method of pure oxygen or fuel staging uses exclusively the furnace heat release, delayed through the staged supply, for controlling the temperature.

The method of "Controlled fuel/oxygen staging with rich/lean burners", besides employing staged fuel or oxygen supply, respectively, uses the lower adiabatic combustion temperature of rich or lean burners, to control the temperature field in the furnace.

12.4.3.1 Simple Air Staging

In the bottom part of the furnace, combustion is effected substoichiometrically (under deficiency of oxygen); the rest of the oxygen necessary for complete combustion is fed in stage-by-stage in the upper part of the furnace (see Figure 12.2).

Air staging is a simple, well-tried technique, which involves reduced furnace dimensions and hence lower costs. Pulverised coal burners, with $\lambda = 0.3$, have been developed to date only for coal gasifiers. The cross-mixing in the furnace poses a problem.

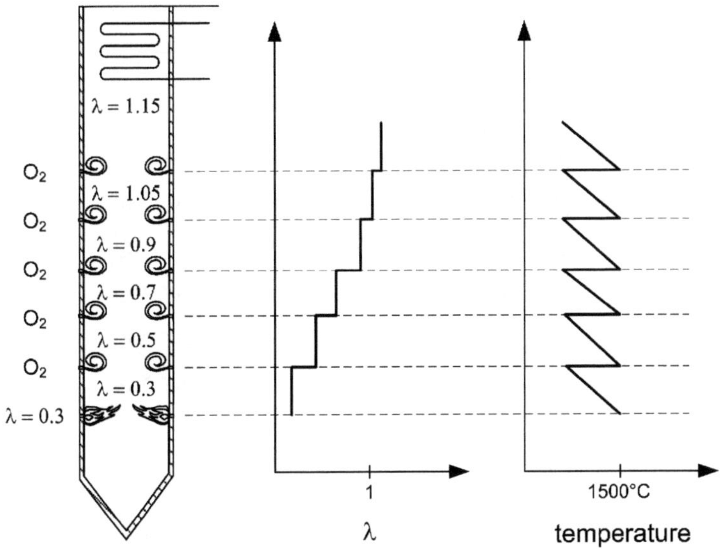

Fig. 12.2 Simple air staging in the furnace.

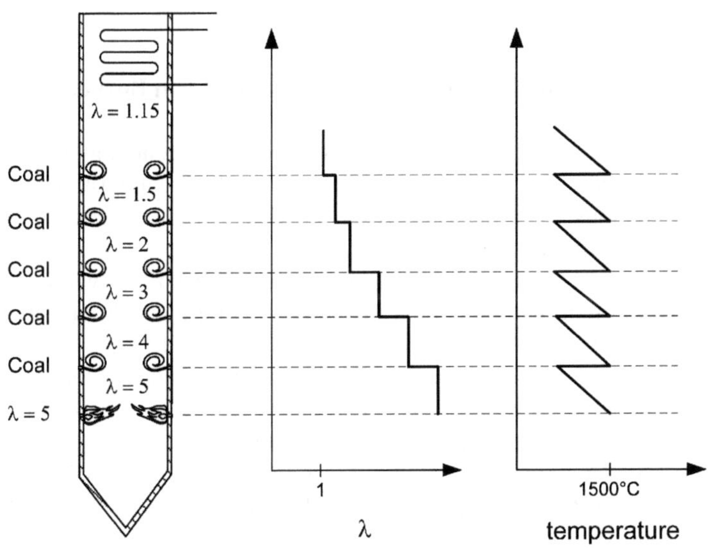

Fig. 12.3 Simple fuel staging in the furnace.

12.4.3.2 Simple Fuel Staging

In this technique, it is the pulverised coal that is fed stage by stage to the furnace (see Figure 12.3). As with air staging, the dimensions of the furnace are small. The problem of the cross-mixing is a central issue here as well.

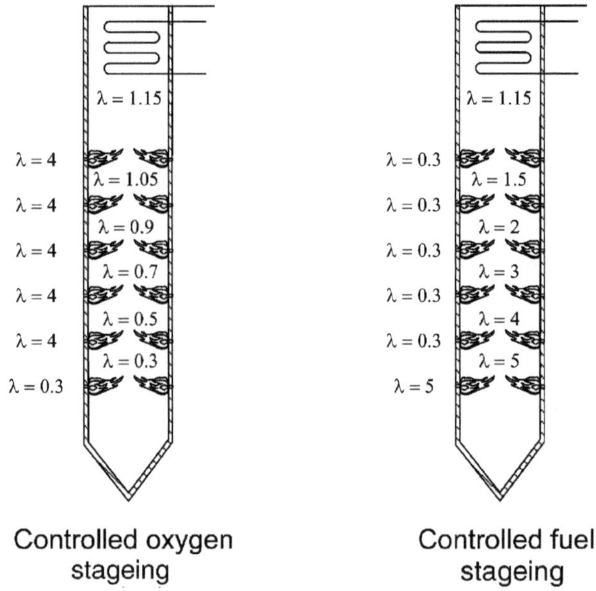

Fig. 12.4 Controlled fuel/oxygen staging in the furnace.

12.4.3.3 Controlled Fuel/Oxygen Staging with Rich/Lean Burners

A special and, for oxy-coal processes new, solution is presented by a staged combustion which makes use of the step-by-step release of the combustion enthalpy by burner combinations operating with excess air and with air deficiency to control temperature in the furnace.

Controlled fuel/oxygen staging distinguishes itself from the two staging techniques above through the application of burners operating with air deficiency or excess air instead of using pure oxygen or coal (see Figure 12.4).

The central problem in realising a firing system with reduced flue gas recirculation lies in the requirement to distribute the heat release over a great part of the furnace in order to avoid improperly high temperatures.

The burners have to be operated with great air deficiency or excess air in order that in the flame itself, too, improperly high temperatures do not develop. Before the flue gas exits the furnace, the air ratio shall be controlled to just above $\lambda = 1$ to ensure complete combustion.

Applying schemes with controlled oxygen-staged firing, the bottom burner levels are operated with air deficiency, the air ratio being raised to the overall air ratio by burners with excess air on the way to the furnace outlet. With controlled fuel-staged combustion using burners with air deficiency and excess air, at first the lower fuel levels are operated with excess oxygen by burners with excess air. By adding fuel through burners with air deficiency, the air ratio in the furnace is lowered step-by-step until the overall air ratio is reached at the furnace outlet. The advantage of the two schemes is the spatial distribution of the heat release. Both concepts avoid the mixing problem of conventional fuel or oxygen staging by means of the higher impulse of the burners operating with air deficiency or excess air, respectively.

12.5 Prospect

Within the framework of a study, the scheme of controlled staged combustion with burners operating with air deficiency or excess air, respectively, was evaluated as an innovative and promising concept for steam generators with combustion using oxygen. The essential advantage lies in the definite reduction of steam generator costs at clearly lower flue gas recirculation rates. At the Chair of Energy Systems, Technical University of Munich, theoretical studies and experiments investigating the concept of controlled staged combustion are being carried out within the framework of an EU-supported project.

References

1. Kather, A. und Plass, L., Drei Technologierouten eröffnen Möglichkeiten zur CO_2-Abtrennung in fossil gefeuerten Kraftwerken, Power Perspektiven 2005, RWE *Power* 2005, http://www.rwe.com/generator.aspx/konzern/fue/strom/co2-freies-kraftwerk/language=de/id=268960/co2-page.html
2. Göttlicher, G., *Energetik der Kohlendioxidrückhaltung in Kraftwerken*, VDI Verlag, Düsseldorf, 1999.
3. Chiesa, P., Kreutz, T. and Lozza, G., CO_2 Sequestration from IGCC power plants by means of metallic membranes, in *Proceedings of ASME Turbo EXPO*, Reno-Tahoe, NV, 2005.
4. Cau, G., Cocco, D. and Montisci, A., Performance of zero emissions integrated gasification hydrogen combustion (ZE-IGHC) power plants with CO_2 removal, in *Proceedings of ASME Turbo EXPO*, New Orleans, LA, 2001.
5. Bozzolo, M., Brandani, M., Traverso, A. and Massardo, A.F., Thermoeconomic analysis of gas turbine plants with fuel decarbonisation and carbon dioxide sequestration, in *Proceedings of ASME Turbo EXPO*, Amsterdam, the Netherlands, 2002.
6. Kvamsdal, H.M., Ertesvag, I.S., Bolland, O. and Tolstad, T., Exergy analysis of gas-turbine combined cycle with CO_2 capture using pre-combustion decarbonisation of natural Gas, in *Proceedings of ASME Turbo EXPO*, Amsterdam, the Netherlands, 2002.
7. Sanz, W., Jericha, H., Moser, M. and Heitmeir, F., Thermodynamic and economic investigation of an improved Graz Cycle power plant for CO_2 capture, in *Proceedings of ASME Turbo EXPO*, Wien, 2004.
8. Bolland, O., Kvamsdal, H. and Boden, J., A thermodynamic comparison of the oxyfuel power cycles Water-Cycle, Graz-Cycle and Matiant-Cycle, in *Proceedings of Power Generation and Sustainable Development*, Liége, Belgium, 2001.
9. Mathieu, P. and Nihart, R., Zero emission MATIANT cycle, in *Proceedings of the 1998 International Gas Turbine & Aeroengine Congress & Exhibition*, Stockholm, 1998.
10. Sundkvist, S.G., Griffin, T. and Thorshaug, N.P., AZEP – Development of an integrated air separation membrane – Gas turbine, in *Proceedings of Second Nordic Minisymposium on Carbon Dioxide Capture and Storage*, Göteborg, Sweden, 2001.

11. Consonni, S., Lozza, G., Pelliccia, G., Rossini, S. and Saviano, F., Chemical-looping combustion for combined cycles with CO_2 capture, in *Proceedings of ASME Turbo Expo*, Wien, 2004.
12. Horn, F.L. and Steinberg, M., Control of carbon dioxide emissions from a power plant (and use in enhanced oil recovery), *Fuel* **61**, 1982, 415–422.
13. Presse Conference of the Vattenfall AB Company, 26 May 2005, chemie.de, Information Service GmbH, http://www.chemie.de/news/d/46247 (accesssed on 04 January 2006).
14. Altmann, H. and Stamatelopoulos, G.-N., Steps towards the minimisation of CO_2 emissions from coal-fired power plants, in *Proceedings of Power Gen*, Milano, Italy, 2005.

Part II
Initiatives in South-East Europe

Chapter 13
CO_2 Mitigation Options for Retrofitting Greek Low-Quality Coal-Fired Power Plants

E. Kakaras, A. Doukelis, D. Giannakopoulos and A. Koumanakos

Abstract. We consider possible application of the state-of-the-art in technological concepts of CO_2 capture and sequestration to retrofitting of low-quality coal-fired power plants. The most promising options, namely the oxy-fuel combustion and the flue gases treatment by amine scrubbing, were evaluated as retrofit options for a typical modern lignite-fired power plant. Results from thermodynamic simulations of the examined cases were used to demonstrate the potential for emissions reduction and evaluate the associated power output and efficiency penalties. Furthermore, an economic assessment of electricity production cost was carried out, in relation to the application of different existing and near future technologies. The economical impact related to fuel prices and CO_2 emission risks was assessed.

Key words: CO_2 sequestration, low-quality brown coal, oxy-fuel, amine scrubbing.

13.1 Introduction

Fossil fuels are the main source of primary energy worldwide, with a share of over 85%. Despite a realistic expectation that other energy options, such as hydro, nuclear, wind, biomass, geothermal and solar will take an increased share in the future, in many countries fossil fuels, especially coal, will remain the primary energy source. The future evolution of the electricity generation sector and meeting the increasing demand for electricity will thus rely on advanced technologies for fossil fuel utilisation and increased use of renewable energy sources (RES), with emphasis on conservation of the available sources and improved environmental performance. Various measures such as reduction of demand, efficiency improvements, substitution of fossil fuels, switching to renewable or nuclear energy, CO_2 sequestration and others can contribute to the above goals, but none of them if applied as a single measure can be a solution towards the stabilisation of CO_2 concentration in the at-

K. Hanjalić et al., Sustainable Energy Technologies: Options and Prospects, 239–250.

mosphere [1]. Instead, according to the Intergovernmental Panel on Climate Change (IPCC), only a combination of actions can have a significant impact on the abatement of the climate change problem.

Taking into account that 1/3 of the CO_2 emitted into the atmosphere is produced by thermoelectric units, the development of novel technological concepts for thermal power production is expected to contribute significantly to CO_2 emissions mitigation. These technologies deal with carbon capture and storage (CCS). The CO_2 sequestration technologies applicable to coal-fired power plants are divided into the following main categories [2]:

- removal of CO_2 from flue gases;
- combustion in oxy-fuel conditions using pure O_2 instead of air; and
- production of a carbon free fuel.

For Greece, lignite is the most significant domestic energy source for electricity production and it will continue to play an important role in the next decades. The two most promising technological concepts for CO_2 sequestration, namely the oxy-fuel process and the MEA scrubbing process are examined for the case of retrofit of low-quality coal-fired power plants.

13.2 Characteristics of the Existing Thermal Power Plants in Enlarged EU

The efficiency of thermoelectric units utilising solid fossil fuels in the enlarged EU ranges from 21.1% up to 44.0% [3]. The efficiency of the power generation plants in EU Member States is higher in comparison to other countries of the enlarged EU, thanks to the implementation of improved technologies, better maintenance and the units' larger capacity. The total efficiency in relation to the commissioning year of the units is presented in Figure 13.1.

The implementation of modern technological solutions during the last decade resulted in an efficiency increase, achieving a range of 38 to 42%. Further improvements incorporate advanced concept designs.

13.3 Greek Lignite-Fired Power Plants

The Greek energy system consists of eight lignite-fired power plants with a total capacity of 5,228 MWel, the majority of which are in the range of 200–300 MWel. More than 11% of the installed capacity was put in operation before 1970, while 35% has been in operation for 20 to 30 years. In comparison to the ageing units in the enlarged EU, of which 50% have been in operation 15 to 22 years, the Greek units can be categorised as relatively young ones [4].

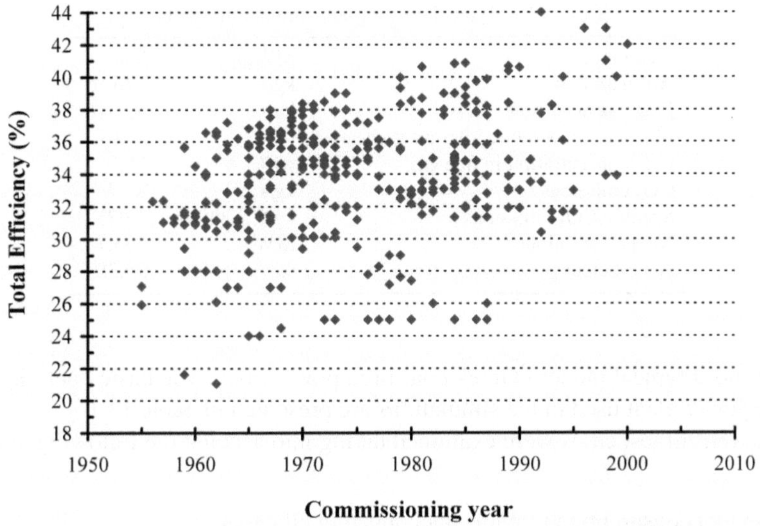

Fig. 13.1 Total efficiency of solid fuel thermoelectric units in the enlarged EU in relation to their commissioning year.

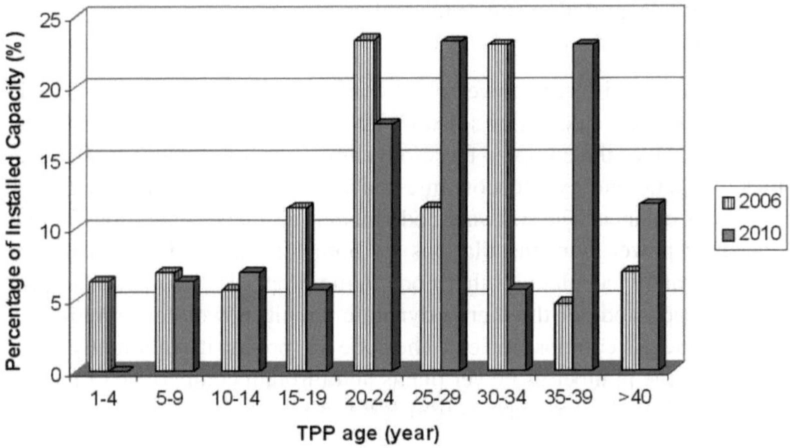

Fig. 13.2 Percentage of lignite-fired power plants capacity in Greece in relation to their age for the years 2006 and 2010.

13.4 Retrofit of a Typical Greek Lignite-Fired Power Plant with CO_2 Sequestration

The plant of 330 MWel gross power capacity consists of a supercritical boiler, an HP-MP-LP steam turbine and eight regenerative feedwater preheating stages. The unit is equipped with a flue gas desulphurisation unit and the examined configuration

Table 13.1 Typical power plant simulation results.

Fuel input	MWth	830
Air mass flow	kg/s	346.46
Flue gas mass flow	kg/s	423.46
Flue gas temperature after air preheater	°C	144.1
CO_2 concentration in flue gas	w %	20.7
CO_2 emissions	kg/s	87.7
Specific CO_2 emissions	kg CO_2/ kWh	1.075
Net power output	MW	293.7
Net efficiency	%	35.38

represents a typical modern Greek coal-fired power plant. The basic operating data of the power plant used in the simulations are presented in Table 13.1.

The retrofit test cases were examined taking into account the following assumptions:

- The fuel consumption remains unchanged in all cases.
- Minor modifications are foreseen in the water/steam cycle for the various cases examined.
- The produced CO_2-rich stream is compressed in a multi-stage compressor to 110 bar for transportation and sequestration purposes.
- The final CO_2 stream purity is 98%.

Simulations of the different power plant configurations were performed with the thermodynamic cycle calculation software ENBIPRO (**EN**ergie-**BI**llanz-**PRO**gram) [5]. This tool solves the complex thermodynamic circuits, computes efficiency and provides exergetic and exergoeconomic analysis of power plants while satisfying heat and mass balance. The software code models all pieces of equipment that usually appear in power plant installations for both full load and part load operation and can accurately calculate all thermodynamic properties (temperature, pressure, enthalpy) at each node of the thermodynamic circuit, power consumption of each component, flue gas composition etc. The code has proven its validity by accurately simulating a large number of power plants and through comparison of results with other commercial software, such as IPSEPRO [6].

13.4.1 Oxy-Fuel Firing Retrofit Option

The main components of the Oxy-fuel retrofit power plant used in the simulation are illustrated in Figure 13.3. Oxygen of high purity is used for combustion. Pure oxygen utilisation results in alteration of the temperature field in the combustion chamber, which has a significant effect on the boiler's design and materials. In order to preserve the temperature level inside the boiler at the same range as in the conventional power plant, flue gas recirculation is implemented. Under the O_2/CO_2

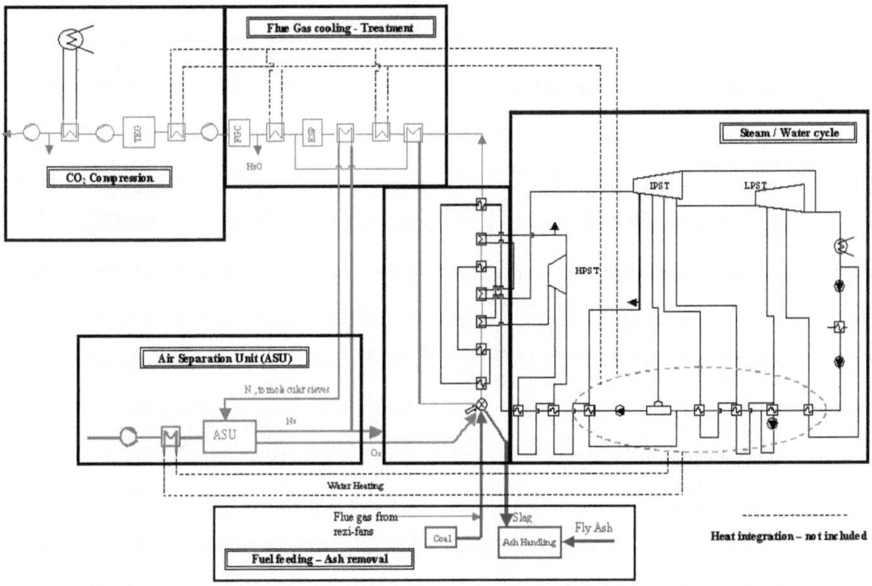

Fig. 13.3 Oxy-fuel firing retrofit configuration of a 330 MWel lignite-fired unit.

recycle configuration, air preheating is not used, which increases the flue gas exit temperature to approximately 310°C.

The high purity (95%) oxygen used for combustion is produced by an Air Separation Unit (ASU). The remaining 5% of the flow is mainly Argon. The ASU process is based on cryogenic air separation, which is the most efficient commercially available technique [7]. The air cleaning process is carried out for the removal of water vapour, CO_2 and any other impurities, by adsorption with molecular sieves. The molecular sieves regeneration is achieved with dry nitrogen, which is also a by-product of the ASU [8]. For that purpose, heaters utilising the heat of flue gases leaving the boiler are used for nitrogen heating. The air inlet pressure in the ASU is 5.5 bar, achieved by a four-stage compression with an isentropic efficiency of 0.83. In order to increase the efficiency of the process, in each compression step, namely 1.013–1.53 bar, 1.53–2.35 bar, 2.35–3.6 bar and 3.6–5.5 bar, an air cooling process is foreseen (up to a final air temperature of 20°C) for compression energy reduction. [8]. The excess oxygen taken into account is 1.5% [9].

The final CO_2 concentration is affected by the boiler's air leakage as well as the level of oxygen purity. Assuming 95% oxygen purity and 3% air infiltration, oxy-fuel combustion can achieve a CO_2 capture of 79% of the produced CO_2 in conventional coal combustion techniques [10].

The simulation results of the power plant retrofit to oxy-fuel operation are presented in Table 13.2. The major findings are the significant losses both in efficiency and power output, but also the achievement of significant reduction in CO_2 emissions compared to the conventional power plant. Taking into account that the flue gases

Table 13.2 Oxy-fuel power plant simulation results.

Air mass flow in ASU	kg/s	315.6
Air compression for the ASU	bar	5.5
O_2 mass flow	kg/s	72.78
Flue gas mass flow at boiler's exit	kg/s	416.5
Recirculated flue gas mass flow	kg/s	270.0
CO_2 concentration (wet)	w. %	58.84
CO_2 concentration (dry)	w. %	96.97
ASU power consumption ($n_{is} = 0.83$)	MW	58.1
Compression of the final product for transportation and sequestration	bar	110
CO_2 compression power consumption ($n_{is} = 0.85$)	MW	22.4
Cooling water pumps consumption for flue gas cooling and air inter-cooling at the ASU	MW	1.5
CO_2 emissions	kg/s	18.11
Specific CO_2 emissions	kg CO_2/kWh	0.31
Net power output	MW	211.0
Net efficiency	%	25.42

are cooled down to 20°C before compression of the final product, an important amount of heat could be provided to the steam cycle by replacing some feed-water steam preheaters with flue gas preheaters, thus lowering the efficiency penalty.

13.4.2 Amine Scrubbing Option

The retrofit plant configuration was based on the adoption of the commercially available MEA process. The CO_2 content of flue gases is absorbed by the liquid solvent. For that purpose, an absorption tower is used where flue gases and solvent are used in counter-current flow. In a second phase, the produced CO_2-rich stream is regenerated in a second column where the CO_2 content is stripped off. The regeneration process is achieved by heating of the amine solution, the required heat supplied by a low-pressure steam extraction at 5.96 bar and 285°C from the steam cycle. The steam provides its latent heat for regeneration of the solvent and the condensate is compressed and returned to the feed-water tank. The stripper column thermal consumption is about 4 MJ/kg of CO_2 removed [11]. In order to increase the flue gas pressure to overcome the system pressure drop, a flue gas blower is used. In addition, pumping of the amine solution and pumping of the absorber wash water is required, resulting in an increase of the unit's electricity self-consumption.

The flue gases at the outlet of the boiler are cooled down to 40°C in order to condense the water vapour and then compressed up to about 1.3 bar. CO_2 recovery from the flue gases is assumed to be 90%.

Table 13.3 summarises the simulation results of the amine scrubbing retrofit. Both efficiency and power output losses are foreseen in comparison to conventional combustion technologies application, which are higher in relation to the oxy-fuel

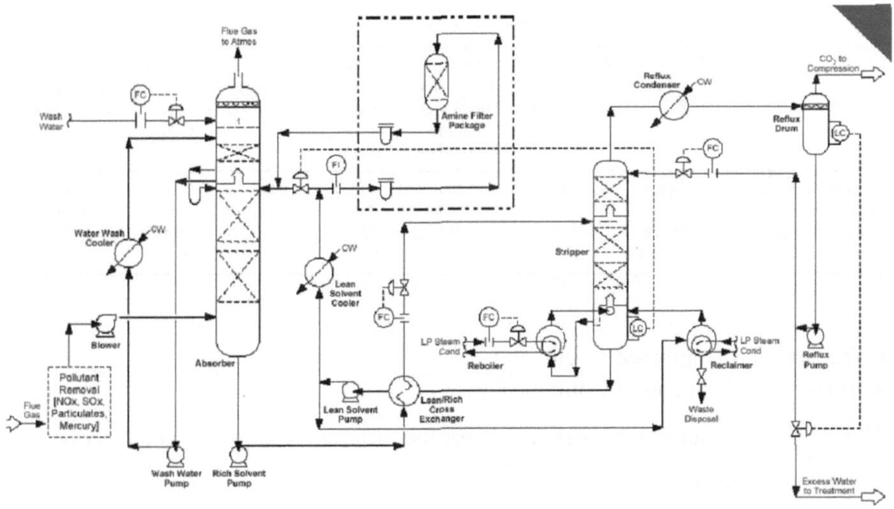

Fig. 13.4 Amine flue gas processing operation [12].

Table 13.3 Power plant with amine scrubbing simulation results.

Steam extraction	kg/s	134.9
Thermal consumption for solvent regeneration	MWth	315.7
CO_2 compression power consumption (nis= 0.85)	MW	20.5
CO_2 absorption unit power consumption (flue gas fan, solution pumps)	MW	8.7
Cooling water pumps consumption for flue gas cooling	MW	0.7
CO_2 emissions	kg/s	8.76
Specific CO_2 emissions	kg CO_2/kWh	0.17
Net power output	MW	189.6
Net efficiency	%	22.84

retrofit option. The estimated CO_2 emission reduction potential is the best among the examined cases.

13.5 Assessment of Electricity Production Cost for Various Technological Options

For the economic analysis of the electricity generation cost, the following technologies were examined:

- conventional lignite-fired power plant;
- lignite-fired power plant with amine scrubbing for CO_2 capture;
- oxy-fuel lignite-fired power plant;
- Clean Coal Technology power plant (supercritical steam characteristics);

Table 13.4 Specific assumptions.

		Conventional lignite-fired power plant	Lignite-fired power plant with amine scrubbing	Oxy-fuel lignite-fired power plant	Clean Coal Technology power plant	Natural Gas Combined Cycle	Integrated Gasification Combined Cycle
Power capacity	MWel	300	200	211	300	380	766
Efficiency	%	35.4	24.2	25.4	44.0	56.5	43.0
Investment cost	€/kW	1100	1900	1570	1150	600	1370
Specific CO_2 emissions	kg/kWh	1.075	0.17	0.17	0.865	0.37	0.76

- Natural Gas Combined Cycle; and
- Integrated Gasification Combine Cycle.

The following assumptions were made for the different technological concepts:

- rate of return: 8%;
- inflation: 3%;
- fuel cost: lignite 1.8 €/GJ/natural gas 5.5 €/GJ;
- investment life for solid fuel unit: 25 years;
- investment life for NGCC and IGCC: 15 years;
- operating and maintenance costs: 3% of investment cost annually and variable cost 0.01 €/kWh for lignite units/0.005 €/kWh for natural gas unit;
- annual operating hours: 7500 h;
- operation in full load;
- cost of transportation is not included in the cost of CO_2 removal; and
- cost of CO_2 emitted (CO_2 market prices): 18 €/t.

Furthermore, specific assumptions were made according to the different technologies, which are presented in Table 13.4 [13].

The different technologies were evaluated in relation to the electricity production cost. Fixed cost was calculated according to the investment amortisation and the maintenance and operating costs. In relation to the fixed cost, the NGCC unit has the lowest cost due to the fact that the investment cost is lower. The high investment and operating cost of CO_2 capture technologies, as well as of the IGCC plant, results in a significantly higher kWh production cost. The variable cost, in which fuel cost is included, is lower for lignite-fired and lignite IGCC units in comparison to natural gas units, due to low fuel costs. The efficiency reduction due to adoption of CO_2 capture schemes impacts the variable cost with a significant increase.

The significant variation of natural gas prices, linked with oil prices, contributes to risk on electricity generation costs. This variation can be up to 40% of the total cost of produced kWh from a NGCC unit [14]. The lignite market is not related to oil price evolution and therefore no risk is accounted for.

For the evaluation of the impact of CO_2 emissions on electricity generation prices, the technologies are divided in two categories: currently commercially available technologies are included in the first category and future technologies in the second. In the first category, conventional lignite-fired power plant, Clean Coal

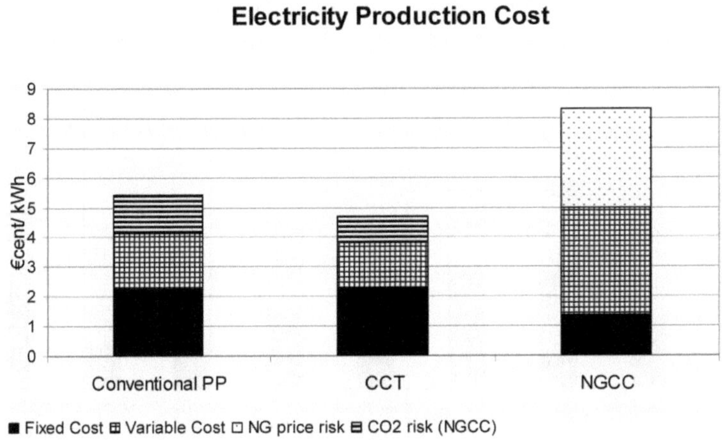

Fig. 13.5 Electricity production cost with CO_2 risk assessment based on Natural Gas Combined Cycle Unit.

Technologies and Natural Gas Combined Cycle are included. In the second category, Integrated Gasification Combined Cycle, Clean Coal Technologies and CO_2 capture technologies are taken into account. The reference case for the first category is NGCC and for the second, CO_2 capture technologies.

The difference in specific CO_2 emissions between the examined technology and the reference case multiplied by the CO_2 emission cost is adopted as an estimation of the risk related to CO_2 emitted, and it is included in the electricity production cost. This cost is low for NGCC due to the low specific CO_2 emissions, in contrast to lignite units.

The utilisation of lignite results in the reduction of total electricity production cost. The higher natural gas price as well as its price risk impact the final production cost. The higher investment cost of CO_2 capture technologies as well as the lower efficiencies, strongly affect the final kWh cost.

Furthermore, a sensitivity analysis based on the fuel price variation was carried out ($\pm 10\%$, $\pm 20\%$, $\pm 30\%$ of fuel cost). The highest sensitivity appeared in the cases of CO_2 capture technologies due to the efficiency reduction, but it was also high for the case of the NGCC unit due to the high fuel prices.

13.6 Conclusions

The state of the art of technological concepts for the retrofit of low-quality coal-fired power plants for CO_2 capture and sequestration were examined, based on the application of the two most promising options, namely the oxy-fuel process and the MEA scrubbing process. Results from thermodynamic simulations based on the retrofit of a typical modern lignite-fired power plant demonstrate the potential

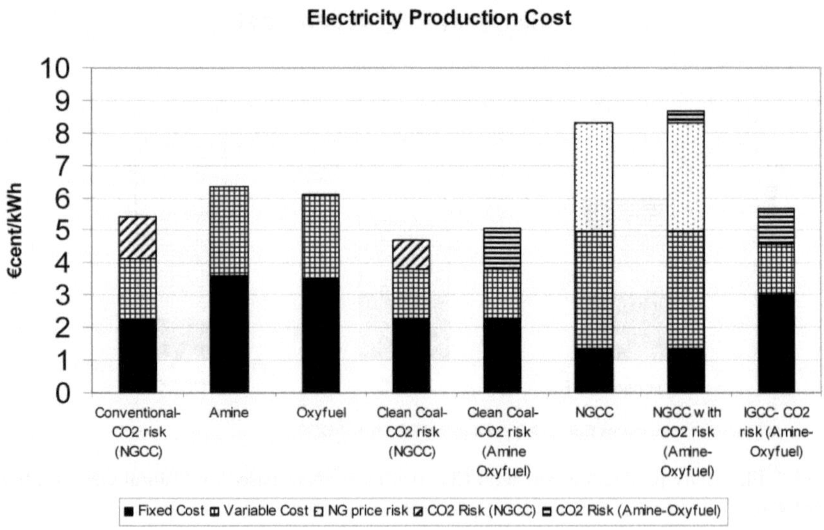

Fig. 13.6 Electricity production cost with CO_2 risk assessment based on CO_2 capture technologies.

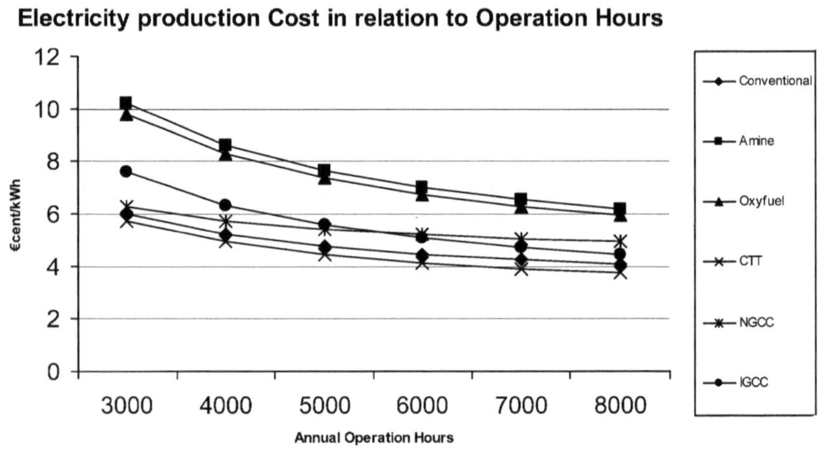

Fig. 13.7 Parametric analysis of electricity production cost in relation to annual operation hours.

of CO_2 emissions reduction of 71% and 84% for the cases of oxy-fuel and amine scrubbing respectively. The associated energy penalties are 82.7 MWel for the power output and 10 percentage points in efficiency concerning the oxy-fuel retrofit and 104.1 MWel and 12.5 percentage points for the amine scrubbing option.

The CO_2 capture technologies also have higher investment costs resulting in higher kWh prices. Furthermore, the purchase of extra emission allowances from the CO_2 market will affect the final electricity cost. From comparison analysis of different available and foreseen technologies for lignite fuel utilisation for electricity generation, it was predicted that lignite units can provide competitive kWh elec-

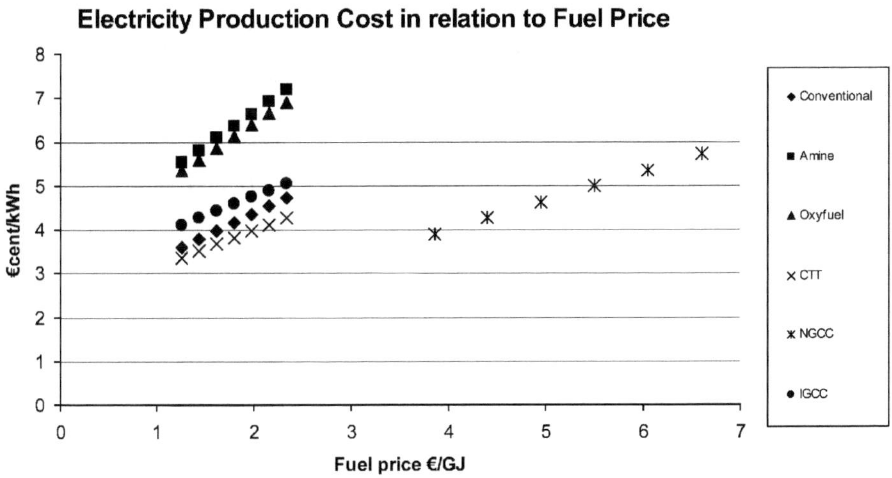

Fig. 13.8 Sensitivity analysis of electricity production cost in relation to fuel prices.

tricity production cost compared to natural gas units, which present higher risks due to fuel price variation.

References

1. Intergovernmental Panel on Climate Change (IPCC), *Climate Change 2001: The Scientific Basis*, Cambridge University Press, 2001.
2. VGB Powertech, Separation and Sequestration of CO_2 from Power Plants – Literature Survey, VGB Project Group CO_2 Separation and Sequestration, Essen, August 2002.
3. CSFTA-VGB, Size and type of existing electricity-generating capacity using solid fuels within an enlarged EU, Final Report (Contract No. 4.1004/D/99-006), Athens, 2000.
4. Kakaras, E., Grammelis, P. and Jacobs, J., Existing electricity generation capacity using solid fuels within the enlarged EU: Market size and renovation options, *VGB PowerTech*, April 2002.
5. Stamatelopoulos, G.N., Calculation and optimisation of power plant thermodynamic cycles, VDI-Regulations, Series 6, No. 340, Braunschweig, Mechanical Engineering Department, 1996 [in German].
6. Pogoreutz, M., Giglmayr, I. and Nixdorf, M., Comparison of software for calculation of thermodynamic processes, Report of VGB research project No. 177, VGB, 2000 [in German].
7. Anderson, K., Birkestad, H., Maksinen, P., Johnson, F., Strömberg, L. and Lyngfeld, A., An 865 MW lignite fired CO_2 free power plant – A technical feasibility study, in *Proceedings of Sixth International Conference on Greenhouse Gas Control Technologies*, Kyoto, Japan, October 1–4, 2002.
8. Andersson, K. and Maksinen, P., Process evaluation of CO_2 free combustion in an O_2/CO_2 power plant, Master Thesis, Chalmers University of Technology, Sweden, 2002.
9. Birkestad, H., Separation and compression of CO_2 in an O_2/CO_2-fired power plant, Master of Science Degree Thesis, Chalmers University of Technology, Sweden, 2002.
10. Singh, D., Croiset, E., Douglas, P.L. and Douglas, M.A., Techno-economic study of CO_2 capture from an existing coal-fired power plant: MEA scrubbing vs. O_2/CO_2 recycle combustion, *Energy Conversion and Management* **44**, 2003, 3073–3091.

11. Bolland, O. and Mathieu, P., Comparison of two CO_2 removal options in combined cycle power plants, *Energy Conversion and Management* **39**(16/18), 1998, 1653–1663.
12. European Technology Platform, Zero Emission Fossil Fuel Power Plants (ZEP), Working Group 1, Power Plant and Carbon Dioxide Capture.
13. VGB Powertech, *CO_2 Capture and Storage*, A VGB Report on the State of the Art, VGB Powertech, Essen, August 2004.
14. Lambertz, J., *Efficient Power Generation in Coal- and Gas-Fired Power Plants*, VGB, 2005.

Chapter 14
Renewable Energy Sources in Slovenia: Facts and Plans

Urška Lavrenčič Štangar and Evald Kranjčevič

Abstract. As a member of the European Union, Slovenia follows a European energy policy and its objectives of a sustainable, competitive and secure supply of energy. Renewable energy sources (RES) have an important role in the global fight against climate change and are at the same time an important factor for a long-term security of energy supply. With about 11% RES share in the national primary energy balance, Slovenia is still far away from meeting its renewable energy targets. This share is slightly decreasing (12% in the year 2000) in spite of increasing use of RES. The main reason is that RES utilisation is growing slower than the primary energy (electricity) demand. However, with over 54% of the country's land covered with forest, the potential for use of solid biomass is relatively high, but it also depends on application of new utilisation technologies. Nevertheless, this type of renewable energy source has started to penetrate the market, despite many barriers. The most significant RES for electricity production is still hydro power, which in part relies on a large amount of old, small hydro power plants. Their renovation as well as the installation of additional new units is a principal part of Slovenian strategy for use of renewable energy sources.

Key words: Renewable energy sources, environment, electricity from RES, biofuels, climate change.

14.1 Introduction

An increased awareness of pollution problems that present a serious threat to our environment initiated a number of actions towards broader and more efficient use of renewable energy sources all over the world. This paper roughly covers such attempts and measures for better use of renewable energy sources, RES, in Slovenia. The data presented here are based on the following most relevant documents: Slovenia – Renewable Energy Fact Sheet [1], Resolution on the National Energy

K. Hanjalić et al., Sustainable Energy Technologies: Options and Prospects, 251–259.

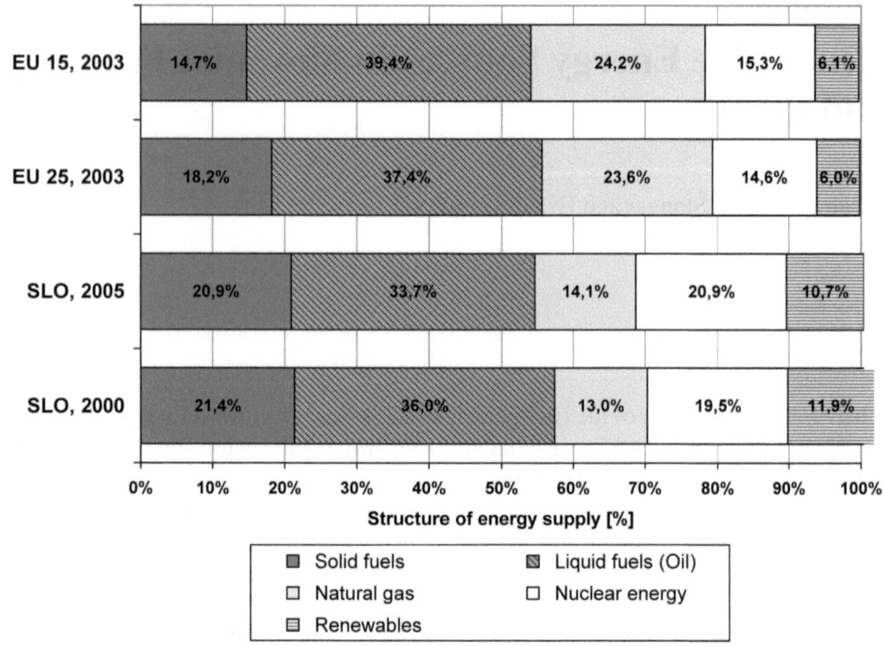

Fig. 14.1 Comparison of structure of energy supply in Slovenia and EU.

Programme (ReNEP) [2], Final Energy Review for the year 2005 [3] and The Report of Republic of Slovenia to the European Commission on Implementation of the Directive 2001/77/EC [4], which concerns the stimulation of electricity production from RES. The first one is available also in English, whereas the others are only in the Slovene language.

Slovenia is obliged to reach relatively difficult environmental targets that are set by the Kyoto Protocol (Slovenia made ratification in June 2002), an NEC directive and other environmental legal frameworks. Energy policy in the context of RES is essential for an acceptable and sustainable energy future.

Utilisation of RES is important for several reasons: lower dependence on (fossil) fuels imports, preventing climate changes and creating new working places. Slovenia is however late in carrying out corresponding measures to reach the targets set by ReNEP, which are comparable to the targets of European Directive 32/2006/EC. The target to increase the share of RES in total (gross) primary energy consumption from 9% in 2001 to 12% by 2010 has been set, in accordance with the White Paper [5] on renewable sources of energy. The target of electricity production from RES to be achieved in 2010 is 33.6% for Slovenia. The ReNEP has set the goal of increasing the share of renewable energy for heating from 22% in 2002 to 25% in 2010. The targets for biofuels have been set as follows: 0.65% in 2005, 1.2% in 2006, 2% in 2007, 3% in 2008, 4% in 2009 and 5% in 2010. The 2010 target is slightly below the guideline in the EU Biofuels Directive, which is 5.75% in 2010.

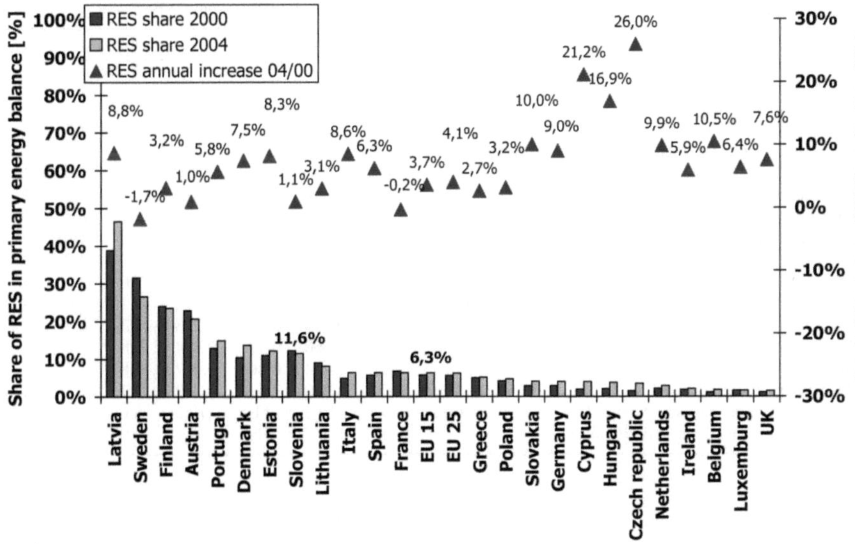

Fig. 14.2 Comparison of RES share in primary energy balance in EU countries for years 2000 and 2004, and annual average growth of energy production from RES in period 2000 2004.

14.2 RES in Primary Energy Balance

The share of RES in the primary energy balance was 10.5% in 2005, which is 1% less compared to the previous year. With respect to this value in 2000 (11.9%) (Figure 14.1), there is an obvious trend in decreasing the share of RES in the long run, as well as in the primary balance, in spite of increasing use of RES (0.3% average yearly increase of RES against 2.9% increase of primary energy consumption in the period 2000–2005). The share of RES places Slovenia in the first half of the EU countries (Figure 14.2) and a negative trend of the share of RES can be noticed also for several other European states.

In the paragraph above, the contributions of geothermal and solar energy are not considered. According to the EurObserve'ER, the use of geothermal energy in 2005 was 0.016 mio toe, while the use of solar energy 0.006 toe in Slovenia. With this correction, the share of RES in the primary energy balance increases by 0.3%, which means that the share of RES in 2005 was 10.8%.

14.3 Electricity from RES

Green electricity in Slovenia is predominantly generated by means of large-scale hydro energy. In 2005, this sector contributed 3078 GWh out of a total amount of 3574 GWh (Figure 14.3, Table 14.1). Small-scale hydro power plants generated

Table 14.1 RES for electricity production and in the primary energy balance.

	2000	2001	2002	2003	2004	2005	05/04	05/03	05/00
							Average annual growth [%]		
Electricity production from RES [GWh]	3,905	3,865	3,415	3,079	4,218	3,574	−15.3	7.7	−1.8
Hydro power plants up to 10 MW	340	371	327	268	436	384	−11.9	19.7	2.5
Hydro power plants over 10 MW	3,495	3,425	2,986	2,690	3,658	3,078	−15.9	7.0	−2.5
Wood and other solid biomass	58	54	84	98	93	82	−11.8	−8.5	7.2
Landfill gas	9	12	16	22	28	30	7.1	16.8	27.2
Gas from waste plants	3	3	2	1	3	2	−33.3	41.4	−6.6
Gross use of electricity [GWh]	12,303	12,694	13,466	13,985	14,492	14,793	2.1	2.8	3.8
Share of electricity from RES [%]	31.7	30.4	25.4	22.0	29.1	24.2	−17.0	1.1	−5.3
Domestic RES production [mtoe]	0.760	0.776	0.716	0.714	0.822	0.772	−6.1	4.0	0.3
Hydro and wind energy	0.330	0.326	0.285	0.254	0.352	0.298	−15.3	8.3	−2.0
Other renewable energy sources	0.430	0.450	0.431	0.460	0.470	0.474	0.8	1.6	2.0
– share wood and other solid biomass			0.426	0.454	0.463	0.467	0.8	1.5	
Total primary energy [mtoe]	6.360	6.749	6.855	6.930	7.128	7.323	2.7	2.8	2.9
Share of RES in primary en. balance [%]	11.9	11.5	10.4	10.3	11.5	10.5	−8.6	0.1	−2.5
– share of wood and other solid biomass			6.2	6.5	6.5	6.4	−1.9	−0.1	

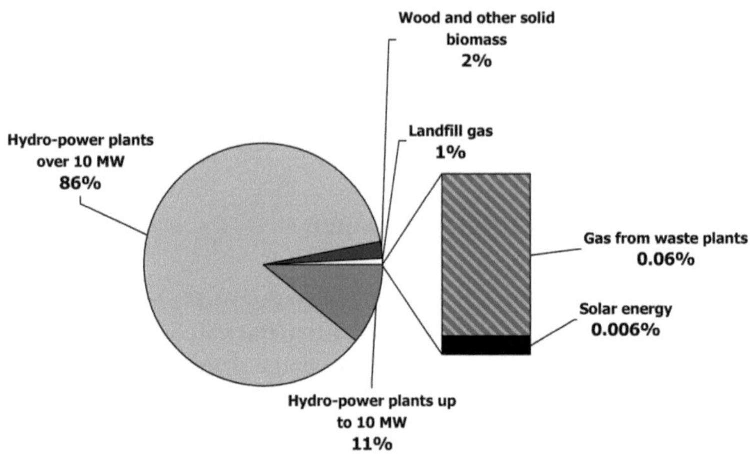

Fig. 14.3 The share of different RES in green electricity production (year 2005).

384 GWh, and recorded an annual increase of 17% according to the average between the years 1977 and 2004, when the highest production was achieved (436 GWh). Bio-electricity (e.g., landfill gas plants, biogas plants, etc.) covered the remaining part of the electricity market by generating 82 GWh (through solid biomass) and 30 GWh (through biogas) in 2005.

The fact is that the portion of electricity produced from RES is generally decreasing (Figure 14.4). For example, in years 2000 and 2001 we were not below the values from the European Directive, but in 2005 only a 24.1% share was achieved, meaning that Slovenia is 7.5% behind the target of 33.6% set for the year 2010. There are several reasons for this shortfall, one of them being the fact that electricity production from RES in Slovenia increases very slowly. Another fact that makes reaching the target even more difficult is dependence of hydro power plants

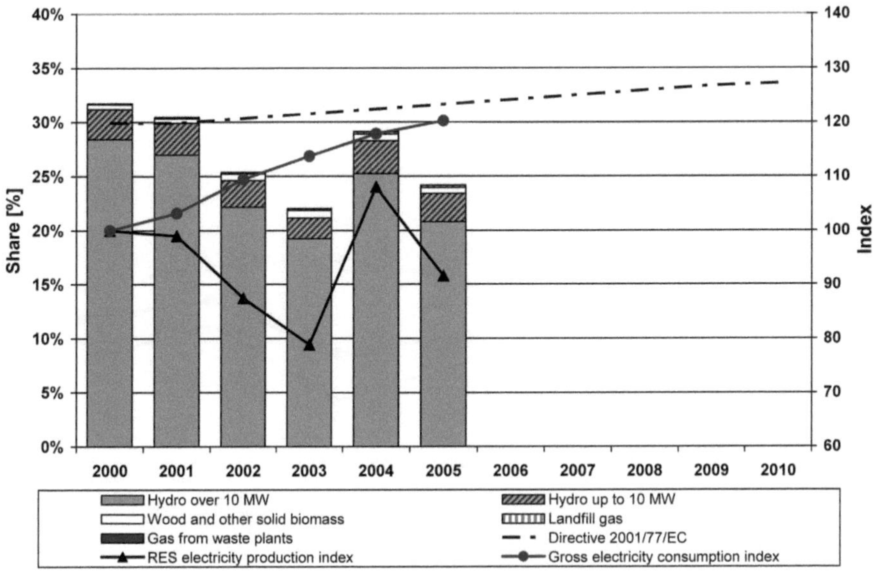

Fig. 14.4 Share of electricity production from RES in gross energy consumption.

(the main source of green electricity in Slovenia) on annual rainfall (and quantities of snow in the Alps). Nevertheless, the biggest obstacle to reaching the target is the fast (and continuous) growth of electric energy consumption in recent years, on average 4% a year, which exceeds the growth of electric energy production from RES (and other conventional sources as well).

An increase of electricity production from biogas (Figure 14.5) is the consequence of some investments in communal companies and animal (especially pig) farms. Among all different RES, the most significant increase in capacities in 2005 were attained by the electricity production from solar power plants. The growth of production capacities in 2005 was the following: biomass 0.55 MW, solar energy 0.12 MW, large-scale hydro power plants 5 MW. This progress is however very small with respect to the aims set by European legislation and the national energy programme.

14.4 Biofuels

The year 2005 represented a big step forward in the regular use of biofuels in vehicle traffic, according to the Directive 2003/30/EC and ReNEP. The appropriate national regulations that were adopted in 2005 determine also the types of biofuels in vehicle traffic and the minimum content of biofuel in the fuel for automotive applications. The measures of agricultural policy for the year 2006 stimulate the harvesting of

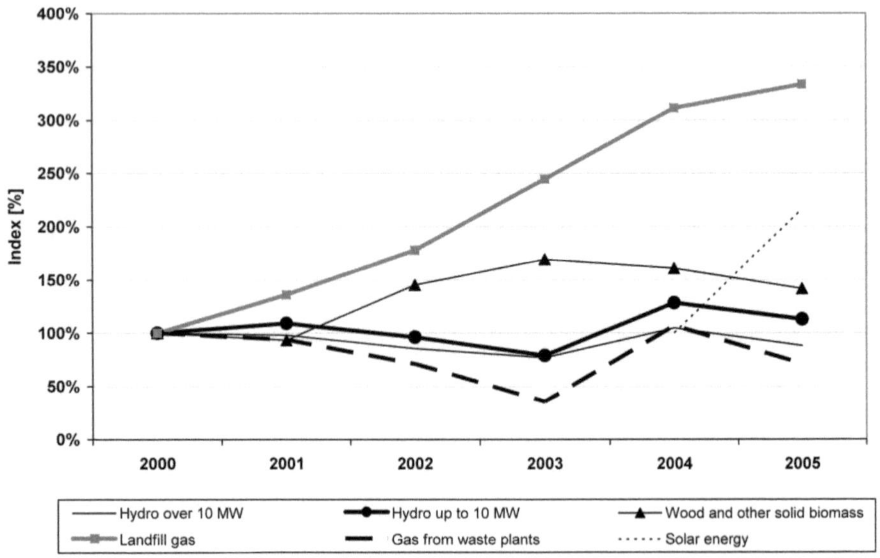

Fig. 14.5 Index of electricity production for different RES in the period 2000–2005.

appropriate crops for biofuels production. The biofuels were first implemented in diesel fuels on an experimental basis already in 2004. The admixed biodiesel was partially imported from third countries or other EU member states and partially produced at Slovene industrial plants for vegetable oils. A biofuel share of 0.35% by energy content was reached in 2005 (7000 tonnes), compared to the target of 0.65% (10,000 tonnes).

The main source for biodiesel production, which also thrives in Slovenia, is the rapeseed plant. It is expected that by the year 2010 approximately 35 km^2 of land will be set aside for rapeseed harvesting. This arable surface would be sufficient for the production of 2080 tonnes of rapeseed plant in an ecologically conservative way or 4200 tonnes in the classical way.

The biggest industrial plant for biodiesel production in Slovenia (and one of the largest in Europe) is planned to be built in the northeast of the country by the Nafta Lendava company together with the Austrian partner CMB Maschinenbau & Handels. It will begin production in 2008. There are however some other producers that will start operating next year as well. The annual production is estimated to be 60,000 tonnes of biodiesel and 6000 tonnes of methanol, which represents 88% of total production in Slovenia. The raw materials will be imported at the beginning, but later on the plant will be more oriented to national suppliers, in accordance with the stimulations and measures of agricultural policy. In 2005, biofuels were also tested in Ljubljana public transportation (Figure 14.6), in the frame of the European project Civitas II Mobilis that incorporates the cities Toulouse, Debrecen, Venice, Odense and Ljubljana. At first the mixture of 20% of biodiesel and 80% of petrochemical diesel was used in the fuel, but due to some storing problems of such

Fig. 14.6 Ljubljana city bus using biodiesel.

Table 14.2 Contributions of biomass, solar thermal and geothermal RES to production of heat in Slovenia.

	Penetration 1997 (ktoe)	Penetration 2004 (ktoe)	Av. Annual growth [%]
Biomass heat	260	430	7%
Solar thermal heat	3	4	6%
Geothermal heat incl. heat pumps	12	15	4%

mixtures, they went over to pure biodiesel in February 2006 [6]. Until June 2006, nearly 35,000 km were traversed in Ljubljana by the two city buses. The results of appropriate measurements showed that the quantity of solid particles in exhaust diminished by 58%, the motor power decreased by 6% and the fuel consumption increased by circa 9% with regard to the petrochemical diesel D2. More reliable results will be available after a higher number of kilometres has been traversed (factors of coincidence will be then excluded).

14.5 RES Heating and Cooling

In Slovenia, biomass heat is the main category of heat obtained from RES with a production of 430 ktoe in 2004. Its growth exceeds that of solar thermal heat and geothermal heat (Table 14.2). The contribution of RES in fuels for heat production was 23% in 2005.

In the year 2005, the Ministry of Environment and Spatial Planning supported the exploitation of RES with financial incentives to individual families to the extent of about 460,000 € and to public and private sectors to the extent of about 1.2 mio €. Within the framework of the GEF project "Elimination of hindrances for the enlarged exploitation of wood biomass as an energy source", which aims to stimulate and accomplish three to five remote systems for heating with wood biomass, two projects of remote heating with wood biomass in the communities Vransko and

Table 14.3 District heating plants using wood biomass, accomplished in the frame of the GEF project.

Location	Investment (Mio EUR)	Installed capacity Wood biomass (MW)	Installed capacity Other (MW)	Heat demand (GWh/a)	Number of consumers (envisaged)	Network distance (m)	CO_2 emission reduction (t/year)
Vransko	1.93	3.2	1.5	6.6	126	3,040	1,670
Kočevje	2.17	4.5	0	14.6	1,359	3,605	3,350
TOTAL	4.10	7.7	1.5	21.2	1,485	6,645	5,020

Kočevje were already finished in 2005. Both systems are already in operation with the total installed power of 7.7 MW. It is expected that the operation of these systems will reduce the emissions of CO_2 by 5 kt a year (Table 14.3).

Among the most important events in this field recently is also the acceptance of the Action plan for biomass by the European Commission, which assures uniform measures for stimulating energy production from biomass through development of the market. The action plan deals with energy generation by use of biomass from wood, wastes and crops for the production of heat, electricity and transport.

14.6 Conclusion

Slovenia has reserves of poor quality coal, but no oil and gas. Therefore, Slovenia is completely dependent on liquid and gaseous fuel imports. Coal, almost exclusively used for electricity production, is mostly obtained from domestic resources. Therefore, the energy policy, through appropriate measures, shall give priority to other energy sources, especially renewable energy sources whose impact on the environment is generally considered as positive, notably concerning CO_2 emissions limitations.

At the end, let us introduce some examples of new initiatives for promoting utilisation of RES in Slovenia. The first one is a part of the European SARA project, which aims to construct seven sustainable, high energy performance, public-access eco-buildings that demonstrate the cost-effective potential for including renewable energies and improving the energy performance of Europe's buildings. The supermarket in Ljubljana will be equipped with advanced energy technologies (photovoltaic and natural ventilation) integrated by an innovative architectural approach (natural lighting and heat recovery) and combined monitoring and building management systems [1]. The second one concerns a very recent initiative to establish the Centre for renewable energy sources and environment protection (COVEVO), which will be situated in the southwestern part of Slovenia. The founders of COVEVO are: Ministry of the Environment and Spatial Planning, Ministry of Higher Education, Science and Technology, University of Nova Gorica, Pivka Community and the Regional Development Agency. By organising workshops, demonstration activities in pilot laboratories and various other initiatives, it will promote the importance of using RES in order to protect our environment. The third initiative is a project to establish an applied technology institute at the coastal region, where one of the tasks will be biofuels research and production. It also plans to offer a graduate study programme in the field of environmental technologies and renewable energy sources.

The new projects described above are still at their initial stage. The present stage and some milestones of future plans for RES utilisation are presented within this paper. Although the current situation is not yet at the desired level, there is still a significant potential left for further development in use of renewable energy sources in Slovenia.

Acknowledgement

The authors wish to thank Mr. Silvo Škornik, MSc., from the Ministry of Economy of the Republic of Slovenia for his help in supplying appropriate literature references.

References

1. Slovenia – Renewable Energy Fact Sheet, available online at http://ec.europa.eu/energy/energy_policy/facts_en.htm.
2. Resolucija o Nacionalnem energetskem programu (ReNEP), Uradni list Republike Slovenije, Uradni list št. 57, str. 7390, 27 May 2004 [Resolution on the National Energy Programme (ReNEP), The Official Gazette of the Republic of Slovenia, No. 75/2004, p. 7390, 27 May 2004].
3. Letni energetski pregled za leto 2005, Končno poročilo, Institut "Jožef Stefan", Center za energetsko učinkovitost, IJS-DP-9467, Ljubljana, 2007 [Annual Energy Overview 2005, Final Report, Jožef Stefan Institute, Energy Efficiency Centre, IJS-DP-9467, Ljubljana, 2007].
4. Poročilo Republike Slovenije Evropski Komisiji o implementaciji Direktive 2001/77/ES Evropskega parlamenta in Sveta o spodbujanju proizvodnje električne energije iz obnovljivih virov energije, Ministrstvo za gospodarstvo Republike Slovenije, Direktorat za energijo, Ljubljana, February 2006 [Slovenia's Report to EC about Implementation of the Directive 2001/77/EC, Ministry of Economy, Directorate for Energy, Ljubljana, February 2006].
5. *Energy for the Future – Renewable Sources of Energy*, White Paper, COM(97) 599 final.
6. Šmajdek, J., Presentation of CIVITAS MOBILIS project, Ljubljana, 8 June 2006 (available in Slovene online at http://www.slovenija-co2.si/aktualno/civitas.pdf).

Chapter 15
Energy Wood Chains in Bulgaria

P. Stankov, D. Mladenov and K. Stanchev

Abstract. The present status of energy wood production and use in Bulgaria is considered in the framework of the current energy policy of the country, existing laws and the respective regulations, the energy market and entrepreneurship in this sector of the Bulgarian economy. We begin with an overview of Bulgarian forests and forestry and wood processing industry and then present an analysis of the structure of wooden biomass and its potential for energy production. Current status of the technology for harvesting, transportation and storage of wooden biomass is considered, followed by an overview of production of pellets and briquettes and other ways of biomass utilisation for energy production. The chapter closes with a perspective on future development of the wood chain industry in Bulgaria.

Key words: Energy wood chains, wood biomass, renewable energy, forest, pellets, brackets, afforestation.

15.1 Introduction

The use of renewable energy sources is becoming ever more important as it concerns several aspects of modern society, such as human-induced climate change – the greenhouse effect, the long-term effect on ecosystems, decreasing reserves of fossil fuels, etc. The Kyoto Protocol was signed in order to limit the rate and level of global climate change. The resulting action plan in the European Union's White Paper (1997) "Energy for the future: renewable energy sources" gives a directive of 12% market share of renewable energy sources by 2010. The published Green Paper as well as other EU directives, such as the Directive on Biofuels for Transportation and the Emission Trading Scheme Directive, strengthen the EU policy towards the use of bioenergy on a sustainable basis. Development and consolidation of European markets and harmonisation of regulations is an important part of EU policy. Nowadays bioenergy is incorporated in the energy policies of all EU and many other countries.

K. Hanjalić et al., Sustainable Energy Technologies: Options and Prospects, 261–280.

The largest projected potential for renewables in the EU is for biomass. The global potential of wood biomass in particular is estimated to be 20×10^9 t. However, only about 10% of this potential is used, even though it has been the main fuel used by people throughout the ages. Market wood fuels are forest industry by-products, logging residuals, upgraded materials such as briquettes, pellets and powder as well as demolition wood. There is quite significant variety in the quality of wood biomass fuels – depending on their storage, handling and the combustion technology.

Bioenergy use has socio-economic aspects. There are socio-economic and environmental drivers, as well as obstacles to the use of bioenergy. The main economic drivers are: additional economic activity; establishment of a bioenergy market; creation of new permanent local jobs; reduction of dependency on energy inputs; reinforcement of the energy supply for local communities; preserved areas and green tourism; contribution to the local research and development potential etc. There are also clear environmental drivers, such as reducing greenhouse gas emissions, emissions trading and better use of local resources. The social drivers concern long-term local employment including new employment, mitigation of rural depopulation and, finally, self-sufficiency.

All these drivers, taken in their entirety, presuppose the sustainable use of biomass fuel. At the same time however, there are several obstacles, such as community indifference, lack of consistency and coordination in government bodies, lack of understanding of the market, unexpected maintenance costs, insufficient demand, etc.

Bioenergy-related policy is concerned with a set of issues such as climate change mitigation, air and water pollution, solid waste and landfill management, rural economic development, fossil fuels supply, prices, etc.

Bulgaria is a country with very good potential for sustainable development in biomass energy use, as far as its geographic and climatic conditions are concerned. About 33% of the land is forestry and most of the rest is agricultural land. Forest has good potential for wood biomass as it is a direct source of biomass, from logging operations (logging residuals) and industrial by-products generated by the forest industry during the processing of timber, plywood, particleboard, etc.

This chapter focuses on energy wood chains in Bulgaria. An energy wood chain is defined as a chain of related operations from the energy wood source to the end use. It is a holistic approach to the energy wood problem, taking into account both energy wood production and utilisation chains.

15.2 Energy Wood Production Chains

15.2.1 Description of the Bulgarian Forests and Forestry

The forests of Bulgaria are of priceless importance for the country, and are a part of European and world wealth. Besides their key function of biodiversity conservation

Table 15.1 Dynamics of total forest area of Bulgaria.

	2000	2001	2002	2003	2004	Units
Total	3,914,000	3,980,000	4,003,000	4,015,000	4,064,000	ha
Coniferous	1,282,000	1,295,000	1,291,000	1,289,000	1,288,000	ha
Non-coniferous:	2,632,000	2,685,000	2,712,000	272,600	2,776,000	ha
• high-stemmed	1,535,000	1,541,000	1,525,000	1,501,000	1,478,000	ha
• low-stemmed	1,097,000	1,144,000	1,187,000	1,225,000	1,298,000	ha

Table 15.2 Dynamics of wooded forest area of Bulgaria.

	2000	2001	2002	2003	2004	Units
Total	86.2	86.5	87.2	87.8	89.2	%
Coniferous	28.5	28.2	28.0	28.0	27.8	%
Non-coniferous:	57.7	58.3	59.2	59.8	61.5	%
• high-stemmed	31.6	31.5	31.4	31.2	31.4	%
• low-stemmed	26.1	26.8	27.8	28.6	30.0	%

and provision of quality natural resources, the forests have global, national, regional, economic, ecological and social importance.

Table 15.1 provides information about the dynamics of the total forest area of Bulgaria for the period 2000–2004, according to the Bulgarian Ministry of Agriculture and Forestry. The total afforested area is 3.4 million hectares or 86% of Bulgarian forest fund. Wooded forest area is land under natural or planted stands of trees and brushwood. The area covered with pine-scrub, varying between 22,000 ha and 23,000 ha, is excluded from the wooded area.

Table 15.2 provides information about the dynamics of wooded forest area of Bulgaria for the same period.

Spruce, black pine and white pine are referred to as coniferous. High forest stands of oak, beech, hornbeam, elm, ash, lime, chestnut and others of seedling origin which are natural or artificially established (through afforestation by planting or seeding) are referred to as non-coniferous.

Information about the distribution of the Bulgarian forestland by above sea-level heights and terrain slope is presented in Tables 15.3 and 15.4, respectively. The distribution of the forestland by above sea-level heights is shown on the map of Bulgaria in Figure 15.1.

The flat forestland species are mainly low stems with low productivity. In addition to this the lower percentage of this terrain (11.9%) preconditions its negligible contribution to the total wood harvesting in Bulgaria (around 14%).

The majority of Bulgarian forestland (59%) is located in the mountain regions, where the terrain is steep or very steep. A large part of Bulgarian hilly forestland is located on sloping terrain.

A huge part of the Bulgarian forestland – 1,334,000 ha (38.75%) is special-function forests – protective, recreational (around sanatoriums) and protected bio-

Table 15.3 Forestland distribution by above sea-level height.

Forestland	Height, m	Area, %
Flat	0–200	11.9
Hilly	201–600	29.1
Mountainous	over 600	59.0
• low mountainous	601–1000	31.1
• medium mountainous	1001–1600	23.3
• high mountainous	over 1600	4.6

Table 15.4 Forestland distribution by terrain slope.

Flat	Slanting	Sloping	Steep	Very steep
0–4°	5–10°	11–20°	21–30°	over 30°
5.2%	12.0%	28.6%	40.6%	13.6%

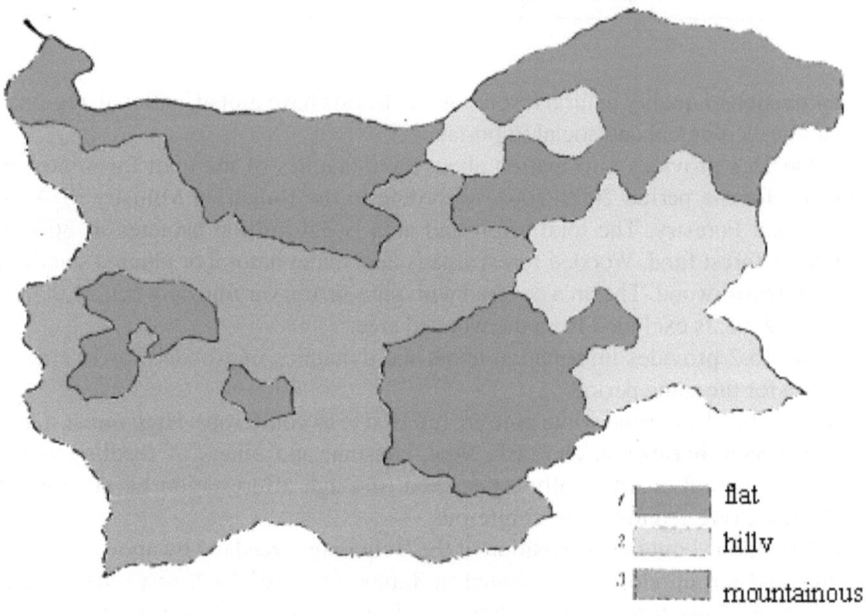

Fig. 15.1 Distribution of the Bulgarian forestland by above sea-level height.

sphere reserves. Three national parks, nine natural parks and 46 biosphere reserves are established on Bulgarian forestland.

The average age of Bulgarian forests is 49 years. Around 55% of Bulgarian forests are young (less than 40 years). The total annual increment of the biomass within the Bulgarian forest fund is 12.3 million m^3 and its expected use is about 4.6 million m^3. The annual increment of the biomass is 3.3% per hectare.

The area of high-stem deciduous forests is decreasing. The regeneration of high-stem deciduous tree vegetation, in places where it has been inadequately replaced by coniferous or transformed into coppice vegetation, is a major objective of the forest management in the country. In order to achieve a faster wood reproduction and to enhance the forests' special functions, the tree composition in some stands, especially in those of small density, has to be reconfigured. Taking into account the situation of the forestry sector, it can be assumed that the rate of forest reconstruction will be slowed down. The large area of the coppice forests intended for conversion into seed forests, and the deterioration of their condition, remains a major challenge to forest management in Bulgaria.

In the wood industry production in Bulgarian forests, felling and pruning are completely mechanised and done with motor saws. A predominant part of wood production projects is in mountainous conditions, in steep areas with a slope of over 25%. The transportation of wood is by tractors, ropeways and animals (horses and mules), depending on the concrete area conditions. When transport technology is determined, the ecological requirements for the protection of surroundings are taken into consideration.

The reason for decreasing wood production is poor density of the forest road network, which amounts to only 7.9 m/ha. That is why, even today, free access to some forest areas is not possible with any means of transport.

To clarify trends in the use of logging residues and primary processing of wood materials, it is necessary to know their structure, quantity, concentration and distribution by area. More information is given in Table 15.5.

The place where residues are produced depends on the technology applied during the harvesting operations. In our conditions larger waste quantities are obtained in clearings and smaller ones in temporary stores. More residues are obtained in temporary stores when whole trees and whole stems are transported using tractors. The adverse state of coppice for conversion and low-stem forests shows that a large part of the wood in these forests could be lost. The reason is natural thinning and felling without the benefit of utilisation. The annual increment of coppice for conversion and low-stem forests is over 3.2 million m^3 and the expected use is 1 million m^3 per year.

The utilisation of coppice for conversion and low-stem forests will increase by 2.2 million m^3 and thus it will not exceed the natural wood growth. Ten percent of this amount consists of logs, roots and leaves which are not used for economic and ecological reasons. The remaining 2 million m^3, about 3.7 million Gcal or 390,000 toe, could be used for energy needs.

Forest fires have been the greatest problem over the past years. In the year 2000 alone, 13,924 ha of coniferous forests and 37,391 ha of deciduous forests were destroyed by fire. This required the organisation of immediate medical felling in the areas destroyed by fire and in forests damaged by natural calamity. The diverse residues, which in many cases are used for power purposes, are obtained by processing timber in wood processing and furniture factories.

Table 15.5 Fallings and removals of main cutting and thinning and cleanings.

	Thousand m³ over bark				
	2000	2001	2002	2003	2004
	Total				
Fallings	4629	4134	5284	6570	7180
Removals	3755	3363	4332	5388	5892
Industrial wood	1593	1447	1771	2172	2459
Firewood	2162	1916	2561	3216	3443
Wood wastes	874	771	952	1182	1288
%	18.88	18.65	18.02	17.99	17.94
	Coniferous forest				
Fallings	1638	1626	1887	2368	2694
Removals	1200	1199	1404	1761	2031
Industrial wood	862	814	993	1317	1554
Firewood	338	385	411	444	477
Wood wastes	438	433	483	607	663
%	26.74	26.63	25.60	25.63	24.61
	Non-coniferous forest				
Fallings	2992	2508	3397	2368	2694
Removals	2555	2164	2928	1761	2031
Industrial wood	731	633	778	1317	1554
Firewood	1824	1531	2150	444	477
Wood wastes	437	334	469	607	663
%	14.60	13.32	13.80	25.63	24.61

Afforestation

Information on the dynamics of afforestation for the period 1997–2004 is presented in Table 15.6. These data are based on the inventory of forests. Substitution of forest-trees is the cutting of unacceptable tree species (out of condition) and the afforestation of new ones of good genetic and technical quality.

Afforestation in mature non-regenerated forest means the afforestation carried out in mature forests (of an age for main cutting) for which natural regeneration is not sufficient. Afforestation of sparse forests is carried out in areas with a small number of trees per unit of area.

Establishment of two-storeyed plantations means afforestation of given forested areas by planting saplings of appropriate tree species, which is designed to increase their productivity. Establishment of intensive plantations covers the afforestation of areas with very good soil and climatic conditions with appropriate tree species, aiming at a speedy accumulation of wood mass.

Reforestation of crops is the afforestation of artificially established forest areas where the density of saplings is lower than expected. The reforestation of the crops area of artificial forests is excluded from the total afforestation area.

Table 15.6 Dynamics of afforestation in ha.

	2000	2001	2002	2003	2004
Preparation of area	6056	5475	8295	8105	6414
Afforestation	6313	5031	7134	8377	7532
Of which:					
Erosion area	128	81	–	14	77
• afforestation	2302	2513	4198	4894	4068
• Substitution of forest-trees	1610	1193	1057	1260	1272
• Mature non-regenerated rests	2047	1056	1744	2038	1987
• Sparse forests	339	256	135	191	199
• Establishment of two-storeyed plantations	15	13	–	9	6
Establishment of intensive plantations	952	634	881	967	989
Reforestation of artificial forests	2086	2344	2733	2352	2562

The distribution trends of afforested area by forest type are the following:

- Afforested area increases by 3.75% compared to 1995 due to the inclusion of forests which have not been uprooted or newly organised.
- Non-wooded area suitable for afforestation increases to 17.74% due to burned down forest, which increased from 6952 ha during 1995 to 22,049 ha during 2000 (three times as much). The renovated cutting area increases by nearly 60%.
- Coniferous forests – percentage decreases for the period 1996–2000 from 34.8% to 33%.
- Non-coniferous forests – percentage increases for the same period from 65.2% to 67%.

Forest Certification

There are two main systems for forest certification: (1) FSC (Forest Stewardship Council) and (2) PEFC (Pan European Forest Certification). FSC is widely used in Europe and the USA as an international forest standard which guarantees the ecological origin of the wood. The use of certified wood stimulates both the preservation of the environment and biodiversity. It also guarantees the interest of the local population. In fact, forest certification concerns the energy wood chain from the source to the end use. In Bulgaria today, 22,000 ha are certified in the Smolyan region and another 100,000 ha are in the process of certification. However 1,000,000 ha are planned to be certified in the next five years.

The national FSC system is being developed by the Bulgarian Association for Forest Certification. The development and implementation of a national methodology for the monitoring of high conservation value forests, HCVF, is also foreseen, as it concerns the maintenance of the balance between a sustainable development of the economy and preservation of the most valuable forestry.

Table 15.7 Wood production of wood processing industry.

No.	Type of material	Measure	2003	2004
1	Broad-leaved wooden materials	Thousand m^3	240	267
2	Oak parquet	Thousand m^3	228	104
3	Shaped details	Thousand m^3	792	803
4	Plywood	m^3	6701	6487
5	Coniferous wooden materials	Thousand m^3	331	227
6	Plates of wooden fractions	Thousand m^3	189	138
7	Window frames	Thousand m^3	38	46
8	Doors and door frames	Thousand m^3	295	307
9	Pallets	Thousand m^3	1583	1952
10	Crates, cases, etc.	t	1869	1688

Table 15.8 Wood residue volumes in % of production ranges volumes.

	Production ranges	Residues, %
1.	Boards and details of coniferous trees	39
2.	Slabs of wood fragments	20
3.	Slabs of wood fibres	20
4.	Plywood	11

15.2.2 Wood Processing Industry

The wood industry in Bulgaria plays an important role in the sustainable development of the country's economy. It is preconditioned by both good forestry as a resource and good traditions, experience and qualification. In the year 2000, 99% of the operating units of the wood industry were already privatised. Practically all of them (99.5%) are SMEs (98% small, 1.5% medium size) and only 0.5% of them are large companies. Information on wood production is given in Table 15.7.

The content of wood waste in the percentage of wood processing industry materials is shown in Table 15.8. The largest percentage of residues is obtained in sawmill and plywood production.

The following residues are obtained from the production of boards and other ranges of round wood:

- Slabs: 5–10%;
- Small pieces: 2–5%;
- Laths: 10–15%;
- Scrap: 2–5%;
- Sawdust: 10–15%.

Sawdust remains almost unused. It represents a big technical and ecological problem for the wood processing industry. It is more suitable for briquette production and is a large reserve of raw-material.

In many cases bark content determines the use of the timber in the production of a diverse range of items. Unused wood-processing industry waste, including branches

Table 15.9 Distribution of wooden biomass.

	Coniferous	Broad-leaved
Branches – leaves	12%	2%
Branches – loppings	10%	15%
Sawn and round firewood	5%	15%
Pulp-wood for boards	15%	10%
Logs	40%	30%
Bark	10%	8%
Stumps	8%	5%

and lopping, is left on clearings. They are obtained by felling and primary processing of timber, which is not used for construction wood or firewood. Currently the felling waste is left on site. The average amount of waste is 10% of coniferous standing wood and 12% of deciduous standing wood. Logs, roots, leaves and other felling waste are not included in that quantity. Total unused wood processing industry waste is about 525,000 m^3 per year. The weight of that dry wood is about 210,000 t, with 1 million Gcal equivalent to 110,000 toe.

15.3 Wood Production Chains

15.3.1 Structure of Total Wooden Biomass

The total structure in coniferous and broad-leaved tree species is of great importance in the choice of process technologies during the harvesting and transportation of wooden biomass and its utilisation for industrial and energy purposes.

The distribution of the forest biomass waste for energy use, for average natural and production conditions, is given in Table 15.9 and in Figures 15.2 and 15.3. Up to 38% of the coniferous tree species biomass is suitable for energy use. For the broad-leaved tree species this percentage is up to 45%.

15.3.2 Factors Influencing the Harvesting and Transportation of Wooden Biomass

The conditions under which harvesting and transportation of wooden biomass are carried out in Bulgaria vary with the above-sea-level height and terrain slope. About 60% of the forests are situated in mountain regions (see Table 15.3). A considerable part of the forests is situated in hilly regions (see Table 15.4). These peculiarities of our forests determine the factors that influence the harvesting and transportation of wooden biomass. The main factors are:

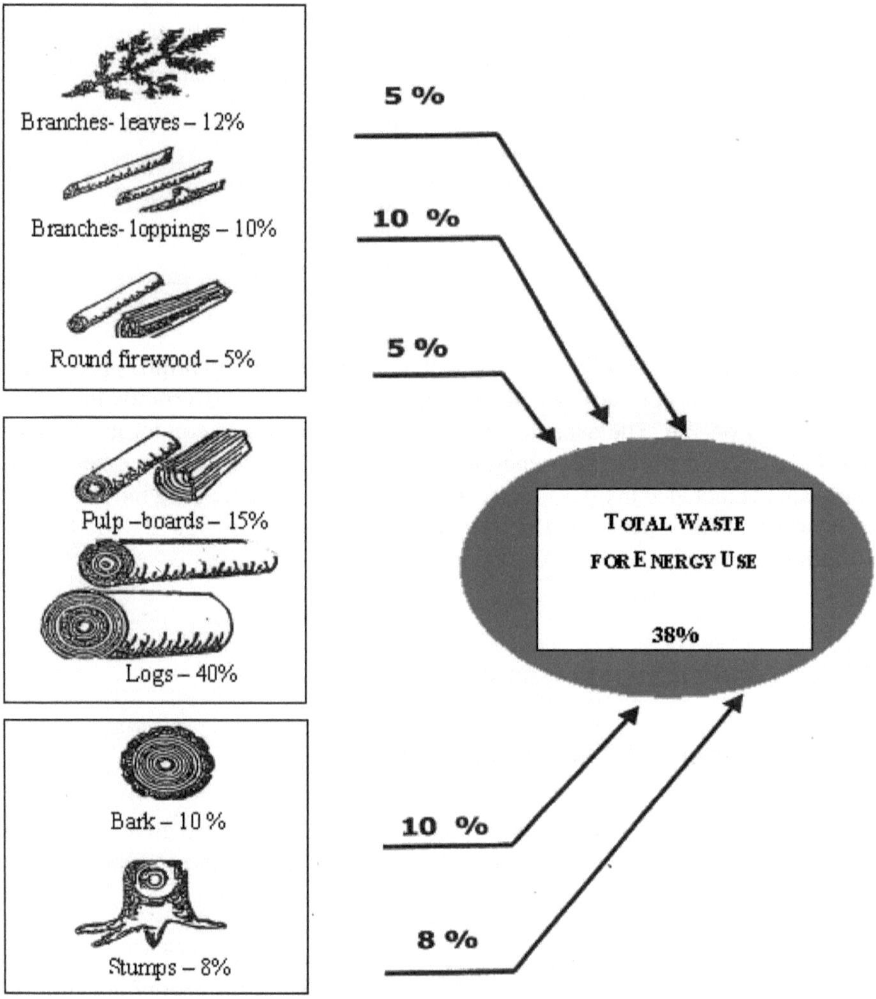

Fig. 15.2 Coniferous wooden biomass waste for energy use.

- Forest use conditions, characterised by:
  - relief and inclination of the locality;
  - type and taxation characteristics of forest stands;
  - concentration of materials harvested; and
  - type of fellings performed.
- State and density of road network in forests.
- Type and number of machines and facilities used.

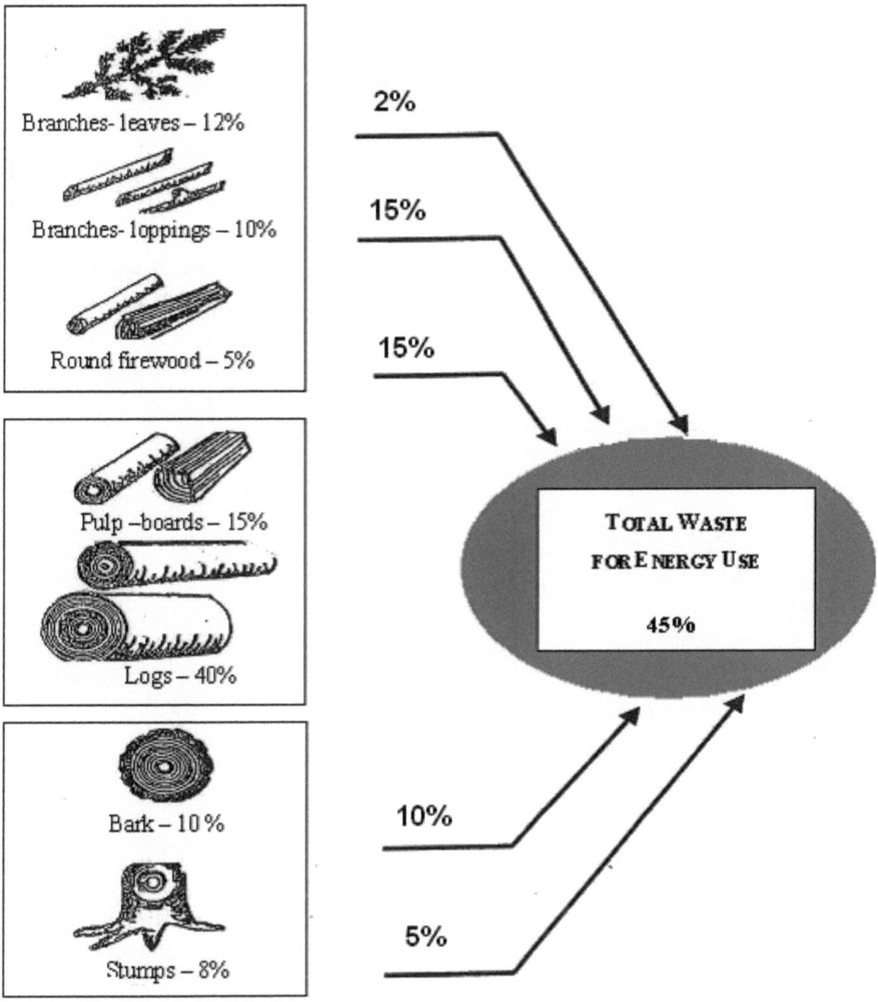

Fig. 15.3 Deciduous wooden biomass waste for energy use.

15.3.3 Technological Process during Harvesting and Transportation of Wooden Biomass

The technological process is a part of the forestry production process and is an aggregate of individual operations related to each other in a certain sequence. The technological process includes three phases of production: works in the cutting area, works related to transportation, and works in timber-yards.

Three series of technological processes are mostly used in Bulgaria, depending on the type of production transported from the cutting area:

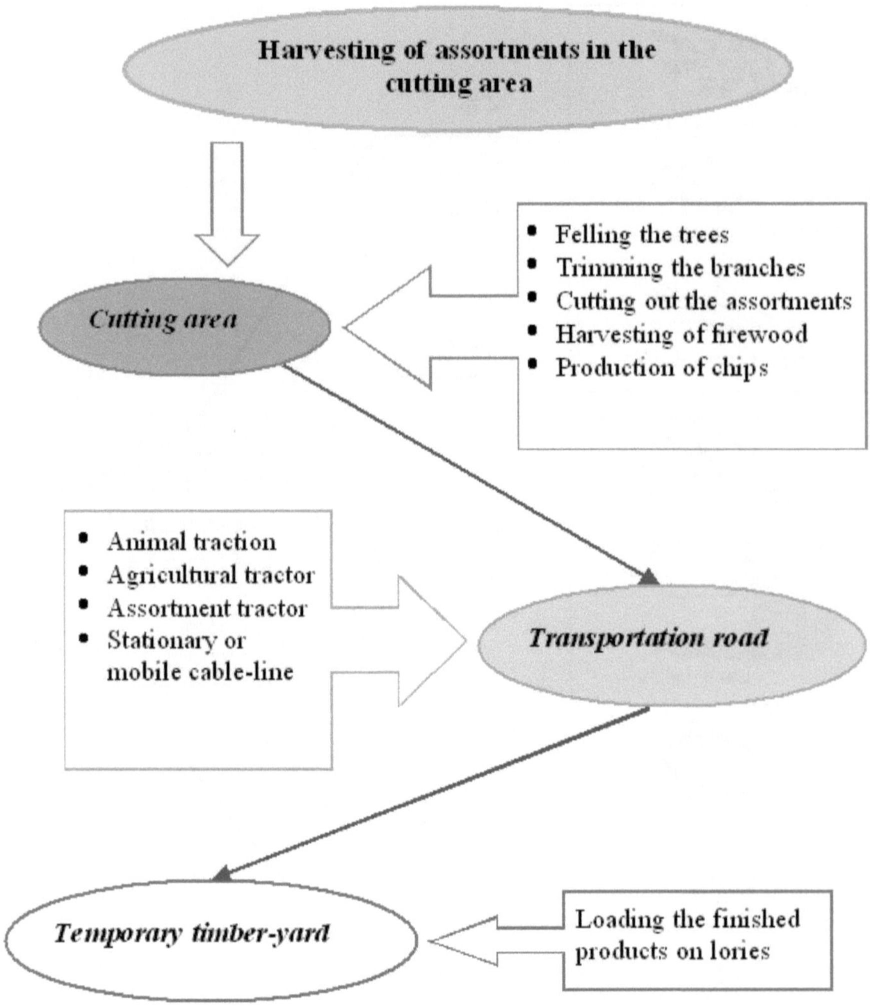

Fig. 15.4 Harvesting of assortments in the cutting area.

- *Harvesting of assortments in the cutting area.* The technological operations performed in this process are shown in Figure 15.4.

This is the most frequently used process in practice. It is characterised by the fact that a great part of the operations are performed in the cutting area under difficult working conditions, especially in mountain conditions.

- *Harvesting of assortments in a temporary timber-yard.* In this process, the technological operations performed are as shown in Figure 15.5.

This technology is characterised by a reduced number of operations in the cutting area and the transfer of a greater part of the operations to a temporary timber-yard.

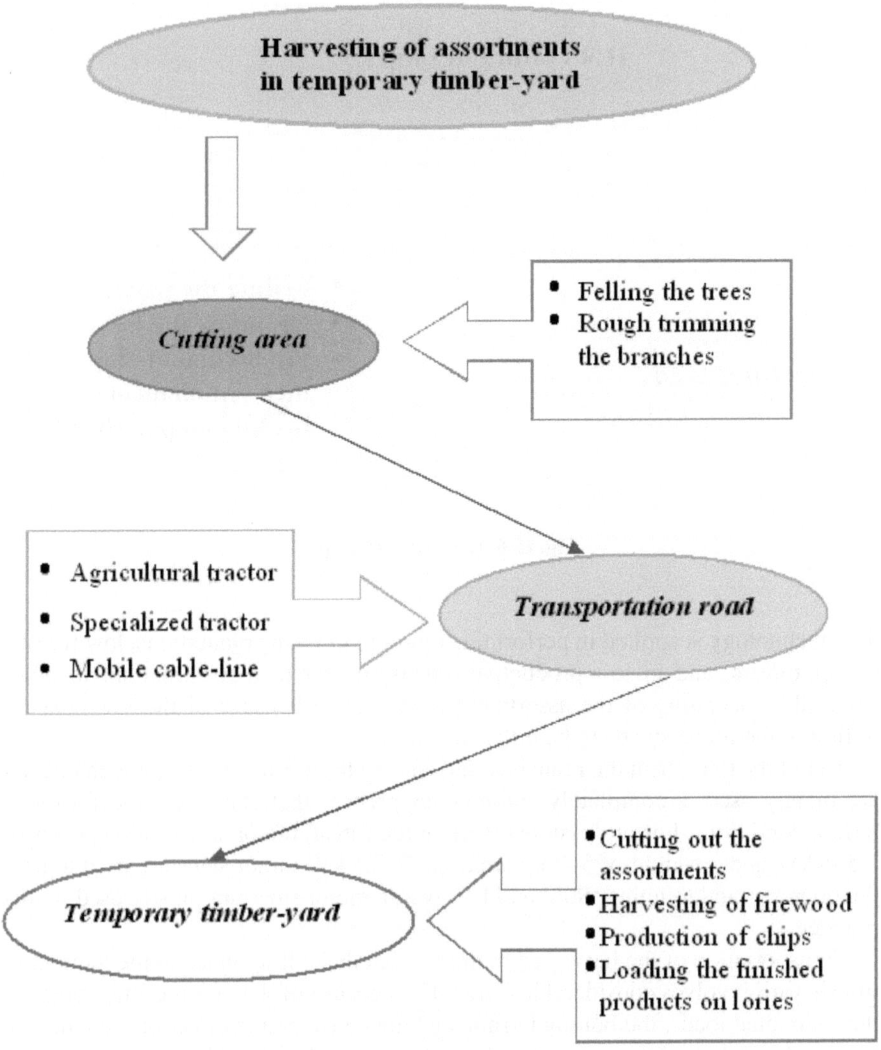

Fig. 15.5 Harvesting of assortments in a temporary timber-yard.

This enables a higher quality of processed materials. The concentration of waste from harvesting of biomass contributes to its more complete utilisation by processing it to chips. Transportation of whole stems considerably increases the productivity of tractors and cable-lines.

- *Harvesting of chips*. In this process, the technological operations performed are as shown in Figure 15.6.

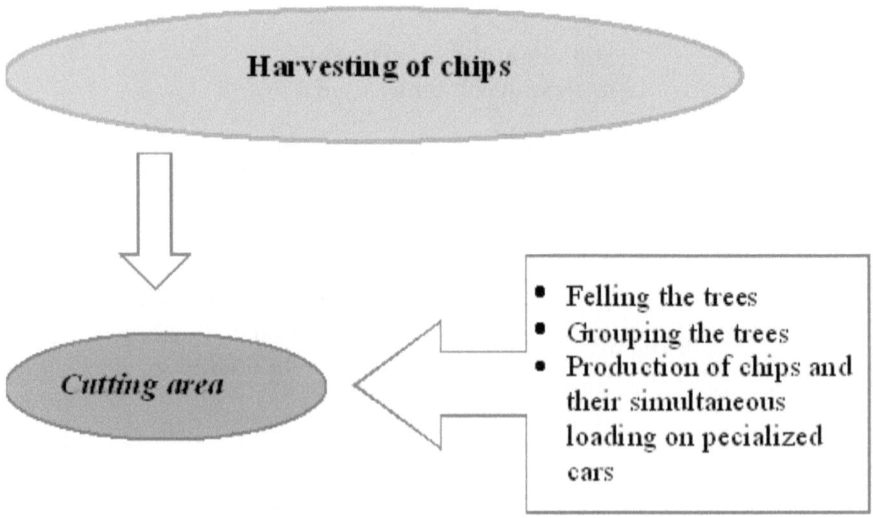

Fig. 15.6 Harvesting of chips.

This technology is applied in performing thinning in young plantations, low forests, sprout forests, and in low-productive forests subject to reconstruction. In those cases, the harvesting of the assortment is inefficient because of the low concentration of the forest stands to be utilised.

To fell the trees, trim the branches, and harvest firewood, petrol-motor chainsaws are mainly used, a completely autonomous process that can work under various terrain conditions. For professional work in the forest, motor saws with a power of 3.5–5 kW and a weight of 5–7 kg are used. 3–3.5 kW motor saws are used to trim the branches and thinning. Stihl and Husqvarna motor saws are mostly used in this practice.

Transportation of the harvested biomass from the cutting place to the temporary timber yard involves considerable difficulties because of the volume and weight of the individual loads, the bumpy terrain, and the considerable effect of climatic and meteorological conditions. For small-sized stems and firewood and for sparse and small quantities of materials, animal traction is used. It is most rational to use animal traction for small distances – up to 300–400 m.

For transportation of both the assortment and whole stems, agricultural wheel tractors of the power class of 50–60 kW are most widely used. They have four-wheel drive, an impact-safe cabin, and are equipped with a single-drum or double-drum winch. They are used mainly on flat and hilly terrains and on mountain terrains with a well-developed tractor road network. Some of them are equipped with bulldozer devices for light repairs of the transportation road and for handling of materials in temporary timber-yards (Figure 15.7).

For large-sized stem transportation on bumpy terrains without roads, specialised forest wheel tractors with an articulated frame of the power class of 60–80 kW

Fig. 15.7 Stem dragging by wheel agricultural tractor in Rhodopa mountain.

Fig. 15.8 Forest wheel tractor in action on a temporal timber yard in Rhodopa mountain.

are used (Figure 15.8). They have four wheel drive and wide-section low-pressure tyres. They are equipped with an impact-safe cabin, a double-drum winch of high traction, and a bulldozer device. They have very good stability, manoeuvrability, and high performance.

Fig. 15.9 Stem extraction from the cutting place by mobile cable-line.

A small transportation machine, the "iron horse", is used for performing thinning and for works in plantations with thin forest stands. It has a 6–8 kW motor and is operated by one operator who performs all the operations during the harvesting and transportation of biomass.

In the case of steep inclinations of the terrain and highly broken terrains, stationary and mobile cable-lines can be used (Figure 15.9). The stationary cable-lines are up to 2000–2200 m long and have a load capacity of 1500–2000 kg.

They are equipped with fully automated hydraulic trolleys whereby loads may be taken along the whole route of the cable-line. The mobile cable-lines are mounted on a wheel tractor. They are up to 300–400 m long and have a load capacity of 1000 kg. They are equipped with automatic hydraulic trolleys. They are used for the transportation of the assortments and whole stems.

The waste biomass is a valuable raw material for energy utilisation. To be able to efficiently transport and use it, it should be processed in the form of chips. For this purpose, mobile machines mounted on a wheel tractor are used. The power of the tractor engine is 60–70 kW. Their performance reaches 3–3.5 m^3/h. Chips are transported with specialised vehicles to users. They may be burned directly in boilers using biomass, or be used as a raw material for the production of pellets and briquettes.

The loading of the finished products is carried out by means of hydraulic jib manipulators and jaw loaders (Figure 15.10). The transportation of the products is carried out by specialised cars with a load capacity of 10–12 tonnes.

Fig. 15.10 Stem loading by jib manipulator with yaw loader on a specialised 12-tonne lorry.

Table 15.10 Storage time for chips.

Summer season	Winter season
From softwood 7–8 months	From softwood – 8–9 months
From hardwood – 3–4 months	From hardwood – 4–5 months

15.3.4 Storage of Chips

Storage of chips is done in the open in the form of piles with a height of 7–8 m and a volume of 2000–5000 m^3. If they are stored in production enterprises on a concrete apron, the volume reaches 13,000–15,000 m^3. The storage time depends on the season and tree species, see Table 15.10.

If terms shown are not observed, putrefactive processes that deteriorate the physicomechanical properties and the calorific value of chips develop in the surface layer.

15.4 Production of Pellets and Briquettes

Pellets are a refined biomass fuel. They are made from sawdust and wood shavings compacted under high pressure into small round objects 6–12 mm in diameter. Pellets are a renewable biomass fuel that does not add anything to the forest and soil. Pellets do not contribute to the greenhouse effect, as oil and coal do, for example.

The production technology of pellets and briquettes consists of pressing the fragmented biomass and obtaining pellets and briquettes of a certain form, where a maximum quantity of pellets and briquette substance is concentrated in minimum volume. The process of pressing into pellets and briquettes is performed under certain conditions: fractional composition of chipped biomass, pressure and temperature of pressing, without using adhesive substances. The technological process of pellets and briquette production depends on the characteristics of the raw material and the perspective usage of pellets and briquettes.

When using waste biomass from logging as a source, pellets and briquetting consist of the following operations:

- gathering of waste branches, brushwood, irregular oval materials, etc.;
- chipping of these wastes with mobile woodcutting machine into technological slivers;
- loading and transporting the technological slivers to the briquette workshop;
- unloading the technological slivers;
- complete milling of technological slivers to size, appropriate for production of pellets and briquettes;
- sorting out the obtained material and removal of the inappropriate in size particles;
- drying the raw material;
- pressing (modeling) the pellets and briquettes;
- cooling the pellets and briquettes ;
- cutting the briquettes with certain length;
- packing them up in thermo shrinkable folio or cardboard boxes;
- storing in storehouse for finished products;
- transporting to end users.

When using waste biomass from wood-processing as a source, pellets and briquettes pressing consist of the following operations:

- transporting the raw material to the workshop;
- unloading the raw material;
- chipping of large-size wastes to size, appropriate for pellets and briquetting;
- sorting out the received material and removal of the inappropriate in size particles;
- drying the raw material; pressing (modeling) the pellets and briquettes;
- cooling the pellets and briquettes;
- cutting the briquettes with certain length;
- packing them up in thermo shrinkable folio or cardboard boxes;
- storing in storehouse for finished products; and
- transporting to end users.

The described technological operations are not always being executed in the technological process of pellets and briquetting. Transporting of waste biomass is included

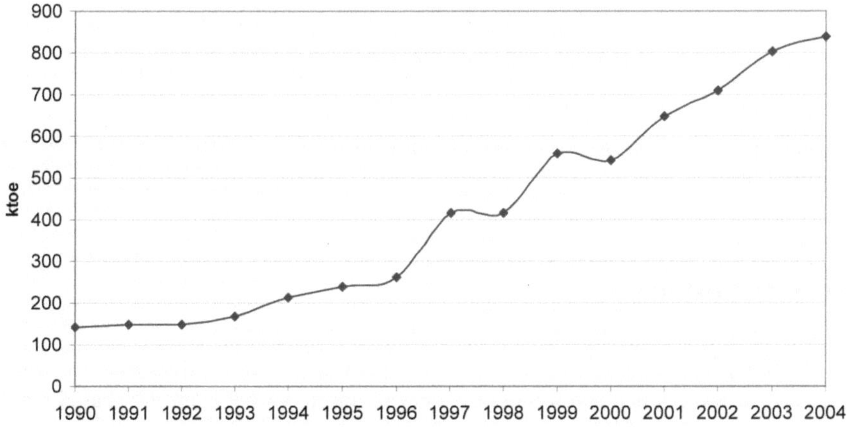

Fig. 15.11 The energy utilisation of biomass in 1990–2004.

in all technological processes. It is done with road or railway transport, pneumatic transport, with rubber belt transporters, etc. Chipping is applied when technological slivers and large-size wastes are used as a raw material. Stationary cutting machines and mills are used. It is desirable that sorting out of the chipped material is always done. It helps to obtain more even final humidity of the source. Raw material with humidity greater than 15% is subject to drying. All the other operations are compulsory in order to guarantee the high quality of wood pellets and briquettes.

15.5 Biomass Present Utilisation

The use of biomass in Bulgaria is continuously increasing. This is illustrated by Figure 15.11, taken from a report of the Bulgarian Energy Efficiency Agency (EEA). In fact, the dynamic of biomass utilisation depends on several factors. The energy policy in Bulgaria plays probably the most important role, but the international energy market, with the increasing prices of oil, gas and electricity, and heat respectively, act in the same direction. The low firewood price, together with the low investment for biomass stoves, are other important factors for the observed remarkable increase of biomass utilisation.

However, the actual use of biomass is higher than the one reported by the Bulgarian Energy Efficiency Agency, as no data concerning unauthorised and illegally cut firewood are recorded in the national statistics – some experts estimate 1050 ktoe in 2004 instead of 840 ktoes as given in Figure 15.11.

From the technological point of view there are two main schemes of converting biomass energy:

- Dry burnable waste (e.g., agricultural waste, straw, steams, etc., or timber and wood working industry, as well as dry waste of stock-breeding).

- Wet waste (e.g., liquid dung, agricultural crops in wax ripeness).

The methods involved in biomass energy transformation are as follows:

- Direct burning and using heat for household needs.
- Burning biomass or biomass products for heating and electro energy in cogeneration units.

15.6 Conclusion

The chapter gives the present status of energy wood chains in Bulgaria. However, the energy wood chains development is very dynamic which is an integrated result of the energy policy of the country, laws and the respective regulations, the energy market and, of course, the activity of many companies in this sector of the economy.

Presentation of companies, producing wood fuel (e.g., pellets, briquettes), specialised machines and equipment as well as heat units (starting from stoves to large-scale units of several MW) is beyond the aim of the present review. However, it should be mentioned that their role for biomass utilisation is of particular importance to create the necessary industrial structure, which could guarantee the practical implementation of all EU and national regulations and satisfy the needs of the local population.

References

1. Stankov, P. (Ed.), ECHAINE Project "Energy Wood Production Chain in Europe", Contract No. ENK5-CT-2002-00623, Reports, http://www.echaine.org/.
2. Iosifov, N., *Briquettes and Pellets form Plant Biomass*, University of Sofia, 2005, ISBN 954-07-2209-8.
3. Lamb, A., European Union energy policies and energy situation, in *Proceedings of the International Seminar "Wood Fuel – An Energy Source for Tomorrow's Europe"*, P. Stankov (Ed.), Pamporovo, Bulgaria, 2004, ISBN 954-91425-5-8.
4. Mladenov, D., Harvesting, transportation and storage of wooden biomass, *Annuals of University of Forestry – Sofia*, 2002.
5. Mladenov, D., Opportunities of wood residues utilization for briquette and pellet production in Bulgaria, in *Proceedings of International Seminar "Low Pollutant Combustion Processes"*, Sofia, 2003, ISBN 954-91425-2-3.
6. Mladenov, D., Small scale production of wood fuel – Examples from Bulgaria, in *Proceedings of International Seminar "Wood Fuel – An Energy Source for Tomorrow's Europe"*, P. Stankov (Ed.), Pamporovo, Bulgaria, 2004, ISBN 954-91425-5-8.
7. National Long Term Program for Renewable Energy Utilization for the Period 2005–2015, Energy Efficiency Agency, Sofia, October 2005.
8. Stanchev, K., Bulgaria energy situation and biofuel market, in *Proceedings of International Seminar "Wood Fuel – An Energy Source for Tomorrow's Europe"*, P. Stankov (Ed.), Pamporovo, Bulgaria, 2004, ISBN 954-91425-5-8.
9. *Statistical Yearbook of Bulgaria*, National Statistical Institute, Sofia, 2000–2005.

Chapter 16
Energy Efficiency in Serbia: Research and Development Activities

Simeon N. Oka

Abstract. The energy system in Serbia is faced with a number of challenges and inefficiencies. The technologies currently used are outdated, the equipment is old and needs to be replaced or refurbished, industrial energy consumption is high, and the energy strategy and price policies need to be revised. The current status of the energy system in Serbia is presented in this chapter, with special emphasis on energy efficiency, the use of renewable energy sources (RES) and the country's dependencc on imported energy. The activities of the National Energy Efficiency Programme (NEEP), founded by the Ministry of Science and Environmental Protection, is presented in detail. Within the framework of the NEEP, 231 research, development and demonstration (R&D&D) projects and 44 feasibility studies have been financed up to now. The main aim of NEEP is to combine science and engineering in order to research, develop, demonstrate and implement efficient modern technologies and equipment, engineering and management methods and to promote the use of renewable energy sources. The Board of Directors is responsible for the choice of priorities, and the selection and control of the projects financed. NEEP has engaged the majority of the most experienced researchers and engineers in Serbia previously working in energy research. In spite of very applicable results achieved, implementation has not been successful due to lack of investment money and an unfavourable environment for the rational and efficient behaviour of the public and the companies concerned.

Key words: Energy efficiency, renewable energy sources, energy policy, research, development and demonstration (R&D&D) projects in energy efficiency.

16.1 Present Status of Energy Sector in Serbia

The inefficiencies in Serbia's energy system can be attributed to several factors. In the past, efficient use of energy has not been a major factor in decision making pro-

K. Hanjalić et al., Sustainable Energy Technologies: Options and Prospects, 281–301.

Table 16.1 Energy reserves in Serbia (not including Kosovo and Metohia), Mtoe.

	Geological reserves		Exploitable reserves	
	Mtoe	%	Mtoe	%
Coal, total	1270	77.58	772.25	87.96
• Possible new mines	82	5.01	41.6	4.47
• Coal mine Aleksinac	12	0.73	5.4	0.62
• Small mines exploitation	105	6.41	65.25	7.43
• Lignite (open pits)	1071	65.42	660.00	75.18
Oil shale	140	8.55	84[a]	9.57
Oil and natural gas	60	3.67	20	2.28
Hydro power potential[b]	167	10.2	1.67	0.21
• >10 MW	152	9.29	1.52	0.19
• <10 MW	15	0.92	0.15	0.02
Total	1637	100.00	877.92	100.00

[a] Estimated as 60% of geological reserves.
[b] For comparison with other energy resources, the hydro power potential was estimated as hydro energy available during 100 years for geological and 1 year for exploitable reserves.

cesses. The industrial infrastructure, developed after World War II, has also contributed to high specific energy consumption. The inefficiencies of the energy system are in fact so high that improvement of energy efficiency is the quickest method to increase security of the energy supply, competitiveness of industrial production, and to reduce dependence on imported energy and fuels. With very small investments, a better mix of energy and fuels, and improved organisation of the energy sector, it is possible to achieve significant energy savings and economic growth.

16.1.1 Structure of the Conventional Energy Resources

The greatest geological energy resource in Serbia is coal (about 80%), in particular low-rank lignite (about 65%). The share of hydro energy in large rivers is less than 10%. Reserves which are not in exploitation but can be activated in a short time (small coal mines, oil shale, and small rivers) amount to about 15%. Of the exploitable energy reserves, the share of coal is about 88%, and other reserves (small rivers, small coal mines and oil shale) make up about 10% (Table 16.1) [1]. Lignite in the open pit mines of Kolubara and Kostolac (35 Mt/year), by far the largest natural energy source in Serbia (not counting lignite resources in Kosovo), is already reserved for existing power plants or for power plants planned to be built in the near future. Power produced in large power plants presently enables Serbia not to import electric energy, but it is questionable whether this will remain possible within the next 10-year period, considering the rapid economic growth (it is estimated that a GDP growth of 7%/year is possible up to 2012).

It is also not possible to count on lignite as a substitute for imported fuels. However, reserves of bituminous and sub-bituminous coal and lignite in small deposits

Table 16.2 Different coal deposits – comparison of energy reserves.

Deposit	Geological reserves			LHV, MJ/kg
	Mt	Mtoe	%	
Lignite – open pit mines in exploitation	5980	1071	85.1	7.5
Sub-bituminous and bituminous coal – small mines, underground mining	275	105	8.5	16.0
Possible future new small mines	297	82	6.4	10–13
Total	6552	1258	100	

Table 16.3 Technically usable hydro power potential in Serbia, GWh/year.

Technically usable potential				Presently exploited	Not exploited
>10 MW	<10 MW	Total	Shared with other countries		>10 MW
17597	1770	19447	6393	10133	7515

can greatly contribute to the reduction of oil and natural gas consumption in district heating and industrial energy production, i.e., in distributed energy production.

Geological reserves of coal in different deposits are given in Table 16.2. According to recent estimates [2, 3], the total geological reserves in 23 small coal deposits amount to about 230 Mtoe (million tonnes of oil equivalent), with about 150 Mtoe of exploitable reserves. Coal production in newly opened small mines can reach 5 Mt/year (mainly high-rank coal). Crude oil and natural gas reserves are significantly smaller, which is the main reason for Serbia's high dependence on imported fuels.

The technically usable hydro power potential is about 20,000 GWh/year, which amounts to 62% of the theoretical potential. About 10% of this can be generated with small hydro power plants ($\leq$10 MW), which amounts to $\approx$1900 GWh/year based on a realizable installed capacity of $\approx$600 MW. About 50% (about 10,000 GWh/year) of the total technically usable hydro power potential is already being exploited (Table 16.3). More than 25% of the installed technical potential is in small rivers, for power plants with an installed capacity <10 MW.

16.1.2 Renewable Energy Sources

Energy resources ready to be used for substitution of imported fuels in the short term must include biomass and geothermal energy sources, as these can be exploited with conventional and cheap technologies, and can be constructed in a relatively short time (wind energy can only be used in specific regions). The available energy potential of biomass waste is given in Table 16.4. This is a very conservative estimation of wood and agriculture wastes, as only 1/3 of the available amount of waste biomass is accounted for.

Table 16.4 Energy potential of waste biomass in Serbia [4].

Type	Energy potential	
	TJ/year	Mtoe/year
Firewood in logs (suitable for small domestic ovens)	17,000	0.4
Wood waste	26,000	0.60
Wood – in total	43,000	1.00
Agriculture wastes – in total	65,000	1.56
Biomass – total	108,000	2.56

The energy potential of geothermal energy in hot waters is estimated to be about 0.185 Mtoe, which is comparable with the hydro power potential in small rivers, but much smaller than the energy potential in waste biomass. Including solar energy has no significant effect on the general picture of available energy sources in Serbia.

16.1.3 Dependence on Imported Fuel and Energy

The importation of fuel and energy has increased over the past several years, and presently amounts to more than 40% of the total primary energy consumption, with a trend of further increase [1, 5]. More than 80% of the required amounts of oil and natural gas have to be imported (3.5 Mtoe/year of oil and 1.9 Mtoe/year of natural gas). Imported coal amounts to about 0.37 Mtoe/year (coke 0.30 Mtoe/year and sub-bituminous coal only 0.07 Mtoe/year). Unless measures are taken to improve energy efficiency and the use of renewable energy sources, importing of oil, gas and electricity will further increase as industrial activities grow and living standards continue to increase.

Figure 16.1 shows that Serbian dependence on imported fuels and energy has permanently increased since 2000, and reached 40% of the total primary energy consumption in 2004 despite the prediction made in [5] that the share of imported energy will reach 40% only in 2015.

According to the projected energy balance for 2006, up to 46.6% of the required energy may have to be imported, making Serbia a highly energy-dependent country. The highest import growth is expected for natural gas, about 13.17% in 2006. The distribution of imported fuels is approximately: oil 47.5%, natural gas 27.9%, and high-rank coal 22.9%. The highest national priority of the Serbian energy policy has to be a reduction in dependence on imported energy by reducing energy consumption, increasing energy efficiency, substituting imported fuels by domestic fuels, exploiting co-generation, and increasing the use of renewable energy sources and wastes.

The distribution of gas and liquid fuel consumption in the last three years is given in Table 16.5 [1]. More than half of the total energy obtained from oil derivatives is used in transport and to drive machines in agriculture, and this part can not be substi-

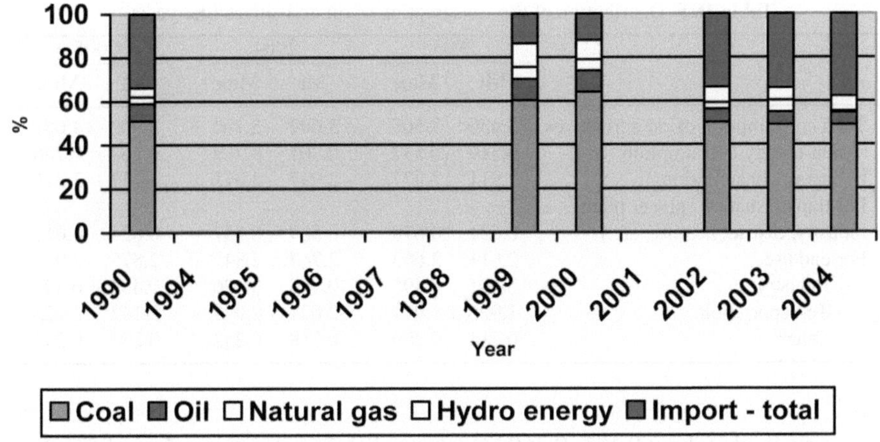

Fig. 16.1 Distribution of primary energy consumption in Serbia.

tuted with other energy resources. In energy production in industry the share of coal is 33% and that of natural gas and liquid fuels is 67%. Central and district heating plants mainly burn natural gas (~70%), liquid fuels (~16%) and coal (~18%), making the dependence on imports in this sub-sector higher than 80%. The dependence on imported fuels for industry and for district heating must be reduced for economic and strategic reasons. About 0.9 Mtoe/year of imported liquid fuel can be substituted by domestic energy sources for heating in households, in district and central heating plants, and for energy production in industry. The energy consumed in transport and agriculture (2.0–2.5 Mtoe/year) can be provided by domestic energy sources (not counting domestic production of oil and natural gas) only if fossil fuels are mixed with biodiesel and bio-ethanol. According to an analysis made in 2004 [6], Serbia has enough natural resources and production capacity to produce 150,000 to 200,000 tonnes/year of biofuel, enough to substitute about 5–6% (0.15 Mtoe/year) of the fossil fuel currently consumed in the transport sector (following EU energy policy directives). In addition, about 0.07 Mtoe/year of imported bituminous and sub-bituminous coal can be substituted by an increased production of coal in small mines or by renewable sources. In total, a substitution of about 1.12 Mtoe/year of imported fuel is possible, which amounts to about 30% of the oil and gas imported for district heating and industry.

Depending on the type of energy needed, the following presently unused energy sources are at our disposal: biomass, geothermal, wind, small rivers, coal, waste coal (small-size classes) of the underground coal production in small mines, and several other minor sources. For electricity production and production of high and low temperature heat the following sources are available: waste biomass (about 2.56 Mtoe/year [4]), waste coal from active small mines (about 0.119 Mtoe/year) and coal from increased production in active new small mines (about 0.250–1.250 Mtoe/year) [3]. For production of low temperature heat, geothermal energy

Table 16.5 Distribution of the consumption of oil derivatives (per year).

	2003		2004		2005	
	Mt	Mtoe	Mt	Mtoe	Mt	Mtoe
Total consumption of oil derivatives	3.430	3.508	3.697	3.790	3.902	4.001
a) non-energy consumption	0.519	0.532	0.701	0.719	0.764	0.786
b) energy consumption	2.911	2.977	2.997	3.071	3.137	3.215
For transformation (power plants, industry, district heating, etc.)	0.691	0.678	0.569	0.557	0.624	0.612
For end use	2.611	2.683	2.767	2.847	2.875	2.958
Industry	0.491	0.495	0.488	0.496	0.483	0.492
Transportation	1.906	1.979	2.021	2.099	2.142	2.223
Other	0.214	0.209	0.258	0.252	0.250	0.243

is also available (about 0.185–0.360 Mtoe/year) [7]. Small rivers and wind for electricity production can give about 0.150 Mtoe/year [8]. In total, a presently unused domestic energy potential of about 3–4 Mtoe/year is available for reducing Serbia's dependence on imported energy (or 1.0–1.5 Mtoe/year for end use). Activation of this energy potential depends on new investments in mines, power and heat plants, as well as on governmental support, laws and directives in the field of concessions, energy prices, ecological taxes and implementation of modern, efficient and environmentally acceptable technologies.

Due to the lack of investment capital, significant results in the reduction of import dependence in the near future can only be achieved by improving energy efficiency. The experience of other countries in transition, and some estimates made in Serbia's economic development strategy up to 2010 [2], show that it is possible to increase energy efficiency by about 20% in a short time (5–10 years) and with relatively small investments. Also, Serbia has the possibility to introduce co-generation, and should do so immediately in order to save primary energy resources. The share of co-generation in the Serbian energy system amounts to only 1% of the total electricity production, compared with an average of 10% in the EU (82% in Germany, 40% in Holland and 50% in Denmark). The introduction of co-generation in existing coal burning power plants can save a large amount of imported natural gas which is presently used in district heating plants.

16.2 Energy Efficiency – Specific Energy Consumption in Serbia

16.2.1 Energy Efficiency

Two facts follow from the presented data: (1) Serbia has limited and practically exhausted energy potentials, and (2) its dependence on the import of energy is high, and will drastically increase in the future. Inefficiency, the most significant characteristic of the whole Serbian energy system, cannot be seen from the presented facts.

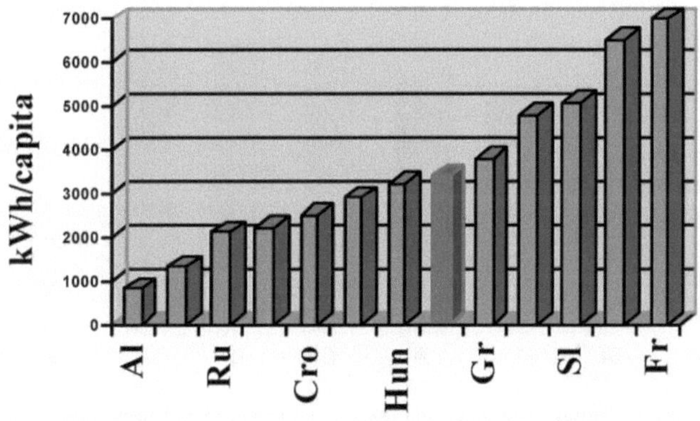

Fig. 16.2 Electric energy consumption, per capita.

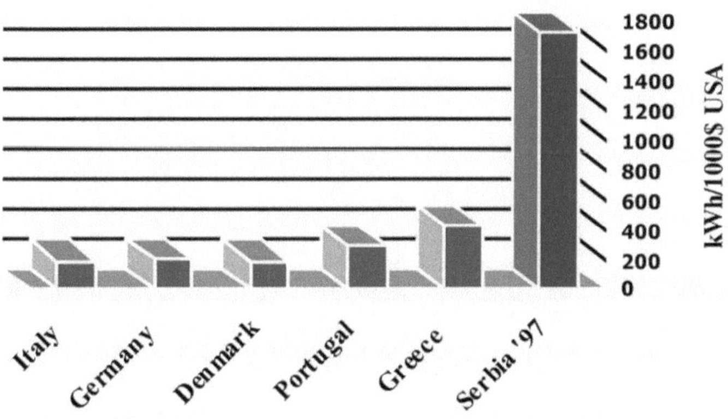

Fig. 16.3 Electric energy consumption, per unit GDP.

Significant efforts have been made in the past five years to refurbish equipment used in coal mines and in power plants, oil and natural gas production, and to increase energy conversion efficiencies in all energy sub-sectors. However, the overall efficiency of the energy sector is still low as a result of (a) old technologies still in use, (b) inadequate management, (c) inappropriate energy strategy in the past, (d) unsuitable energy price policy, (e) late revitalisation of equipment, (f) lack of or neglect of standards and regulations, etc.

From 2001 until 2004 Serbia's GDP increased at an average rate of 5.1%/year (8.6%/year in 2004, 5.2%/year in the first quarter of 2005). Consumption of electric energy in 2002 was 3400 kWh/capita, i.e., at the level of European countries at an intermediate state of development (Figure 16.2). However, the energy consumption normalised to the GDP is the highest in Europe, i.e., 1900 kWh per US$1000 of GDP, due to the specific structure of the industry and energy sector, as well as to low energy prices and decreased industrial activity since 1990 (Figure 16.3). High

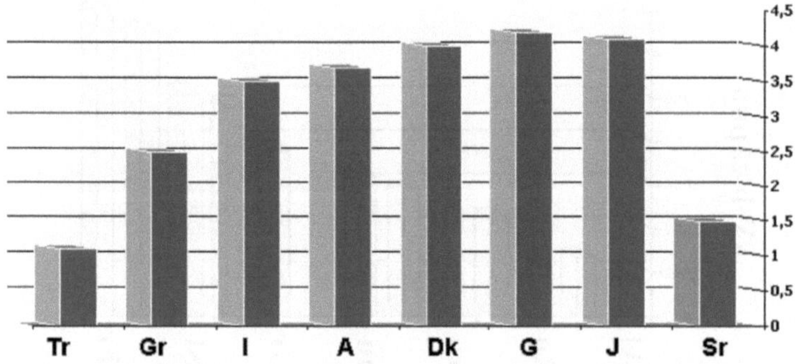

Fig. 16.4 Primary energy consumption, energy consumption, toe/capita.

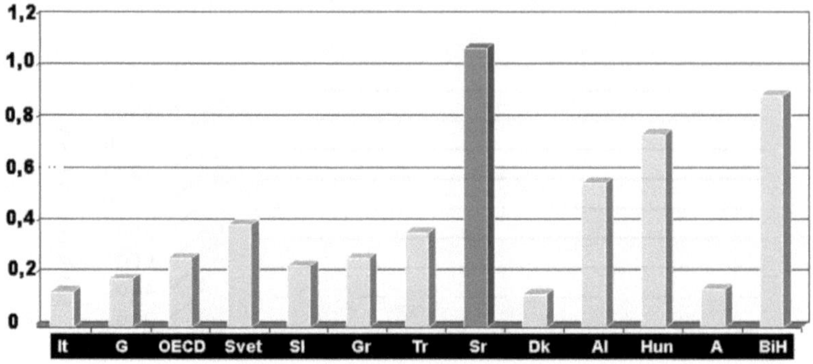

Fig. 16.5 Primary energy consumption, toe/1000 $ of GDP.

electric energy consumption per capita, combined with low prices of electric energy (low living standard) and reduced industrial activity, indicate a non-rational use of electric energy for heating in households. The GDP increase after the year 2000 indicates that specific energy consumption per unit of GDP will be reduced in the future.

Non-rational use of natural energy resources and the inefficiency of the Serbian energy system can also be illustrated with primary energy consumption data. Primary energy consumption per capita in Serbia is 2.5 times smaller than in developed countries (Figure 16.4). In contrast, the primary energy consumption per unit of GDP is very high (Figure 16.5).

Another sub-sector in which inefficient energy consumption is evident, is heating of houses and buildings. In buildings constructed before 1945 the average installed capacity of heating systems is 200 W/m2, in those constructed after 1960 it is about 145 W/m^2, which means that average annual consumption of heat in buildings is 200–300 kWh/m^2 for an average cold year. Keeping in mind the maintenance condition of the buildings, the average annual consumption in buildings connected to district heating systems is 420 kWh/m^2, and in buildings with central heating systems up to 600 kWh/m^2. In European countries with similar climate conditions, buildings are constructed today with an annual energy consumption for heating, hot water, air-conditioning and lighting less than 100 kWh/m^2. The three to four times higher energy consumption for buildings in Serbia indicates an opportunity for significant energy savings.

16.2.2 Activities for Increasing Energy Efficiency in Serbia

To address the problems in the energy sector, the government of Serbia undertook two parallel activities. The National Energy Efficiency Programme (NEEP) was founded on March 1st 2001 with the aim to stimulate fundamental and applied research, as well as development and demonstration activities for increasing energy efficiency and to increase use of renewable energy sources. It is financed by the Ministry of Science and Environmental Protection (MoSEP) with about 2 M€/year, and engages the most prominent experts and researchers in the field.

Furthermore, the Serbian Energy Efficiency Agency (SEEA) was founded in the beginning of 2002. SEEA aims to create a favourable environment for activities aimed to improve energy efficiencies by developing technical standards, regulations and laws, and by educating engineers, managers and other professionals in the energy industry. SEEA was initially financed by the European Agency for Reconstruction (EAR), but since 2004 its activities fall under the Energy Law. Funds for the stimulation and financial support of energy efficiency projects and implementation of RES will soon be established under the Energy Law.

The main objective of NEEP is to unite research, engineering and legislative activities in order to study, develop and implement new technologies and methods, including feasibility studies, and energy and economic analysis in real industrial applications. NEEP also aims to overcome the problem of insufficient funds in industry, and to financially support transfer, research, and implementation of energy-efficient technologies. It also provides funds for transferring methods and equipment already proven in developed countries, and for demonstrating them in the specific Serbian natural and energy environment in order to promote their wide dissemination.

Responsibilities of the SEEA were defined as improvement of the environment for energy efficient and rational behaviour, development of economic and financial instruments for support and stimulation of energy efficiency and use of RES, technical standards, pricing policy, taxes and tolls, etc. Responsibilities also include the education of citizens, professionals, experts and engineers, the transfer of knowledge and best practices from developed countries, the use of donations, and the realisation of demonstration projects (case studies). In addition, SEEA was tasked with increasing energy efficiency in buildings, industry, municipal systems, transport, and households, and to making use of renewable energy sources. Up until now SEEA has received about 5 million Euros for project realisation from EAR, US$21 million from the World Bank, and about 300,000 Euros/year from the Norwegian

Government. So far, 16 projects for energy saving have been realised (in public buildings, district heating, public lighting and for the implementation of renewable energy sources) as well as 5 courses for the education of professionals and engineers in industry (150 persons, 35 companies) and in local authorities (46 persons, 46 communities).

The main deficiencies in activities in the fields of energy efficiency and the use of renewable energy sources are a lack of coordination between the different ministries, an unfavourable environment for making rational decisions, and lack of stimulation and financial support from the government. The only ministries actively involved in energy efficiency and use of renewable energy sources are MoSEP and the Ministry of Mining and Energy (MoME).

16.3 National Energy Efficiency Programme

16.3.1 Foundation of NEEP – Motivation

In the Strategy for the Development of the Energy Sector in Serbia up to 2010 [2], and later in the Strategy of Energy Development in Serbia up to 2015 [5], a 20% increase in energy efficiency by 2010 was set as one of the main objectives, in order to contribute to emission control and at the same time to avoid, for the moment, the need for large investments in new power plants. The founding of NEEP has also demonstrated a specific emphasis on the importance of energy efficiency and the use of RES in the development and reanimation of the Serbian economy. The experience of many countries shows that energy inefficiencies should be attacked along the whole energy transformation chain, from primary energy production (supply side) to the consumption of energy (end use). An increase in energy efficiency also requires a whole set of different measures and activities, from focused fundamental investigations of the physical and chemical processes involved, up to different engineering activities, management and organisation followed by appropriate legislation. These are the reasons it was necessary to join and coordinate in one united national energy efficiency programme a range of activities, starting with definition of objectives and priorities and choice of the best proposed projects, up to the control and follow-up on realisation of the projects financed. NEEP also has an objective to demonstrate the best technologies, and/or to implement proven, commercially available technologies, methods and equipment. It aims to disseminate best practices, and to help companies and local authorities to find and secure financial support to implement these practices and technologies.

In 2001, the Strategy of the Development of the Energy Sector in Serbia up to 2010 set the objective to increase energy efficiency by 20% by 2010. This was based on the assumption that the Serbian Government would, in a short time, adopt a set of laws, legislative measures, technical standards and financial and economic incentives in order to stimulate the implementation of more efficient technologies, equipment and methods, creating a favourable environment for saving energy and to use natural resources in a rational manner.

16.3.2 Basic Principles of NEEP

The strategy of NEEP is based on the following principles [9]:

1. to cover all sectors and sub-sectors of the energy system, from heat and electric energy production to end use in industry, transport, buildings, municipal systems and households;
2. to take care not only of energy efficiency itself, but in general, of the rational use of domestic energy resources, impact on the environment, alternative and renewable energy sources, and imported energy and fuels — in a word, sustainable development;
3. to assure security of supply and a high quality of energy for the end users;
4. to stimulate research, development and demonstration of modern technologies and their use within the context of domestic energy sources and local conditions.

16.3.3 Structure of the NEEP

Since NEEP covers the entire energy chain, it consists of nine Development programmes managed by directors and deputy directors: (1) Energy efficiency in electric power production, (2) Energy efficiency in electric power transmission and distribution, (3) Energy efficiency in industry, (4) Energy efficiency in municipal systems, (5) Energy efficiency in households, (6) Development of domestic ovens and boilers that burn solid fuels, (7) Use of alternative and renewable energy resources, (8) Energy efficiency in buildings, (9) Energy efficiency in transportation.

The Board of Directors, consisting of the directors and deputy directors of the nine development programmes, is responsible for the management and direction of all activities of NEEP including: formulation of the strategy, choice of priorities, announcement of open invitations for financing research, development and demonstration (R&D&D) projects, evaluation and choice of projects for financing, control, follow-up of the realisation of projects and evaluation of the results of projects. However, final decisions are made and approval given by the Minister of Science and Environmental Protection. The Board of Directors is chaired by the director and deputy director of NEEP.

16.3.4 General Objectives of NEEP

In the beginning of 2002, the Board of Directors of NEEP formulated the strategy and priority topics of NEEP as a basis for its future activities [9]. The overall strategy was based on the Strategy of Energy Development in Serbia up to 2010 [1], as well as on the knowledge and experience gained in developed countries. In 2005, new priorities and topics were chosen in close cooperation with the MoME, the Ministry of Agriculture and Waterworks, the Serbian Energy Efficiency Agency and the Standing Conference of Towns and Municipalities.

The activities and realisation of NEEP are expected to contribute to:

1. a safer and more continuous energy supply to consumers (industry, district heating and individuals);
2. reduced use of imported oil and gas;
3. a decrease of the specific energy consumption per unit of GDP and an increase of the competitiveness of the domestic industry in the world market;
4. rational and efficient use of domestic energy resources;
5. rational and efficient use of energy in general and reduced production costs;
6. reduced environmental impact of power generation equipment;
7. implementation of European standards in energy efficiency and environmental protection;
8. reduced impact of energy costs on the prices of industrial products; and
9. a more favourable environment for small, medium and large enterprises.

To achieve these objectives, open invitations for financing research, development and demonstration projects and feasibility and strategic studies were announced.

16.3.5 Specific Objectives and Priorities of NEEP

Projects financed by NEEP have to achieve the following specific goals: increased, but rational use of domestic and local energy resources, in particular those that are renewable; reduced use of imported fuels for power and heat production in industry and in district heating; reduced use of electric heating in households; increased, efficient and economical use of waste biomass; increased use of alternative energy resources, e.g., from municipal and industrial wastes; higher efficiency of energy generation in utilities, industry, and district heating plants; reduced losses in energy transmission and distribution; reduced heat and energy consumption in buildings; higher efficiency in the end use of energy in industry, municipal systems and households; R&D&D, promotion and implementation of efficient and clean energy technologies for heat and electric energy generation; increase of the energy efficiency of domestic appliances and energy labeling; decrease of fuel consumption and increase of energy efficiency in the transport sector, and control of the environmental impact of energy plants and equipment.

It was also the intention of NEEP to supply research organisations and authorised laboratories with modern experimental and measuring equipment, in order to make them qualified for the audit and control of energy efficiency and energy balances that conform to European standards and regulations.

16.3.6 Creation of Favourable Ambient for Energy Efficiency and Rational Use of Energy Sources

Achievement of the objectives formulated by NEEP and implementation of the results of projects financed by NEEP is not possible if the government fails to create a favourable environment by taking appropriate legislative, economic and financial measures. Special attention was paid in NEEP to create a database for devising strategies in different energy sub-sectors [4, 10–15], and for setting up regulations, technical standards, and economic, financial and technical measures to stimulate, promote and support energy efficiency projects and implementation of renewable energy sources. Creation of a favourable environment that stimulates and supports the implementation of new, modern, energy efficient, environmentally acceptable technologies and the use of alternative and renewable energy sources are the responsibility of the MoME and other ministries. Realisation of the objectives formulated by NEEP depends greatly on coordination between different ministries, agencies and other governmental bodies. NEEP has initiated cooperation of the MoSEP with the MoME and other relevant ministries, i.e., the Ministry of Agriculture and Water Management, the Ministry of Capital Investments and the Ministry of Finance.

In a number of studies financed by NEEP, the adoption of new laws in the energy field, technical standards, regulations and other measures necessary to create a favourable ambient for the rational behaviour of people and enterprises were proposed. Examples include new regulations and measures for controlling the quality and energy efficiency of equipment that burns different types of fuels, industrial equipment and domestic appliances; supply of testing equipment for authorised research laboratories; regulations and economical and financial measures to support the use of renewable energy sources; regulations for testing energy equipment in exploitation; standards and regulations in the field of energy transmission and distribution, and regulations and standards for energy consumption in buildings, for insulation materials and heat insulation of buildings.

In order to reduce electric energy consumption for heating in households, financial and economic support was proposed for: (1) production of ovens and boilers burning solid fuels and biomass, waste and liquid and gaseous bio-fuels; and (2) to reduce the price of equipment burning solid fuels and biomass for heating, cooking and other purposes in households. This also included subsidies for solid fuels (coal, briquettes from coal and biomass), favourable loans for solid fuels and equipment, a tax reduction for manufacturers producing ovens and boilers burning solid fuels, and financial support for starting mass production. Furthermore, long-term "soft" loans were proposed to enable the start-up of mass production of equipment using solid fuels and biomass, and the production of briquettes made from coal wastes and biomass. Financial support and tax reductions were also proposed to enterprises that use renewable energy sources or waste from their own production. Long-term soft loans were made available for constructing small power plants using local resources and renewable energy sources, such as coal, biomass, small river hydro power, municipal waste, wind, geothermal and solar energy. Finally, tax reductions, loans and financial support were proposed for better heat insulation of buildings and other projects increasing energy efficiency.

16.3.7 Types of Projects Financed by NEEP

Since the intention of NEEP is to harmonise short-term and long-term objectives, three types of projects are financed:

- *Research, development and demonstration projects (R&D&D)* aim to create new knowledge of the processes, and new technologies, equipment, products and services contributing to increased energy efficiency and use of renewable energy sources. Results of those projects are databases and conceptual designs, construction and testing of pilot or demonstration units, new prototypes and production activities for starting mass production of new equipment, and demonstration of their energy efficiency. R&D&D projects are planned for a maximum duration of three years.
- *Demonstration projects* aim to use the existing knowledge and experience gained in developed countries, or domestic knowledge accumulated in the past years, in order to demonstrate and implement proven modern, energy efficient and environmentally acceptable technologies, including equipment and/or engineering procedures and methods. Realisation of the demonstration projects must include either construction of pilot or demonstration plants and their testing under conditions of exploitation, or the design and construction of prototypes and start-up of production of new products. Demonstration projects have to either verify the viability of technologies and equipment, or to start up new production followed by an analysis of the optimal conditions for mass application. The duration of demonstration projects is limited to 12 months.
- *Strategic and feasibility studies* aim to prepare foundations for new strategies, planning and choice of priorities in the fields of energy efficiency, renewable energy sources, or the energy system in general. Project results include databases of the availability and regional distribution of different renewable and non-renewable energy resources, and the characteristics, optimal conditions and fields of application of those energy sources. Also included are the validation and optimal fields of application of different modern technologies, in particular technologies for the use of renewable energy sources, possibilities for increased energy efficiency in different industrial and public sectors, etc. [4, 10–15]. The duration of strategic and feasibility studies is 3–12 months.

R&D&D projects have to unite focused fundamental and applied research and development with all the necessary engineering activities for the design, construction and testing of the new technologies and equipment. This also includes testing of the quality of new products, integration of the activities of research and engineering organisations, and active involvement of professional staff in the industry for realisation of the projects. Research has to be focused on topics relevant for the available domestic energy sources – coal, biomass, industrial and municipal wastes – while taking into account local natural and social conditions. Demonstration projects have to integrate the activities of the research organisations with those of industry and local authorities, and use available government funds in the construction and testing of pilot and/or demonstration units, and in validation of the feasibility of mass implementation of modern technologies and equipment, methods or procedures.

16.3.8 Financing

Projects are financed jointly by scientific funds and funds provided by project users – industry, public organisations, local authorities, and others.

The Ministry of Science and Environmental Protection (MoSEP) financially supports the following activities:

1. objective focused fundamental and applied research and feasibility studies: up to 100%,
2. research and development activities – 50–100%,
3. conceptual as well as practical design studies – up to 50%,
4. construction and manufacturing of prototypes and laboratory equipment – up to 50%,
5. technical and economic analyses and feasibility studies – up to 50%,
6. testing of pilot or demonstration units, and start-up of new production facilities – up to 80%,
7. energy audits and measures for increasing the energy efficiency – up to 50%, and
8. environmental impact studies – 50–100%.

Since funds obtained from the MoSEP are not sufficient for the realisation of projects accepted by NEEP, particularly if pilot and/or demonstration units have to be constructed and tested, end users must give financial support, and must be engaged by delivering the necessary materials and personnel (engineers and technicians). Practically all project activities (except fundamental and applied research) have to be supported by the users. These activities include conceptual designs, construction and testing of pilot and demonstration units, development of both prototype and production equipment, management, and technical and economic analyses.

16.4 Activity of the NEEP in the Period 2002–2005

In the period from 2002 to 2005, five open invitations for submission of project proposals for all nine development programmes were announced. In total, 140 projects (both R&D and Demonstration), and 25 feasibility or strategic studies have been selected for financing. According to the new strategy and priorities of NEEP for the next period, 2006–2008, two open invitations were announced in 2006. As a result, 91 projects and 19 studies were selected for financing. By the end of 2005, 120 projects and 30 studies were successfully concluded. Results of the projects and studies have been regularly presented to scientific, engineering, professional and public communities.

16.4.1 Priorities of the Specific Development Programmes

Considering the general and particular objectives defined in the strategy of the NEEP, the following priorities were announced in open invitations for specific development programmes.

Energy Efficiency in Electric Power Generation

The following priorities were defined in coal mining, thermal and hydro power plants:

- Implementation of (a) selective mining and homogenisation of the lignite in open pit mines, and (b) water draining systems at open pit mines;
- Increase of energy efficiency in: (a) mining, preparation and transport of lignite from open pit mines, ash and slag transport, and (b) boiler units, turbines, generators, cooling systems, etc.
- Increase of energy efficiency, availability and power level: (a) of the hydro power plants, (b) by introduction of co-generation in thermal power plants, and (c) by balancing energy production and consumption.
- Study of (a) the advantages of co-generation or combined cycles in electricity generation, (b) hydro power plant operation on multipurpose water accumulations, and (c) improvement of the joint operation of hydro power and thermal power plants.

Energy Efficiency in Electric Power Transmission and Distribution

Priorities were defined in both in transmission and distribution networks:

- Estimation and decrease of energy losses in transmission, distribution and industrial networks, by reconstruction of the grids and optimal management.

- Increase of energy efficiency in transmission, distribution and industrial networks by using compensation of the reactive power and high capacitance condensers.
- Estimation and decrease of energy losses during transit of electric energy through the Serbian grid to different countries in the region.
- Improvement of the application software for: (a) load levelling and consumption management in distributive and industrial networks, (b) analysis of the static and dynamic stability of electric transmission networks, (c) limiting transmission capabilities of the network's interconnections, and (d) analysis of the deregulated electric energy market.
- Implementation of (a) full analysis and diagnostics of the energy efficiency in transmission and distribution networks, including data acquisition and storage, (b) temperature measurements of the high voltage equipment, (c) measurements, analysis and equipment for improvement of the electric energy quality, and (d) standards for electric energy quality.

Energy Efficiency in Industry

Priorities were defined as follows:

- Energy management and control of energy systems in industry.
- Implementation of co-generation in existing power plants in industry.
- Development of (a) equipment for measurement and control, (b) industrial control systems, and (c) energy efficient equipment and technologies.

Energy Efficiency in Municipal Systems

Priorities were defined in district heating systems, gas distribution and use, and water supply systems:

- Use of local fuels, distant control and heat flow meters in district heating systems.
- Implementation of (a) gas engines for combined electricity and heat production, leading to lower return-water temperatures in district heating systems, and (b) optimal combined implementation of district heating and gas distribution systems in towns.
- Efficient use of (a) natural gas in households, (b) natural gas in energy systems in large buildings, hospitals, sport centers, etc., (c) gas engines in the main control stations of gas pipe lines, and (d) natural gas from small fields for local energy production.
- Improvement of the efficiency, optimisation and re-engineering of water pumping systems in towns, and implementation of small hydro power plants in water supply systems.

Energy Efficiency in Households

Priorities were defined as follows:

- Development and application of a method for the evaluation of (a) energy efficiency indicators in households, and (b) economic aspects of energy efficiency in households.
- Reduce energy consumption by improvement of thermal isolation, building design according to modern climate control principles; substitution of electric heating by other energy sources; energy management in apartment buildings connected to district heating systems; responsible use of natural gas in households.
- Development and implementation of (a) a more efficient system for hot water supply, (b) use of renewable energy sources for heating, and (c) heat pumps for heating and cooling in households.
- Reduction of energy consumption of air conditioning systems and lighting.
- Reduction of energy consumption of electronic equipment in stand-by mode.
- Saving of energy in households by: (a) stimulation and support measures, (b) new methods for energy consumption management, (c) improvement of standards for energy efficiency, and (d) education of people using electronic and written media.

Development of Domestic Ovens and Boilers Burning Solid Fuels

The following priorities were set with regard to the burning of solid fuels, in particular biomass:

- Development of technology and equipment for (a) production of briquettes and pellets from waste biomass, and (b) distribution and transportation of the briquettes and pellets.
- Development of (a) highly efficient ovens with natural and forced convection suitable for highly volatile fuels, (b) combined oven-boiler designs for combustion of different solid fuels, (c) small boilers for households for the combustion of briquette and/or pellets, and (d) small boilers for local coal combustion.
- Feasibility studies of (a) the energy potential and characteristics of different kinds of waste biomass, and (b) technology and equipment for briquetting and pelleting.

Use of Alternative and Renewable Energy Sources

Priorities were defined for several renewable energy sources – biomass, geothermal, wind, solar and for small river hydro power:

- Feasibility studies for (a) the energy potential of solar and wind energy, (b) the potential of geothermal energy; (c) fuel cells, (d) authorised laboratories for testing equipment for renewable energy technology, and (e) production and application of biodiesel and bio-ethanol.

- Development, design, construction and testing of:
  - solar panels, hybrid solar panels for heat and electric energy production, heating and air-conditioning systems using hybrid active and passive solar energy systems, and photovoltaic systems for lighting;
  - small wind generators for households, wind generator farms, and wind-powered water pumping stations for agriculture;
  - heating and air-conditioning systems with heat pumps using geothermal hot water energy and earth heat, geothermal energy systems in agriculture and industry;
  - mini/micro hydro power plants in multipurpose water reservoirs in mountain regions, mini/micro hydro power plants for irrigation in agriculture;
  - heating and/or co-generation plants using solid biomass or solid wastes in municipal systems, agriculture, paper and pulp industry, wood processing etc.,
  - equipment for biodiesel and bioethanol production, and sustainable, energy efficient and ecologically clean farms.

Energy Efficiency in Buildings

Priorities were defined as follows:

- Determination of (a) actual outdoor design temperatures, and (b) meteorological year for calculations and modeling of the dynamic behaviour of buildings.
- Application and intensification of predictive methods for: (a) passive/natural night cooling in domestic climatic conditions, (b) influence of the "thermal mass" in cooling and heating, and (c) double facades for reduction of heat losses and cooling loads.
- Development of (a) complex control technologies and software for heating and air-conditioning systems in buildings, (b) modern natural lighting (daylight) technology, (c) measurement techniques for testing the energy efficiency of buildings, and (d) joint simulation and optimisation of all energy systems in buildings.
- Development of (a) windows with a heat transfer coefficient less than 2 W/m^2K, (b) special glass with increased reflection and selective absorption and emission abilities, (c) methods to design buildings with an annual energy consumption $<100\ kWh/m^2yr$, and (d) conservation methods for old monumental and historic buildings.
- Application of previously awarded patents: energy-saving construction designs, insulating materials, heating systems, construction equipment which can be produced in large quantities.

Energy Efficiency in Transport

Priorities include all types of transport of people and goods:

- Feasibility studies of (a) European legislation, regulations and technical standards in energy efficiency and environmental protection in transport; (b) a system for monitoring energy efficiency in transport; (c) the use of alternative fuels, such as liquid gas, natural gas, biodiesel and bio-ethanol, and (d) the present state of motor vehicles in Serbia and quality of maintenance.
- Implementation of (a) modern methods for management, monitoring, control and maintenance of vehicles and traffic, (b) integrated transport systems, and (c) new concepts for transport of people.
- Implementation of information technologies, modern technologies in the transport of hazardous materials, and European legislation concerning fuel consumption and emissions;.
- Study of (a) trends in the development of internal combustion engines, vehicles and equipment, (b) modernisation of vehicles and engines from domestic production, and (c) conformation of domestic motors to European standards.

16.5 Conclusions

Although NEEP has engaged the most experienced researchers and engineers, it can not be said that the results have been satisfactory. Although most of the projects realised gave very applicable results, many of these were not implemented in practice due to lack of funding and the unfavourable environment for energy-efficient behaviour. Users of the results, i.e., enterprises from industry and agriculture, and municipal and local authorities, were not, and presently still are not, economically thriving and are not motivated to invest in energy efficient equipment and technologies.

Many very valuable research and engineering results have been obtained and are ready to be applied in improved economic circumstances and a more favourable environment. The results of the strategic and feasibility studies have been extremely valuable, and include databases for planning, definition of strategies, use of renewable energy sources and implementation of new efficient and environmentally acceptable technologies and equipment in the energy sector, agriculture, industry and municipal systems.

NEEP achieved many additional results which form a sound basis for further activities in the fields of energy efficiency and the use of renewable energy sources. The importance of R&D&D has been realised, and a large number of young researchers and engineers gained experience and knowledge in new modern technologies, equipment and methods. Furthermore, new approaches in the preparation of project proposals and in the supervision and follow-up of completed projects have been explored.

References

1. Oka, S. and Mesarović, M., Development of efficient infrastructure – Energy system, in *National Strategy of Economic Development of Serbia up to 2012*, Belgrade, 2006, pp. 100–108 [in Serbian].
2. Oka, S. and Mesarović, M., Electric energy production and coal mining – Present status, development potentials, strategy, expected effects, necessary institutions, governmental policy and measures, in *Strategy of Economic Development of Serbia up to 2010*, Book II, Ministry of Science and Environmental Protection, Belgrade, 2002, ISBN 86-7282-052-6, pp. 251–266 [in Serbian].
3. Oka, S., Gagić, D., Mesarović, M. and Blečić, N., Use of coal from small mines for energy production, in *Strategy of Economic Development of Serbia up to 2010*, Book II, Ministry of Science and Environmental Protection, Belgrade, 2002, ISBN 86-7282-054-1, pp. 505–520 [in Serbian].
4. Ilić, M., et al., Energy potential and characteristics of waste biomass in Serbia and technologies for biomass preparation and energy production, Study financed in the frame of NEEP, Belgrade, 2003 [in Serbian].
5. *Strategy of Development of Energy System in the Republic Serbia up to 2015*, Belgrade, Ministry of Mining and Energy, Belgrade, 2005 [in Serbian].
6. Furman, T., et al., Production and use of biodiesel – Alternative and ecological fuel for diesel engines, Study financed in the frame of NEEP, Novi Sad, 2004 [in Serbian].
7. Bašić, Đ., et al.: Energy potential and possible use of geothermal energy in Serbia, Study financed in the frame of NEEP, Novi Sad, 2005 [in Serbian].
8. Gburčik, P., et al., Solar and wind energy potential in Serbia, Study financed in the frame of NEEP, Belgrade, 2004 [in Serbian].
9. Oka, S., National energy efficiency program – Strategy and priorities, *Termotehnika* **XXVIII** (1/4), 2002, 1–110 [in Serbian].
10. Mićević, Z., et al., Increase of energy efficiency by reconstruction of thermal power plants in exploitation, Study financed in the frame of NEEP, Belgrade, 2004 [in Serbian].
11. Bašić, Đ., et al., Implementation of combined heat and electric energy production in existing industrial power systems, Study financed in the frame of NEEP, Novi Sad, 2005 [in Serbian].
12. Pejčinović, J., et al., Economical analysis of exploitation and use of oil shale for district heating and electric energy production for towns Nis and Alekcsinac, Study financed in the frame of NEEP, Belgrade, 2003 [in Serbian].
13. Marković, B., et al., Development and implementation of the indicators for energy efficiency for individual houses in region of Nis, Study financed in the frame of NEEP, Niš, 2006 [in Serbian].
14. Radovanović, M., et al., Available technologies and equipment for production and use of briquettes and pellets, as substitution for electric energy in heating households, Study financed in the frame of NEEP, Belgrade, 2003 [in Serbian].
15. Nemoda, S., et al.: State of the art and use of fuel cell technologies and possibility of use in Serbia, Study financed in the frame of NEEP, Belgrade, 2005 [in Serbian].
16. Georgijević, V., et al.: Determination of outside design temperature and humidity for designing of heating and air conditioning systems, Study financed in the frame of NEEP, Belgrade, 2004 [in Serbian].

Chapter 17
Energy Sector in Macedonia: Current Status and Plans

Bozin Donevski

Abstract. A survey of the developments in the energy sector of Macedonia is presented, including past and current activities as well as perspectives for further development. Considered are the connection between the development of the energy sector and industrial systems, the infrastructure for energy resources and perspectives on its utilisation. Attention is focused on basic problems in the development of the energy sector in Macedonia, the present energy balance, problems with the energy supply, costs and obstacles associated with imported energy, and challenges with the underdeveloped energy infrastructure and connections with neighboring countries. The review aims to provide a perspective on continued development of the energy sector as a part of the total economic development of the country.

Key words: Energy, resources, reserves, infrastructure, consumption, indicators.

17.1 Background

Supplying energy is a basic requirement for the development of any country and for the improvement of its social and economic standard. Development of the energy sector is a continuous and ongoing process that should be actively supported by other industrial sectors.

The energy resources available for use in Macedonia are limited. Therefore, it is necessary to achieve industrial development of the country with as little consumption of energy as possible. A reduction in the consumption of energy also has important environmental advantages, and will help Macedonia to meet its own pollution targets and meet the Kyoto Protocol requirements, which went into action on February 16, 2005. In order to achieve this, several initiatives were started at appropriate institutions in Macedonia.

Until 1991, the energy infrastructure in Macedonia was an integral part of the former Yugoslav energy system. After the separation of Macedonia in the period

K. Hanjalić et al., Sustainable Energy Technologies: Options and Prospects, 303–320.

1991–1995, many unexpected political problems arose, such as the Greek embargo on Macedonia, and the sanctions imposed by the UN against the former Federal Republic of Yugoslavia. These developments had a significant effect on energy exploitation and on working conditions in the energy sector in Macedonia. The crisis in the region of Kosovo has also influenced the Macedonian energy system. Since the Macedonian and Kosovarian energy systems are directly connected, the Macedonian system was also affected by a complete collapse of the energy system in Kosovo as a result of the NATO air strikes.

The energy strategy for Macedonia can be analysed by distinguishing two main periods, i.e., before and after 1991. For the period prior to 1991, the energy strategy for Macedonia is described in a study that was financially supported by the former government of SR Macedonia [1]. The problems in the energy sector following 1991 are presented in the National Development Strategy for Macedonia [2]. This study was financially supported by the Macedonian government of the new independent state. In addition, another new strategy for developing the energy sector of Macedonia, published in 2000, has been jointly supported by the Macedonian government and the US Agency for International Development (USAID) [3].

The first study for Energy Development in Macedonia [1], covering the period from 1986 to 2005, was conducted in 1986. The objective of this study was to propose a new direction for the complicated development of the energy sector in the Republic of Macedonia. The assumptions used in this study were based on the energy sector being part of the former Yugoslav energy infrastructure, in anticipation of long-term social and economic progress. The study suggested possibilities for both the autonomous production of primary and converted energy, and for the import of energy from other former Yugoslav republics and abroad. Two variants for the strategic direction of energy development in Macedonia were proposed, indicated as Variant I and Variant II. In this study, priority was given to Variant I as it could be realised at minimum cost. Specifically, the criteria were to minimize expenses for the production of electric energy and in investments for an electric power system, and minimal foreign currency expenses for the import of various forms of energy. Variant II is defined as the development of an energy system which uses imported natural gas in the total energy balance. In its directive for 2001 to 2005, this variant suggested that a nuclear power plant of 300 MW installed power was to be built in Macedonia. Due to the split of the former Yugoslavia in 1991, the long war in Bosnia and Herzegovina, the crisis in the region of Kosovo, the blockade of Macedonia by Greece and the UN embargo on the former Federal Republic of Yugoslavia, the effects of the proposed variants in this study on the energy sector in Macedonia are quite different than originally anticipated.

17.2 Energy Supply

17.2.1 National Energy Infrastructure

The basic energy infrastructure in the Republic of Macedonia includes the electrical power system, lignite coal mines, oil refineries, oil pipelines, a gas pipeline system, district heating systems, small geothermal systems, and biomass (firewood). According to the Energy balance in 2006 [4, 5], the average total primary consumption of energy in the Republic of Macedonia is around 2.6 Mtoe per year. One toe represents a tonne of oil equivalent. For comparison, the oil equivalents of several other energy carriers are given: coal (1 tonne = 0.667 toe), gas (1000 m^3 = 0.900 toe), and electricity (1 MWh = 0.083 toe). Within the primary energy supply, the share of coal is 56%, of crude oil 30%, of natural gas 2.3%, and the remainder of 14% consists of hydro energy, biomass (fire-wood) and geothermal energy.

17.2.2 Total Installed Generation Capacity of the Electrical Power System

Until 2005, Elektrostopanstvo na Makedonija – ESM (the Public Electricity Supply Company of Macedonia) was the national power utility company. It had a vertically integrated state monopoly on electricity generation, and on the transmission and distribution of energy throughout the country. The total installed generation capacity for electricity production is 1524 MW, with an annual production of ∼6.5 GWh. Of this capacity, 1010 MW are from thermal power plants with an annual production of 5 GWh, and 514 MW are from hydro power plants with annual production of ∼1.5 GWh.

The re-structuring of the electricity sector started in 2004 with the separation of power generation, distribution and supply from power transmission and grid management within ESM. As a consequence, ESM was split into separate companies in 2005 for generation and distribution:

- AD ESM, a generation and distribution company which also owns 11 mini and small hydro power plants with a total capacity of 38.15 MW [6],
- AD MEPSO,[1] for the transmission of power,
- AD ELEM,[2] the largest generation company, and
- AD TEC Negotino is a one-plant generation company.

In December 2005, an international tender for the sale of 90% of the capital shares of AD ESM was issued. In spring 2006, the Austrian utility company EVN[3] won

[1] http://www.mepso.com

[2] http://www.elem.com.mk

[3] http://www.evn.at

Table 17.1 Transmission connection capacities with neighbouring countries [6].

Transmission interconnection capacities of Republic of Macedonia with neighbouring countries (MW)	Serbia and Montenegro	Greece	Bulgaria	Albania
Current situation	1200	1200	0	0
Period 2006–2009	1200	2400	1200	0
Period 2010–2019	2400	2400	1200	1200
Period 2020–2030	2400	2400	1200	1200

the tender and bought 90% of ESM for 225 million Euro. The planned privatisation of the largest generation company, AD ELEM, is still under discussion. The largest production capacity is the lignite-fired Bitola complex with three units of 225 MW each and net production of about 1434 GWh per unit. Peak demand is met by additional hydro power capacity. AD ESM-EVN has about 0.5 GW of hydro power capacity, which includes pumped storage, run of river and small hydro electric plants. The biggest hydro power plant in the basin of Crn Drim is HPP Spilje, which generates one third of the total energy production (380 GWh) generated by hydro power in the country [6].

17.2.3 Electricity Grid

The main transmission network of the energy infrastructure consists of 400 kV, 220 kV and 110 kV lines. The 400 kV line forms the backbone of the transmission network. It connects the Bitola Thermal Plant to the main load centre at Skopje, and connects Macedonia to the European UCPTE [7] system, as shown in Figure 17.1. Further connections (400 kV) to Greece and Bulgaria, and a 200 kV line to Albania are planned, see Table 17.1 and Figure 17.2.

Macedonia has 28 distribution districts, with a distribution network of 35 kV, 10 kV and 0.4 kV lines. Electricity coverage is extended to almost all populated Macedonian regions. Skopje's consumption dwarfs that of the other distribution districts, accounting for 37% of the total household consumption in Macedonia. Only three other distribution districts account for more than 5% of the total consumption: Tetovo (8.1%), Bitola (6.7%) and Kumanovo (5.9%). On average, electricity consumption in each of the 24 other distribution regions is about 2% of the total electricity consumption. The website of ESM [6] provides detailed information on the grid.

All 400 kV power lines built by Macedonia have a maximum transmission capacity of 1200 MW. To determine the practically achievable transmission capacity to each country separately, international methods and procedures are available (UCTE) [7].

All the investments carried out and planned by Macedonia can significantly contribute to expansion of the transmission capacities for power exchange among coun-

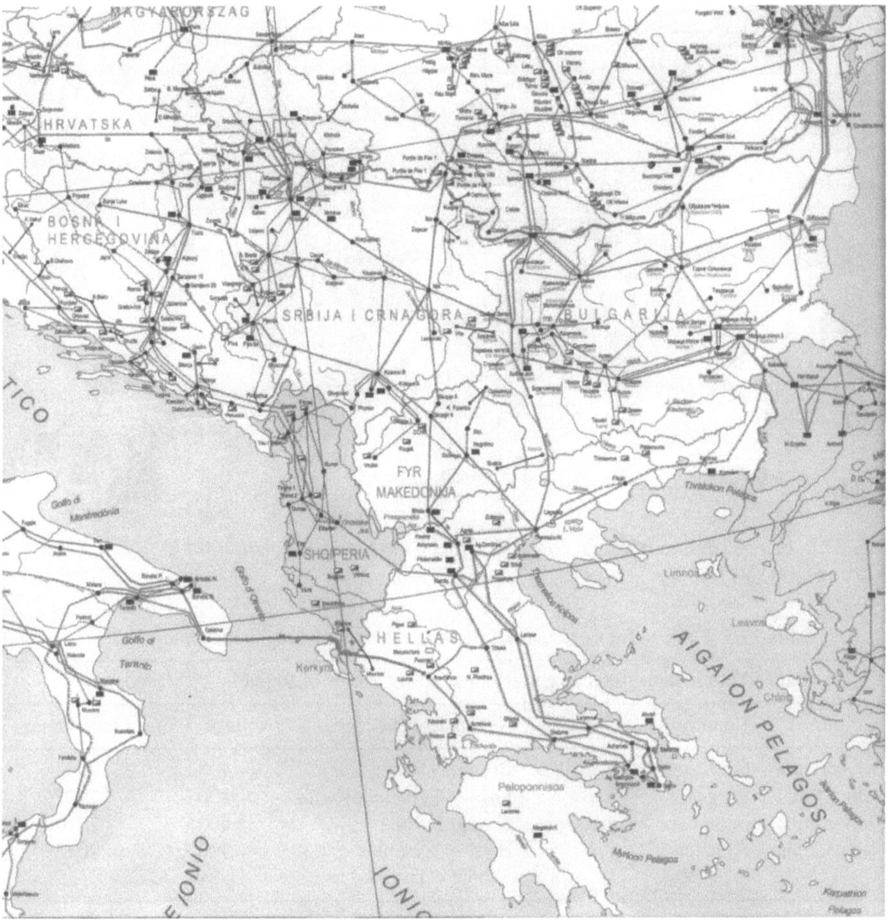

Fig. 17.1 Connections between the energy systems of the Balkan countries (UCPTE system, 22 July 2005) [7].

tries in the Balkan region (Table 17.2) and with the European Union. Furthermore, the power exchange capacity with neighbouring countries is currently close to the peak load capacity of the internal system, which significantly surpasses the criterion of a 20% safety margin that is recommended by the EU. Investments in new interconnections increase the flexibility and security of Macedonia's power exchange with neighbouring systems. Moreover, it will give Macedonia greater regional influence on measures to overcome potential power congestion, and in the trade and exchange of electrical power with EU countries.

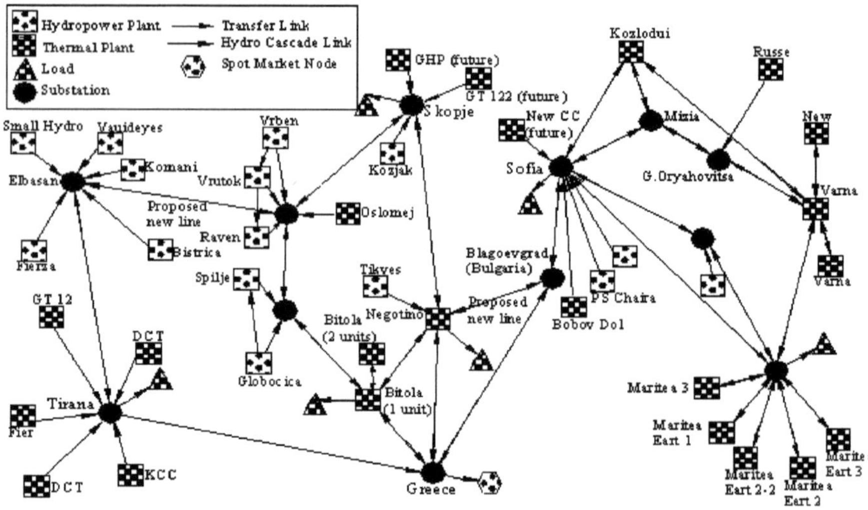

Fig. 17.2 Planned future network connections to Greece and Bulgaria (400 kV), and a line to Albania (200 kV).

Table 17.2 Planned future network interconnection with neighbours countries [6].

Planned interconnection capacities in the power sector of Republic of Macedonia				
Country 1	Country 2	Network connection	Voltage (kV)	Operational
Macedonia	Greece	Bitola – Florina	400	2007
Macedonia	Bulgaria	Štip – Chervena Mogila	400	2007
Macedonia	Serbia and Montenegro	Skopje 5 – Niš	400	2010
Macedonia	Albania	Bitola – Zemblak or Skopje – Tirana	400	2012

17.3 National Energy Resources and Perspectives on Its Utilisation

The total energy reserves in Macedonia are presented in Table 17.3. The table shows that the energy reserves are very modest in terms of diversity and size.

17.3.1 Coal

The total reserves of coal in Macedonia are 941 million tonnes, most of which (about 80%) are proven reserves. Additionally, exploration reserves comprise about 60% of the total balance; approx. 82% of existing reserves are suitable for surface exploitation, while 18% reserves can only be delivered from underground. The

Table 17.3 National energy reserves in Macedonia.

Energy Reserves		
Resources	(Mtonnes)	%
Coal (total discovered reserves)	243.45	72.8
Hydro power (technical useable)	58.96	17.6
Oil shale (not proven reserves)	26.00	7.8
Uranium (not proven reserves)	6.08	1.8
TOTAL	351.62	100.0

recoverable reserves would cover domestic needs for another 25 years. Currently, the annual production of lignite is 7.2 million tonnes, concentrated in four mines: Suvodol (6.3 Mtonnes); Oslomej (1.1 Mtons); Brik (0.1 Mtons) and Priskupstina (0.1 Mtonnes). Two of them are state-owned, and two are privately owned. Most of the coal (between 85 and 90%) is used for electricity generation, mainly from the mines in Suvodol and Oslomej. Around 200,000 tonnes of coal per year are used for industrial purposes and domestic heating. The two bigger mines supply lignite to nearby power plants Bitola I, II and III, and Oslomej. The coal industry does not receive any direct or indirect subsides, as it operates profitably. According to a source at the Ministry of Economy (February 2007), a new coal mine at Brod-Gneotino is planned to start operating in the near future.

The coal is mainly low-rank lignite, i.e., lignite with a low heating value. The average heating value of the lignite is between 6189 and 8290 kJ/kg (with an average of 7238 kJ/kg), and contains 0.5–1.5% of sulphur and 8–25% of ashes. The ratio between not proved reserves *versus* coal is approximately 3.2:1 to 26:1.

Although most local coal is low-grade lignite, it is used extensively for domestic energy production. Higher quality anthracite coal and coke (approximately 130,000 tonnes per year) have to be imported to supplement primary electricity generation and the local metallurgical industry.

Macedonia currently has two thermal lignite-fuelled power plants in operation. REK Bitola consists of three generators of 675 MW (3×225), and REK Oslomej, Kičevo consists of one generator of 125 MW. The characteristics of these plants are given in Table 17.4.

17.3.2 Hydro Power

The potential for hydro-power-based electricity generation is estimated to be much higher than is currently being exploited. The total hydro power potential in Macedonia that can be technically exploited is about 6436 GWh/year. This is consistent with a Master Plan prepared a long time ago (1976) and other studies made at a later time, in which the technically usable hydro power potential of the rivers in the country was estimated to be 5483 GWh/year.

Table 17.4 Basic characteristics of installed thermal power plants in Macedonia. (Source: The Public Electricity Supply Company of Macedonia (ESM). Yearly Report 2000, Skopje 2001 [6].)

Thermal power plant	Capacity (MW)	Net annual production (GWh)	Year of installation	Hours of operation until the year 2000	Basic fuels	Fuel heating values (kJ/kg)
Bitola I	225	376.8	1982	7480	lignite	7238
Bitola II	225	43.2	1984	7867	lignite	7238
Bitola III	225	31.4	1988	8100	lignite	7238
Oslomej	125	178.2	1980	4580	lignite	7667
Negotino	210	128.3	1978	45	heavy oil	40190
TOTAL		5159.0				

This energy can be best utilised by building hydro power plants with a total capacity of 1620 MW installed power (for all plants combined). However, only 24% of the available hydro power resources are used in the existing plants. Taking into consideration the rapid urbanisation of the region combined with the lack of planning and control over water resources, the capacity of the foreseen hydro electric potential is expected to be significantly less than the estimated reserves, i.e., about 4085 GWh/year [8].

In 2007, the Ministry of Economy will announce a public invitation for strategic partnerships for investment in and building of hydro power plants at the locations of Čebren, Galište and Boškov Most. Recently, a public call for proposals was also announced for building the first 60 of a total of 400 small hydro power plants on different locations.

Current Status of Hydro Power

Macedonia currently has 12 hydro power stations in operation with a total power capacity of 480 MW. Approximately 49% of the power produced with these stations is owned by Mavrovo Hydropower Systems. The capacity of existing smaller stations amounts to 50 MW.

Production of electricity from hydro power has fluctuated widely in the period 1980–2000. The generated power was about 1850 GWh (159 ktoe) in 1980, 848 GWh in 1992, 1389 GWh in 1999, 626 GWh in 2001, and 757 GWh in 2002. In 2003, the generated power rose again to 1483 GWh, which corresponds to about 4% of the total primary energy production.

The Kozjak hydro power project (82.5 MW) was due to be completed in 2001, but is still not fully operative. Also, several other hydro-power-related projects are planned, one of which is the storage facility in Lukovo Pole. Other projects are additional reservoirs for the three existing downstream power plants in the Mavrovo system (HPP Vrben, HPP Vrutok and HPP Raven) and HPP Spilje 2 (72.8 MW),

Table 17.5 Basic characteristics of installed hydro power plants in Macedonia [8]. (Source: The Public Electricity Supply Company of Macedonia (ESM), 2004 annual report, Skopje, 2005 [6].)

Power plants	No. of generators	Installed capacity (MW)	Water drop (m)	Water accumulation (10^6 m^3/sec)	Estimated production (GWh/year)	Production in 2000 (GWh)	Year of installation/refurbishment
Vrben	2	12.80	196.00	run of river	45.00	31.4	1959
Vrutok	4	150.00	525.00	277.00	350.00	376.8	1957/1973
Raven	3	19.20	74.00	run of river	40.00	38.3	1959/1973
Globočica	2	42.00	88.11	15.00	250.00	178.2	1965
Špilje	3	84.00	85.76	212.00	350.00	289.9	1970/1998
Tikveš	4	92.00	88.00	272.00	210.00	128.3	1968/1981
Kozjak	2	82.50	94.55	260.00	15.46	0	2001
Smaller plants	22	35.80		run of river-reservoir	169.57	122.2	1938–1993
TOTAL		527.10		1036.00	1572.03	1170.0	

and a new generating unit that will be added to HPP Spilje. The basic characteristics of the hydro power plants in Macedonia are given in Table 17.5.

In the summer of 2006, the Slovenian company "Turbo Institute" and the former Macedonian government made an agreement for Turbo Institute to construct and manage 20 small hydro power plants with a total capacity of 20 MW. However, the new government has questioned this agreement and has cancelled it. In September 2006, the Macedonian government tasked the Ministry of Economy to launch bidding for the construction of small hydro electric power plants in Macedonia.

During the GEF Mini Hydro project that was completed in 2004, it was demonstrated that financially viable opportunities exist for exploiting small hydro power resources, e.g., by incorporating small turbines into existing water supply pipelines. In the follow-up of this project, the Ministry of Economy has developed a portfolio of four similar projects (each in the rage of 0.5 to 5 MW) proposed by public sector entities, and 11 mini hydro projects (each in the range of 0.1 to 1 MW) proposed by private investors. These projects have all been screened for technical and financial viability, but most still face institutional constraints (e.g., the resolution of water rights issues) and all suffer from lack of funding [6].

17.3.3 Oil

Since there are no domestic oil reserves in Macedonia, oil and oil products must be imported to satisfy the basic needs. Imported crude oil is refined in the Power Plant Oil Distillery (OKTA Refinery) near Skopje, which has a capacity of 2.5 millions tonnes per year. The plant has been built between 1978 and 1982, using mainly former Soviet Union equipment. The refinery is small and adds little value to the oil. The crude oil is transported by train in special containers from the port of Thesaloniki in Greece, which is located at a distance of 220 km from the plant).

In 1999, following privatisation, ownership of the OKTA Refinery went to the Greek company Hellenic Petroleum. The refinery is currently uneconomic when compared to the alternative of importing petroleum products directly into Macedonia.

Presently, the refinery is supplied with crude oil through a pipeline jointly owned by Hellenic Petroleum (80%) and the Government of Macedonia (20%). Built by Hellenic, and completed in 2002, the pipeline has capacity of 2.5 million tonnes per year, which is more than adequate to meet domestic demands and for exports to other countries in the region, and connects the refinery to Hellenic's terminal in Thessaloniki. It is the most economical method of supplying crude oil to the OKTA refinery, even though the pipeline access charges used to finance construction were high in 2002–2007. The country's domestic demand for petroleum products is relatively stable at around 700,000–900,000 tonnes per year. The biggest oil derivatives and gas distributor in Macedonia is AD MAKPETROL, established in 1947, and since 1998 a fully private joint-stock company. AD MAKPETROL is the owner of 114 petrol stations and 12 warehouses for oil derivatives. It makes more than 60% of all the oil derivatives in Macedonia, and in 2002 the company placed approximately 410,000 tonnes of oil derivatives on the market. The storage capacities for oil in Macedonia are not sufficient for covering 90 days of consumption.

As can be seen from Table 17.4, one thermal power plant with an installed capacity of 210 MW and fired by heavy oil was built in 1978 in Dubrava, Negotino. Due to the Greek embargo on Macedonia from 1992 to 1996, the thermal power plant was not in operation for four years. Due to increased prices on the global oil market, the electricity production in this power plant is too expensive, and it mainly serves as a backup plant for the Macedonian power system. The plant only operated for 117 hours in 1997, and did not operate at all in 1998. At present (2007), this power plant operates at 50% of its capacity, and most of its power is exported to Greece. Recently, the plant was put up for public sale, and more than 15 foreign companies have shown interest in buying it.

17.3.4 Gas

There are no domestic gas resources in Macedonia. Macedonia imports about 100 million cubic metres of natural gas from Russia through a pipeline through Bulgaria, which was constructed in 1996. The distance of the pipeline from the Bulgarian border near Deve Bair to Skopje is approximately 98 km. So far, 26 km of pipeline has been built to connect various cities to the main line, and a further 31.5 km has been built for further distribution.

Gas is imported from Russia since 1997, and is transported from the Ukraine via Moldavia, Romania, and Bulgaria at a flow rate of 800 million m^3/year at a pressure of 0.4 MPa. The capacity can be increased up to 1200 million m^3/year by increasing the pressure.

Still, less than 10% of the available capacity is used, and gas demand represents only 1% of the total energy demand in Macedonia. Gas is used only in industry, and for district heating (Toplifikacija AD, Skopje). A gasification project to supply the industry surrounding Skopje is under way. Macedonia's national action plan aims for the gasification of urban residential areas around Skopje in order to replace electricity as an energy source for domestic heating. Due to the presence of the gas pipeline and its high population density, Skopje is the most likely candidate for a Macedonia gasification project. The economics of gasifying areas in Skopje currently not served by the district-heating network depend on the relative prices of gas and electricity. A previous study suggested that the development of a Skopje gas distribution network would be economically viable at the then current gas and electricity prices [3]. In order to establish a gas distribution network for Skopje, a number of issues would have to be resolved (e.g., issues relating to ownership of the gas pipeline), and a framework for regulation of the gas industry would have to be developed. These and other issues should be addressed in a feasibility study. Recently, another initiative was taken for developing an infrastructure for the gasification of the small city of Kriva Palanka, located in the eastern part of Macedonia at about 16 km from the Bulgarian border, in a partnership with a private company in the USA.

The government of Macedonia established the company GA-MA as a public enterprise for the supply, transport and distribution of natural gas in October 1996. Since Makpetrol (as a private company) is a shareholder in this enterprise, the enterprise (according to the Law on Trade Companies and Law on Public Enterprises) has been approached with an offer to be transformed into a joint company. However, a legislative decision on how the import and transmission of natural gas should be arranged between these two companies has still not been taken.

Very recently, the Council of the city of Skopje has decided to establish a new public enterprise "Energetika Skopje" for building the gasification infrastructure in Skopje. This enterprise will be responsible for distributing natural gas, and will build and maintain a network of gas pipelines in Skopje.

The first pipeline in Macedonia was built from Obilić (in Kosovo) to Skopje in order to supply the Still Company (Zelezara Skopje) with gas to power their generators. The length of this gas pipeline is about 90 km; however, this system is presently out of order.

A study for building a power plant located in Skopje and driven by a gas turbine generator is still under discussion as an ongoing project.

Finally, it should be noted that there are no gas storage facilities available in Macedonia.

17.3.5 Heat Generation

District heating in Macedonia is largely confined to Skopje. There are only two other small heat systems, located in Kocani and Bitola. The District Heating System of Skopje currently has 487 MW of hot water generation capacity and 26 MW of steam generating capacity. Annual heat production is around 685 GWh, predominantly for residential and commercial consumers. System losses of around 14% are not unduly high by regional standards, although there may be scope for economically viable loss reduction projects.

AD Toplifikacija Skopje,[4] a district heating company whose main activity is the production and distribution of heat, was founded in 1965. The district heating system of Skopje comprises four heat plants and approximately 150 km of distribution grid. In the past 15 years, 50% of the energy delivered to the consumers was metered. Since the year 2000, all of the delivered heat is metered. The total consumption of the district network company for the production of hot water and steam for heating is 71,400 toe (2.9 PJ) of fuel (85% gasoline, 15% natural gas).

There are plans to build a CHP plant for the Skopje District Heating System. For this purpose, a company is to be established together with the National Utility. The proposed plant would have a capacity of 100 MW of electrical power and 150 MW of thermal power [11].

However, electricity is widely used for residential heating in Macedonia, particularly in urban areas [9].

17.3.6 Geothermal Energy

There are some regions in Macedonia (Vinica, Kocani and Gevgelija) in which geothermal energy resources are being exploited. At present, geothermal water is used for agricultural purposes, heating greenhouses, and swimming pools for balneology[5] purposes. No electricity is produced from geothermal energy resources in Macedonia. Geothermal energy has only recently taken its place in the energy balance of the country. In the year 2003, 543 TJ of geothermal heat was produced. An assessment by the World Bank estimates its potential at 22 MW, i.e., 694 TJ/year (assuming continuous operation) [10–12].

Investigations conducted in the above mentioned regions indicate that geothermal energy resources may actually exceed the presently exploited energy sources. As such, it is expected that geothermal energy will play an increasingly important role in the energy sector.

---

[4] http://www.toplif.com.mk/

[5] The study of the medicinal use of hot baths.

17.3.7 Solar Energy

The solar irradiation intensity in Macedonia is one of the highest in Europe. The most favourable areas record a large number of sunshine hours, with the yearly ratio of actual-to-total possible irradiation reaching approximately 50% for former Yugoslavia as a whole, or 45% for the mountainous central regions (including Macedonia), due to the prevailing weather pattern. The annual average daily solar irradiation varies between 3.4 kWh/m^2 in the northern part of the country (Skopje) to 4.2 kWh/m^2 in the southwestern part (Bitola). The total annual solar irradiation varies from a minimum of 1250 kWh/m^2 in the northern part to a maximum of 1530 kWh/m2 in the southwestern part, which leads to an average annual solar irradiation of 1385 kWh/m^2. The climate characteristics – high intensity and number of hours of solar irradiation, and high temperatures and air humidity – provide favourable conditions for the successful development of solar energy. The continental climate with hot and dry summers makes Macedonia a country with a significantly higher potential for the utilisation of solar energy than the average European country.

The main form of solar energy technology used is flat plate solar thermal collectors for heating houses and some commercial and public premises. However, their contribution to the total energy demand is insignificant (less than 1%). Nor is it expected that this figure will increase substantially in the near future, as installation of new solar thermal installations would mainly be limited to new buildings or installations. Likewise, electricity production from solar photovoltaic sources will be restricted to research or remote locations, primarily for telecommunication equipment. PV solar energy is still three to five times more expensive than alternative fossil fuel derived sources.

In the period 2005–2007, experts and professionals were trained to improve the technology and production of solar thermal installations in Macedonia within the framework of a project of the Austrian Development Agency on Solar Water Heaters. The project is being carried out by the Austrian AEE INTEC[6] together with CRES[7] from Greece and SWT[8] from Germany.

17.3.8 Uranium

Little attention has been paid to the presence of uranium reserves in Macedonia, and the use of this energy source within the next 20 years seems unlikely [1]. Exploration should continue in order to determine the quantities of already discovered reserves and to determine if they can be classified as a higher grade. The energy potential of uranium is estimated to be circa 6 million tonnes of (not proven) reserves.

[6] http://www.aee-intec.at

[7] http://www.cres.gr

[8] http://www.swt-technologie.de

Table 17.6 Selected 2003 indicators for the Republic of Macedonia. (Source: IEA Energy Statistics, 2003.)

Key Indicators		Compound Indicators	
Population (million)	2.05	TPES[a]/Population (toe/capita)	1.31
GDP (billion 2000 US$)	3.57	TPES/GDP (toe/thousand – 2000 US$)	0.75
GDP (PPP)[b] (billion 2000 US$)	13.15	TPES/GDP (PPP) (toe/thousand – 2000 US$ PPP)	0.20
Energy Production (Mtoe)	1.56	Electricity Consumption/Population (kWh/capita)	3116
Net Import of Energy (Mtoe)	1.07	CO_2/TPES (tonnes CO_2/toe)	3.07
TPES (Mtoe)	2.68	CO_2/Population (tonnes CO_2/capita)	4.02
Electricity Consumption[c] (TWh)	6.38	CO_2/GDP (kg CO_2/2000 US$)	2.31
CO_2 Emissions[d] (Mt of CO_2)	8.23	CO_2/GDP (PPP) (kg CO_2/2000 US$ PPP)	0.63

[a] TPES = Total Primary Energy Supply
[b] GDP (PPP) = Gross domestic product at purchasing power parity (PPP) per capita
[c] Gross production + imports – exports – transmission/distribution losses
[d] CO_2 Emissions from fuel combustion only. Emissions are calculated using IEA's energy balances and the Revised 1996 IPCC Guidelines.

17.3.9 Oil Shale

With (not proven) reserves of 26 million tonnes, oil shale appears to be a promising source of energy. However, the technology of exploitation and processing is not yet mature, and many technical and economical challenges must first be overcome before it can play a significant role in the energy sector.

17.3.10 Biomass (Firewood)

In the region of Macedonia, wood as biomass is thus far only used for the purpose of heating. According to the results presented in a previous study [1], the average annual use of firewood is 930×10^3 m^3 per year. According to the energy balance for the year 2005 [4], biomass contributes 6.2% to the gross inland energy consumption. Biomass, in the form of wood and charcoal, is almost exclusively used in the domestic sector. Industrial or other uses are negligible and represent less than 1% of the total biomass consumption. There is relatively high potential in the country for utilising biogas from animal manure for energy generation purposes, as well as for growing crops for the production of biofuels.

Table 17.7 Energy Balance for Macedonia in 2006 and adopted for 2007. (Source: Energy Balance of the Republic of Macedonia for the year 2007 [5].)

	2006		2007		Index
Primary source	Quantity	10 TJ	Quantity	10 TJ	Ratio 2007/2006
1. Electricity (10 kWh)					
(a) production	6.557	23.65	6.794	24.50	103.61
– hydro	1.646	5.94	1.270	4.58	77.10
– thermal	4.911	17.72	5.524	19.92	112.48
(b) import	1.923	6.94	2.947	10.63	153.25
(c) export	–	–	–	–	–
(d) consumption ($d = a + b - c$)	8.489	30.62	9.741	35.13	114.73
2. Coal (10^3 tonnes)					
(a) production	6.856	52.50	6.442	48.95	95.62
(b) import	87	0.67	76	0.58	87.88
(c) export	–	–	–	–	–
(d) consumption ($d = a + b - c$)	6.943	53.16	6.518	49.53	93.17
3. Coke (10^3 tonnes)					
(a) production	–	–	–	–	–
(b) import	115	3.19	122	3.38	105.96
(c) export	–	–	–	–	–
(d) consumption ($d = a + b - c$)	115	3.19	122	3.38	105.96
4. Oil products (10^3 tonnes)					
(a) production	999	44.01	1.015	44.72	101.60
(b) import	–	–	–	–	–
(c) export	–	–	–	–	–
(d) consumption ($d = a + b - c$)	824	36.30	1.218	53.66	147.70
5. Natural gas (10^6 mN^3)					
(a) production	–	–	–	–	–
(b) import	82.22	2.75	113	3.78	137.45
(c) export	–	–	–	–	–
(d) consumption ($d = a + b - c$)	82.22	2.75	113	3.78	137.45
6. Biomass (firewood; 10^3 m^3)					
(a) production	650	7.08	700	7.62	107.69
(b) import	–	–	–	–	–
(c) export	–	–	–	–	–
(d) consumption ($d = a + b - c$)	650	7.08	700	7.62	107.69

17.4 Indicators for Energy Consumption and Energy Balance in Macedonia [16]

In Table 17.6, some selected key and compound indicators are presented. Table 17.7 presents the energy balance for 2006 and 2007 as adopted by the government of Macedonia [5].

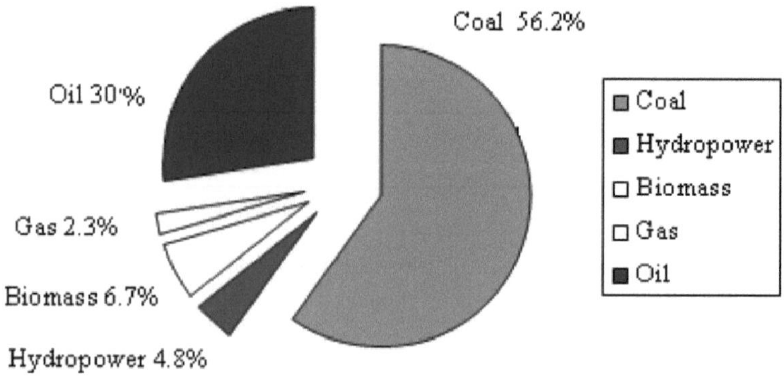

Fig. 17.3 Primary energy consumption for Macedonia in 2004 [9].

17.5 Background of the National Energy Policy, Legislation and Programmes

In December 2005, Macedonia was granted candidate status for membership in the European Union. As a result, Macedonia plans to harmonize its energy policies, including the ones on environment, with those of the EU so as to promote closer integration with other European countries.

In the EU Council Decision of 30 January 2006 on the principles, priorities and conditions for the Republic of Macedonia to become a EU partner, the following priorities were outlined for the energy sector [13]:

- Begin to align legislation on the internal electricity and gas markets, energy efficiency targets and renewable energy ambitions with the aim to gradually open the energy market to competition.
- Strengthen the independence of the Energy Regulatory Commission.
- Start implementing the Energy Community Treaty.
- Enhance administrative capacity in all energy sectors.

The Energy Regulatory Commission of the Republic of Macedonia became operational in 2003. It is engaged in establishing tariff systems and prices, licensing, and customer protection. Following the requirement as set out by a new law, a Macedonian Energy Agency was founded in 2006. The Energy Agency has a mandate to initiate and coordinate energy-related studies in cooperation with domestic and foreign experts and specialised companies. Based on these studies, it can suggest solutions and activities to the government through the Ministry of Economy. This agency is independent in its work and financially accountable to the Ministry of Economy.

In order to further develop the energy sector, the following areas for cooperation within this sector have been identified:

- Formulation and planning of an energy policy, modernisation of the infrastructure, improvement and diversification of supply, and improvement of access to the energy market, including facilitation of energy transit.
- Promotion of energy saving, more efficient use of energy, use of renewable energy, and measures to reduce the environmental impact of energy production and consumption.
- Restructuring of energy utilities and more intense cooperation in this sector.

The Government of Macedonia, together with the governments of Albania, Bosnia and Herzegovina, Bulgaria, Croatia, Greece, Kosovo, Romania, Turkey, and Serbia and Montenegro signed the "Athens Memorandum – 2002" whereby they have agreed to develop a South East Europe Regional Electricity Market (SEEREM). In 2003, the Revision of this protocol was also signed, which expands the regional market to gas.

In order to maintain a healthy energy balance, the Macedonian government has formulated the following long-term goals:

- developing the domestic energy sources (primarily coal mines and hydro power);
- intensifying the gasification of the country;
- developing the energy infrastructure and connecting the energy grid to those of neighboring countries;
- promoting efficient use of energy and increased use of renewable energy sources.

In line with these goals, the energy policy in the country generally contains the following strategic directions: conformation to the legislation of the EU, opening of the energy market, strengthening the role of the Regulatory Commission, making reforms in the transformation of the energy sector, further expansion of the energy infrastructure by building new energy generating facilities and grid connections to neighboring countries, increasing energy efficiency, more intensive use of renewable energy resources in the country, and introduction of adequate environmental standards and measures. These strategic priorities are further developed and strengthened with the new Energy Law, which was adopted by the Macedonian parliament on 11 May 2006 [14].

The Republic of Macedonia ratified the EC Treaty for South East Europe on 3 May 2006. The main goal of this treaty is to provide free energy trade under transparent conditions between the countries from the region and with the EU member states [13].

17.6 Conclusions

In the energy sector in Macedonia, there is significant potential for exploiting renewable energy sources (hydro power, geothermal energy, and solar energy) and Macedonian government is intent on starting its use.

The implementation of new technologies to provide for efficient energy use is taken into consideration in the Energy Efficiency Strategy until 2020, adopted by the Macedonian Government in 2004.

The favourable geo-strategic location of Macedonia provides prospects for becoming a main energy hub in the region. Hence, building additional grid connections to neighboring countries is one of the key elements that will allow Macedonia to establish a stable and modern energy infrastructure, and activities to construct these connections will be further intensified.

Finally, activities to promote the preservation of nature and the protection of the environment are being undertaken, and the authority to take preventive measures and impose sanctions has been delegated to both local and national authorities.

References

1. A study for the development of the electric power industry in SR Macedonia in the period of 1986–2005, Institute for Energy, Skopje, EMO-Ohrid, 1986 [in Macedonian].
2. Chapter 2: Energy, in *National Development Strategy for Macedonia, Development and Modernization, Macedonian Academy of Sciences and Arts*, Skopje, 1997, pp. 259–273.
3. A strategy for development of the energy sector in Macedonia, Information from the Ministry of Economy, Republic of Macedonia, Skopje, May 2000 [in Macedonian].
4. Regular Review 2006, Energy Chapter Protocol on Energy Efficiency and Energy Balance, Ministry of Economy, 2006.
5. Energy balance of the Republic of Macedonia for the year 2007, Ministry of Economy, *Official Gazette of the Republic of Macedonia*, No. 136, December 27, 2006.
6. AD ESM-EVN AG, Electric Power Company of Macedonia Skopje (Elektrostopanstvo na Makedonija AD ESM), http://www.esmak.com.mk.
7. UCPTE, http://www.ucte.org.
8. Donevski, B., A survey of the energy situation in Macedonia, paper presented at International Workshop and Round Table Discussion: Energy Future of Bosnia and Herzegovina, Center for Energy, Faculty of Mechanical Engineering, University of Sarajevo, June 15–16, 2000.
9. World Energy Database 2004, 2005, 2006, ENERDATA.
10. Kennedy, D. and Besant-Jones, J., World Bank framework for development of regional energy trade in South East Europe, Paper No. 12, Energy and Mining Sector Board Discussion Paper, The World Bank, Washington, DC, March 2004.
11. FYR Macedonia energy policy paper, Report No. 29709-MK, Infrastructure and Energy Services Department Europe and Central Asia Region, The World Bank, Washington, DC, July 2004.
12. World Bank, Macedonia energy policy paper, July 23, 2004.
13. European Commission, Council decision, January 2006.
14. Energy Law, Ministry of Economy, *Official Gazette of the Republic of Macedonia*, No. 63, May 23, 2006.
15. Österreichische Energiegentur – Austrian Energy Agency.
16. IEA Energy Statistics, 2003.

Information from the follow websites was used for this study:
http://www.energy-community.org
http://www.energyagency.at
http://www.worldenergy.org
http://www.balkanenergy.com
http://www.ucte.org

Chapter 18
Energy Sector in Bosnia and Herzegovina: Current Status and Plans

Alija Lekić

Abstract. Bosnia and Herzegovina has substantial primary energy resources and its energy industry could be profitable. This chapter presents the current status of the energy sector and the data on primary energy resources, potentials, energy infrastructure, and the production and consumption of energy. A comparison of consumption with neighbouring countries and the world is shown, as well as some energy indicators. There are ambitious plans for electric power generation in Bosnia and Herzegovina and an overview of the capacities under consideration is presented.

Key words: Energy, resources, potential, infrastructure, consumption, indicators.

18.1 Introduction

Bosnia and Herzegovina possess substantial primary energy resources and has a long tradition in energy production. It is believed that its energy industry, especially the electric power production, has still large unexploited potential and could emerge into a prosperous economic segment. This optimism is based on the country's large coal reserves, high hydro power potential and existing power plants on the one hand, and the opportunity to export electricity and limited amounts of coal to neighbouring countries, on the other.

The total balanced coal reserves in Bosnia and Herzegovina are estimated at 5464 million metric tonnes. The major constituents are the brown coal (calorific values between 11,728 and 18,471 kJ/kg) with estimated reserves of 1886 million tonnes, and the lignite (calorific values between 8,002 and 11,439 kJ/kg) with reserves of 3578 million tonnes. The total hydro power potential is estimated at 23,395 GWh annually, with about 6500 MW of installed capacity. Preliminary research surveys of oil and gas, which were interrupted by the war, had indicated the presence of promising deposits at a number of sites in some areas of Bosnia and Herzegovina.

K. Hanjalić et al., Sustainable Energy Technologies: Options and Prospects, 321–329.

The potential for exploitation of geothermal energy, wind energy, solar energy and biomass energy has not been sufficiently explored, and the proportion of these energy sources in overall consumption will most certainly remain modest for some time. However, because of the significant increase of the use of renewable sources of energy in the world, their potential and the feasibility of their use should be seriously investigated.

In spite of its potential, Bosnia and Herzegovina is currently importing energy. Although the consumption of primary energy per capita is lower than the world average, about 34.3% of the consumed primary energy was imported in the year 2004. Most of the liquid fuel and natural gas used is imported.

18.2 Status of the Energy Sector

According to the present constitution, signed in 1995, Bosnia and Herzegovina is divided into two "entities", the Federation and Republika Srpska. Only limited matters are the responsibility of the institutions of Bosnia and Herzegovina as a state, although the possibility remains that Bosnia and Herzegovina "shall assume responsibility for such other matters as are agreed by the Entities". Energy matters are the responsibility of the Entities, but it was expected that "the Entities shall begin negotiations with a view to including in the responsibilities of the institutions of Bosnia and Herzegovina other matters, including utilisation of energy resources and cooperative economic projects".

Nevertheless, at present most of the decisions in the energy sector are the responsibility of the Entities. The initiatives and efforts to include energy resources and the energy sector in the responsibilities of Bosnia and Herzegovina have not yet received a positive response. Neither Bosnia and Herzegovina nor the Entities have accepted the energy strategy. The Action Plan of the Poverty Reduction Strategy Paper (PRSP) included a requirement to prepare an energy strategy, with the objective of forming a plan for the development of the energy sector. However, and in spite of the support from the World Bank, the strategy is not ready – even though it was recommended that the strategy should be defined and adopted by the first half of 2005.

Lately, certain steps in that direction have been undertaken. Some action plans for the energy sector have been prepared, but the implementation of even accepted plans is very slow. In July 2004, laws on an Independent System Operator and a Transmission Company were adopted. The Entities have also adopted the laws on energy generation and distribution. The Independent System Operator was established in 2005 and the process of establishing the Transmission Company is underway. A State Electricity Regulatory Commission has also been set up. However, the Entities have decided to have their own regulators as well. The creation of an Energy Department within the Ministry of Trade and Economic Relations, which is in charge of energy policy in Bosnia and Herzegovina, is being considered and has been given the support of the CARDS programme. An energy strategy for Bosnia and Herzegovina is expected to be developed under this programme as well.

18.3 Energy Infrastructure, Production and Consumption

The main energy resources in Bosnia and Herzegovina are coal (brown and lignite) and hydro power. There are a number of coal mines in Bosnia and Herzegovina; 14 coal mines are now in operation, with 10 in the Federation of Bosnia and Herzegovina: Kakanj, Zenica, Breza, Bila, Gračanica, Kreka, Đurđevik, Banovići, Tušnica and Kamengrad; and 4 in Republika Srpska: Ugljevik, Gacko, Stanari and Miljevina. Most (80–90%) of the produced coal is used for electricity generation.

The total installed capacity for electricity generation in Bosnia and Herzegovina in the year 2004 was 3857 MW: 1780 MW in thermal power plants and 2077 MW in hydro power plants. In addition to that, there are about 25 MW installed capacities in small and mini hydro power plants.

There are, at present, three vertically integrated electric utility companies:

1. Elekroprivreda BiH (Electric Power Utility of Bosnia and Herzegovina – EP-BiH);
2. Elektroprivreda HZ HB (Electric Power Utility of the Croatian Community Herzeg-Bosnia – EPHZHB); and
3. Elektroprivreda Republike Srpske (Electric Power Utility of the Republika Srpska – EPRS).

Elektroprivreda BiH disposes of 1704 MW of the power generation capacity, which represents about 47% of the total available capacity in Bosnia and Herzegovina. The installed capacity in the thermal power plants is 1225 MW and the rest is in hydro power plants (479 MW). In 2003, more than 77% of this electricity was generated in two thermal power plants. The remainder was produced in hydro power plants.

Elektroprivreda HZ HB controls six hydro power plants with a total capacity of 775 MW.

Elektroprivreda Republike Srpske has the power generating capacity of 1378 MW: 555 MW in thermal power plants and 823 MW in hydro power plants.

Some of the thermal power plants supply heat for district heating systems. The unit Tuzla A3 has a heat capacity of 174 MW_t, the unit Tuzla A4 has a capacity of 220 MW_t and the unit Kakanj A5 has a capacity of 150 MW_t. Previously, there were another six units (two in Tuzla and four in Kakanj) with an installed electric capacity of 29 MW each, which are now out of use.

Several units in Tuzla and Kakanj have been reconstructed and one is presently under reconstruction in Tuzla.

A number of capacities (14 in total) with a total installed capacity of 293.6 MW_e were previously in use in industrial facilities, supplying electricity, process steam and heat, but only four of these capacities, with a total output of about 34 MW_e, are now available. Several small hydro power plants have recently been constructed.

Bosnia and Herzegovina does not have its own resources of oil and natural gas. However there is an oil refinery in Bosanski Brod, Republika Srpska, refining imported oil. (Another refinery, Modriča, produces only motor oils and lubricants.) In the last few years the refinery has again been in operation, but with only a small part of its capacity (the currently available capacity is estimated at 2 million tonnes per year) depending on the amount of oil imported and the market requirements. The company Jadranski naftovod d.d. (JANAF – the same pipeline that is used in Croatia) manages the transport of crude oil from a harbor on the Croatian coast to the refinery. Storage capacities for oil are not sufficient to cover 90 days' consumption.

In both entities, there are two major state-owned distributors. Small private distributors cover a large share of the market as well. Considering that the number of private petrol stations is on the increase and has reached approximately 300 stations, objective estimates suggest that the commercial capacities in the B&H market are already oversized.

Natural gas is imported from Russia by a pipeline through the Ukraine, Hungary and Serbia. The connection in Bosnia and Herzegovina is in Zvornik, on the eastern border of the country. This is the only connection for natural gas supply. The companies managing the import and transmission of natural gas are Energoinvest and BH Gas, Sarajevo and Gaspromet Pale (which manages only a part of the transmission line from Karakaj to Zvornik – approximately 20 km). There are two main gas distribution companies: Sarajevogas – Sarajevo (gas distribution in Sarajevo) and Sarajevogas – Lukavica (gas distribution in Eastern Sarajevo). All companies involved in the gas sector are state-owned.

There are no gas storage capacities in Bosnia and Herzegovina, although some studies about the possibilities for this were made.

Energy consumption in Bosnia and Herzegovina was considerably reduced after the war 1991–1995. As the standard of living is, in a way, connected to energy consumption, one of the main social goals in Bosnia and Herzegovina, i.e., to reduce poverty in the coming years, can hardly be achieved without increasing energy consumption. But increasing energy consumption must be followed closely with the reduction of the energy intensity (i.e., increasing energy efficiency), which has increased since the war.

The total primary energy consumption in Bosnia and Herzegovina (in 2004) was 4704×10^3 tonnes of oil equivalent (toe) or 196.95 PJ. The energy consumption per capita (in the same year) was 1.2 toe or 50.2 GJ/cap which is considerably lower than the world average – 1.77 toe/cap or 74.1 GJ/cap – and even lower than the average energy consumption in South-East Europe (SEE) countries – 1.83 toe/cap or 76.6 GJ/cap (Figure 18.1).

More than half of the primary energy consumption is in coal (2667×10^3 toe or 56.6% of total consumption), about 25.9% in petroleum products, 10.8% is hydro power, 6.5% natural gas and the rest from other sources.

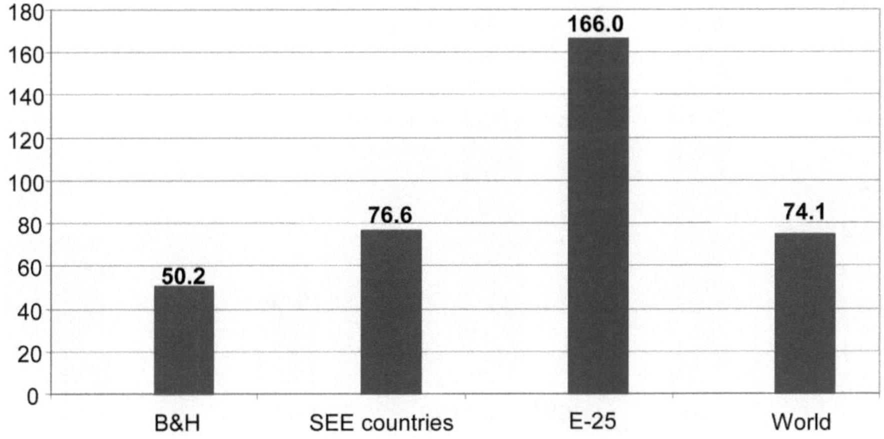

Fig. 18.1 Primary energy consumption per capita [GJ/cap].

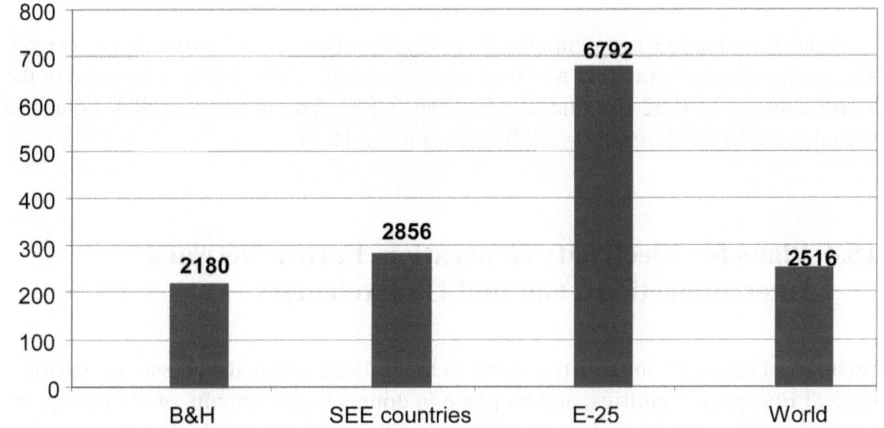

Fig. 18.2 Electricity consumption per capita [kWh/cap].

Crude oil, petroleum products and natural gas are all imported. Coal production in 2004 was 8648×10^3 metric tonnes — 2302×10^3 of lignite and 6346×10^3 tonnes of brown coal.

Electricity generation (in 2004) was 12,743 GWh and consumption was 10,720 GWh; about 15% of the production was exported. Electricity consumption per capita in that year was 2180 kWh, which is lower then the world average – 2516 kWh/cap – and, again, even lower then the average electric energy consumption in SEE countries – 2856 kWh/cap (Figure 18.2).

The majority of the energy infrastructure in Bosnia and Herzegovina, especially electric energy facilities, has been reconstructed, enabling not only an increase in electric energy consumption, but also its export.

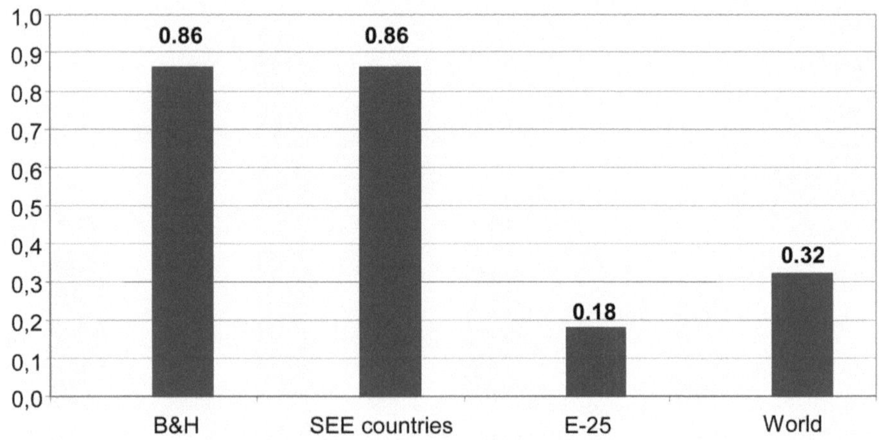

Fig. 18.3 Energy intensity [toe/thousand 2000 US$].

Apart from the low consumption of energy, the efficiency of its use is low as well. The energy intensity in 2004 was 0.86 (toe/thousand – 2000 US$), compared to the world average of 0.32 (toe/thousand – 2000 US$). The average in SEE countries was also 0.86 (toe/thousand – 2000 US$) (Figure 18.3).

18.4 Plans for Electricity Generation, Future Network Interconnections, Gas and Oil Exchanges

Bosnia and Herzegovina is in the center of the Balkans and at the crossroads between several European countries, and its place in connecting segments of the energy infrastructure (electricity and gas) could be significant. Activities in that direction and its association to the regional market should help the development of the energy sector in Bosnia and Herzegovina.

Bosnia and Herzegovina has signed the Memorandum of Understanding on the Regional Electricity Market in South East Europe and its Integration into the European Union Internal Electricity Market, and recently the Treaty establishing the Energy Community in South East Europe.

The integration of Bosnia and Herzegovina into the European Union is a generally accepted goal of all sectors, including the energy sector. In the document "Mid–term Development Strategy of B&H", prepared at state level, a section is devoted to the energy sector. Most of the activities and measures foreseen in the document refer to the establishment of conditions for further development. The concrete plans made by different parts of the energy sector are quite often the product of the wishes of the institutions and enterprises in charge of the planning.

The comprehensive transformation and restructuring of coalmines is proposed in order to lower coal production costs, increase productivity, reduce the number of employees, improve working conditions, raise the living standard of employees, and protect the environment.

The introduction of new production technologies, which will improve profitability and ensure better protection of the environment, is foreseen as well. The required funds for the implementation of the proposed measures have been estimated.

The main activities in the electric power sub sector were the completion of the Power III project, supported by the international community, and the establishment of the Transmission Corporation at the state level, the Independent System Operator, state and Entity regulatory commissions and internal and regional electric power markets. The separate Action Plan for these activities in the Federation of Bosnia and Herzegovina has been prepared and harmonised with the Action Plan of Republika Srpska, but not as yet approved, although some activities are underway.

Parts of the Power III project have enabled the reconstruction of the substations and 400 kV lines and the interconnection of the grid to the UCTE system, which was done in October of 2004.

The plans for new electric power capacities are prepared by each electric utility separately. The basic project for the electricity and heat-supplying unit in Banja Luka, based on the coal from the Stanari mine, has been prepared by Elektroprivreda Republike Srpske (Electric Power Utility of the Republika Srpska – EPRS). The planned capacity is 167.4 MW_e for electricity generation and 256 MW_t for heat supply. The Energy Financing Team from the UK has expressed an interest to build a significant electric power generating capacity based on coal from the Stanari mine, for which it has obtained a concession for the next 30 years.

There is considerable potential in Republika Srpska (covered by Elektroprivreda Republika Srpska) in hydro power. Technically usable potential has been estimated at 10,027.5 GWh per year, out of which about 7040 GWh are not in use. Some plans for hydro power facilities have been prepared, including the Buk Bijela power plant on the river Drina with the capacity of 450 MW (and annual electric energy generation of $\sim$1158 GWh), two power plants on the river Vrbas with the capacity of about 85 MW (and the annual electricity generation of $\sim$327 GWh), and the power plant Dabar in the Trebišnjica river basin with the capacity of 160 MW and annual electricity generation of $\sim$271 GWh. The outcome of these plans is difficult to estimate, bearing in mind the need to find financial support. Recently, the electric utility of Slovenia has expressed an interest to invest in the reconstruction of the existing unit in Ugljevik, and in a new unit of 600 MW.

Elektroprivreda HZ HB (Electric Power Utility of the Croatian Community Herzeg-Bosnia – EPHZHB) has started work on the construction of the new hydro power plant Mostarsko blato. The installed capacity of the power plant will be 61 MW (two Francis type turbines, 30.5 MW each) with a nominal flow rate of 40 m^3/s and expected electric energy generation of 170 GWh per year. Preparation activities for the initial stages of the project of the coal mine and thermal power plant Kongora in the region of Duvanjsko polje are underway in the EPHZHB. The pre-feasibility study has shown the possibility of having two units with a total capacity of 550 MW (2 × 275 MW) and electric energy generation of 3300 GWh per year.

Elekroprivreda BiH (Electric Power Utility of Bosnia and Herzegovina – EPBiH) is in the initial stages of projects for a number of power plants, both thermal power and hydro power. Some of these thermal power plants are planned for existing locations, such as two units in Tuzla with the capacity of 340 MW and 465 MW, and one unit in Kakanj with the capacity of 210 MW. (An alternative for the location in Kakanj is to build two different units with a higher total capacity – 450 MW.) Another new unit is planned for Kamengrad with the capacity of 400 MW. The production of the above mentioned thermal power plants would be 8464 GWh per year.

The capacity of the planned hydro power plants is 464 MW (6 hydro power plants at Glavatičevo, Konjic, Ustikolina, Vranduk, Vrhpolje and Čaplje with capacities ranging from 12 to 172 MW). The production of the hydro power plants would be 1159 GWh per year.

To initiate these activities, the Ministry of Energy, Mining and Industry of the Federation of Bosnia and Herzegovina has announced a public invitation to express interest in strategic partnerships for the following investment projects: Ustikolina (3 × 22 MW), Vranduk (21 MW), Rmanj Manastir (2 × 36 MW), Vrilo (42 MW), Tuzla (370 MW), Kakanj (250 MW), Bugojno (2 × 300 MW) and Kongora (2 × 275 MW).

A number of locations for small hydro power plants have been proposed and preliminary measurements of the wind energy potential in the southwestern part of Bosnia and Herzegovina have been completed.

As mentioned earlier, the connection of the gas supply to Bosnia and Herzegovina is on the eastern border of the country in Zvornik. This is the only connection for natural gas supply. The experience of having a single gas connection supply line has exposed many difficulties, such as in the recent conflict situation in Bosnia and Herzegovina.

Several studies concerning natural gas and the development of a natural gas infrastructure have been performed. The possibilities of connecting lines with the natural gas networks in Croatia, at different places, were analysed. The connection to the network in Croatia (the Memorandum of Understanding between the authorised companies was signed in June 2006) would be very important in opening the possibility for connections with other sources of natural gas and to integrate into the European market. (A connection with Serbia exists.)

Expanding the network in Bosnia and Herzegovina to include new consumers was also considered. A preliminary study on natural gas storage has been performed and the used salt mine near the town of Tuzla with a capacity of 60 million m^3 was considered, although that area is not, as yet, connected to the existing gas network.

The development of the gas sector will to a great extent depend on developments in the region. Specifically, the construction of the "South European Gas Ring" could determine the future of both the gas sector and the energy sector in Bosnia and Herzegovina, and it would considerably improve the position of the natural gas sub sector if it were carried out throughout Bosnia and Herzegovina.

A few months ago, a Memorandum of Understanding was signed between Bosnia and Herzegovina and Croatia in order to construct a gas pipeline connecting Hungary, Croatia and Bosnia and Herzegovina. This would enable Bosnia and Herzegovina to gain access to other sources of gas supply.

The existing Jadranski naftovod (JANAF – the same pipeline that is used in Croatia), with possible extensions, will remain the main oil supply pipeline. Further activities will be focused on increasing the use of the existing refinery capacities and the adoption of appropriate laws and regulations in line with the EU directives. Initial improvements were already made by adoption of the Decision on the Quality of Liquid Oil Fuels by the B&H Council of Ministers in September 2002, stipulating an obligation to import only liquid fuels that correspond to the regulations and meet EU quality standards.

The capacities for oil and oil derivatives storage are not sufficient and some earlier capacities, which were destroyed during the war, should be reconstructed.

18.5 Conclusions

The energy sector in Bosnia and Herzegovina has considerable potential and could become a significant exporter in the region. Some initial steps to unify it at the state level are underway, but the implementation of accepted plans is slow and could reduce the potential effects.

New technologies, especially for electric power generation, should be taken into consideration for the planned new capacities.

Index

MIX
Papier aus verantwortungsvollen Quellen
Paper from responsible sources
FSC® C105338

If you have any concerns about our products,
you can contact us on
ProductSafety@springernature.com

In case Publisher is established outside the EU,
the EU authorized representative is:
Springer Nature Customer Service Center GmbH
Europaplatz 3, 69115 Heidelberg, Germany

Printed by Libri Plureos GmbH
in Hamburg, Germany